*The Star Wars Controversy*

# The Star Wars Controversy

AN *International Security* READER

EDITED BY

*Steven E. Miller*
*and Stephen Van Evera*

PRINCETON UNIVERSITY PRESS
PRINCETON, NEW JERSEY

Published by Princeton University Press, 41 William Street, Princeton, New Jersey 08540
In the United Kingdom: Princeton University Press, Guildford, Surrey

First Princeton Paperback printing, 1986
First hardcover printing, 1986

LCC 86-4280     ISBN 0-691-07713-4     ISBN 0-691-02253-4 (pbk.)

The following materials were first published in *International Security*, a publication of the MIT and International Affairs at Harvard University:

"The SDI in U.S. Nuclear Strategy," Fred S. Hoffman, Senate Testimony
"Rhetoric and Realities in the Star Wars Debate," © 1985 James R. Schlesinger
"Why Even Good Defenses May Be Bad," and "Do We Want the Missile Defenses We Can Build?"
   Charles L. Glaser
"Preserving the ABM Treaty: A Critique of the Reagan SDI," Sidney Drell, Philip Farley and David
   Holloway
"Ballistic Missile Defense and the Atlantic Alliance," David S. Yost.

# Contents

# The Contributors

STEVEN E. MILLER teaches defense studies in the Department of Political Science at the Massachusetts Institute of Technology and is a co-editor of *International Security*.

STEPHEN VAN EVERA is the managing editor of *International Security*.

FRED S. HOFFMAN is Director of Pan Heuristics, a Los Angeles-based research group. He was director of the study group that prepared the report "Ballistic Missile Defenses and U.S. National Security" in October 1983 for the Future Security Strategy Study (generally known as the "Hoffman Report").

JAMES R. SCHLESINGER was U.S. Secretary of Defense 1973-75 and Secretary of the Department of Energy 1977-79. He is presently Counselor at the Center for Strategic and International Studies at Georgetown University.

CHARLES L. GLASER is a Postdoctoral Fellow on the Avoiding Nuclear War Project at the John F. Kennedy School of Government, Harvard University, and a Research Fellow at the School's Center for Science and International Affairs.

SIDNEY DRELL is Deputy Director of the Linear Accelerator Center and Co-Director of the Center for International Security and Arms Control, both at Stanford University. PHILIP FARLEY is Senior Fellow at the Center for International Security and Arms Control at Stanford and former Deputy Director of the Arms Control and Disarmament Agency. DAVID HOLLOWAY is Senior Research Associate at the Center for International Security and Arms Control at Stanford and Reader in Politics at the University of Edinburgh.

DAVID S. YOST is an Assistant Professor at the U.S. Naval Postgraduate School, Monterey, California. He is the author of *European Security and the SALT Process*, Washington Paper, Number 85 (Beverly Hills and London: Sage Publications, 1981) and editor of *NATO's Strategic Options: Arms Control and Defense* (New York: Pergamon Press, 1981).

# Preface | *Stephen Van Evera*

On March 23, 1983 President Ronald Reagan announced in a nationally televised speech that his administration would seek to develop a defense against ballistic missiles. "What if free people could live secure in the knowledge that their security did not rest upon the threat of instant U.S. retaliation to deter a Soviet attack, that we could intercept and destroy strategic ballistic missiles before they reached our own soil or that of our allies?" he asked. The President answered with "a vision of the future which offers hope. It is that we embark on a program to counter the awesome Soviet missile threat with measures that are defensive," and he continued on to announce "a comprehensive and intensive effort to define a long-term research and development program to begin to achieve our ultimate goal of eliminating the threat posed by strategic nuclear missiles." The President acknowledged that he proposed a "formidable, technical task" that could take decades to achieve and that ballistic missile defenses could be provocative: "If paired with offensive systems, they can be viewed as fostering an aggressive policy, and no one wants that." Nevertheless, he suggested that his program "holds the promise of changing the course of human history" and could "free the world from the threat of nuclear war . . ."[1]

The President did not explain how his defensive task would be accomplished, saying nothing of boost phase intercepts, space-based lasers, aerosol clouds, "pop-up" systems, and the like. Nor did he discuss whether aerial defenses against bombers and cruise missiles (without which missile defenses would provide little protection to the American population) would also be developed.

This announcement marked the birth of the President's Strategic Defense Initiative (SDI), soon labelled "Star Wars" by the press. Ten weeks later the President appointed three panels of experts to assess the feasibility and desirability of building strategic defenses. These reports were submitted by October 1983, and in early 1984 Secretary of Defense Caspar Weinberger released unclassified summaries of two of these panel studies: the *Defensive Technologies Study*, or "Fletcher Report" (after panel chairman James C. Fletcher), exploring possible technologies for ballistic missile defense (BMD); and the *Future Security Strategy Study*, or "Hoffman Report" (after panel chairman Fred S. Hoffman), exploring the strategic implications of BMD. (The third panel study, the "Miller Report," remains fully classified at this writ-

---

1. "Appendix A: The Conclusion of President Reagan's March 23, 1983, Speech on Defense Spending and Defensive Technology," in Ashton B. Carter, *Directed Energy Missile Defense in Space* (Washington, D.C.: Congress of the United States, Office of Technology Assessment, 1984), pp. 85-86 (reprinted below [p. 165], and hereinafter cited as "March 23 1983 speech").

ing.)[2] After approving these reports, also in early 1984, the President chartered a Strategic Defense Initiative Organization (SDIO) to carry out his program, and appointed Lieutenant General James A. Abrahamson as its director. The administration then announced that spending on strategic defenses, which had previously stood at roughly $1 billion per year, would be increased fourfold over the next five years.[3] The SDIO also initiated a program to test and demonstrate defensive technologies that, observers noted, might violate the 1972 Soviet-American anti-ballistic missile (ABM) treaty.[4] Observers also noted that even if early SDI tests could be designed to fit within the 1972 Treaty, fuller development of SDI systems would eventually violate the Treaty, thus requiring its amendment or abrogation. The 1972 Treaty banned the deployment of any space-based ballistic missile defenses, or any nationwide BMD, while the deployment of such systems is the apparent goal of the SDI program.

The SDI program was harshly criticized from its inception, stirring an intense controversy that soon dominated discussion of American strategy arms policy. Other issues, such as the perennial MX missile dispute, were pushed aside. Rhetoric on both sides has often been heated: Former Undersecretary of State George Ball termed the President's SDI speech "one of the most irresponsible acts by any head of state in modern times," and the Union of Concerned Scientists declared "the end unattainable, the means harebrained, and the cost staggering," while Secretary of Defense Weinberger answered that the SDI "offers the world more hope than any concept developed since the nuclear era began."[5] Many arguments on both sides have been heard before, during the "first ABM debate" of 1967–72, when a fierce battle developed over a plan developed by the Department of Defense (DoD) to deploy a more primitive BMD system (known as the "Safeguard" system). This battle ended in victory for the anti-ABM forces, with the signing of the 1972 ABM

---

2. The unclassified summary of the Hoffman Report was formally entitled *Ballistic Missile Defenses and U.S. National Security*. It is hereinafter cited as "The Hoffman Report," and the Fletcher study is hereinafter "The Fletcher Report." Both are published below (p. 273 and p. 291).

   On these reports an excellent discussion is Donald L. Hafner, "Assessing the President's Vision: The Fletcher, Miller, and Hoffman Panels," *Daedalus* Vol. 114, No. 2 (Spring 1985), *Weapons in Space, Volume I: Concepts and Technologies*, pp. 91-108.

3. Albert Carnesale, "The Strategic Defense Initiative," in George E. Hudson and Joseph Kruzel, eds., with a foreword by Alexander L. George, *American Defense Annual, 1985-1986* (Lexington, Mass.: Lexington Books, 1985), p. 191.

4. The terms ABM and BMD are usually used interchangeably to indicate defenses against ballistic missiles.

5. George W. Ball, "The War for Star Wars," *New York Review of Books* (April 11, 1985), p. 38; Union of Concerned Scientists, *The Fallacy of Star Wars* (New York: Vintage Books, 1984), p. 176, borrowing language from Senator Arthur Vandenburg; Caspar W. Weinberger, *Annual Report to the Congress, Fiscal Year 1986* (Washington, D.C.: U.S. Government Printing Office, 1985), p. 214.

Treaty. The President's 1983 SDI announcement re-opened many of the same disputes, now cast in a new context, but fought with equal vehemence.

This preface will survey the technical background behind the SDI controversy, outline the principal issues in dispute, and sketch the background to the essays and documents printed in this volume.

### Ballistic Missile Defense Technologies

In principle a ballistic missile can be destroyed during any of four phases of flight: the "boost phase," during which the rocket boosters propel the missile toward outer space; the "post-boost phase," after the rocket motors burn out, during which the post-boost vehicle (known as the "bus") dispenses its warheads to travel their separate trajectories and releases any decoys or other penetration aids; the "midcourse phase," during which these warheads and decoys streak through space toward their targets; and the "terminal phase," during which the warheads enter the atmosphere, and finally strike their targets. A typical thirty-minute intercontinental ballistic missile (ICBM) flight consists of three minutes of boost, five minutes of post-boost, twenty minutes of midcourse, and two minutes of terminal phase.[6]

Killing ICBMs in the boost phase is a tempting method of attack, because the defender can destroy an entire busload of warheads and decoys by killing a single missile, before these objects are dispersed to overwhelm the defense with numbers. However, boost-phase and post-boost phase interception schemes also require exotic defensive systems that are based in space, or that can very quickly "pop up" from ground level to operate from space. This is the case because the boost phase of Soviet ICBMs takes place low over Soviet territory, and the curvature of the earth conceals Soviet missiles at this point from any ground-based weapons located in the United States or other western countries. As a result Soviet missiles in boost phase can only be attacked by weaponry located high above the earth, or directly over Soviet territory.

The American BMD concepts of the 1950s and 1960s conceived of intercepting Soviet warheads only during the last two phases of flight—the midcourse and terminal phases—because schemes for attacking missiles and warheads from space seemed technically implausible. By contrast, the BMD schemes now under discussion in the SDI Office envision intercepting Soviet missiles in all four phases of flight, with a heavy emphasis on boost phase interception accomplished from space. This commitment to boost phase in-

6. Carnesale, "Strategic Defense Initiative," p. 188.

tercept is the chief novelty of the SDI, and the case for the technical feasibility
of SDI rests heavily on proponents' claims that new technologies have finally
made boost and post-boost phase interception feasible.

Schemes for boost-phase intercept are still preliminary, and their effectiveness is quite hypothetical. However, nine possible kill mechanisms are
discussed most often. Four of these nine are varieties of laser beam: so-called
chemical lasers, x-ray lasers, excited dimer (or "excimer") lasers, and free
electron lasers. These laser beams would destroy Soviet missiles by burning
holes in their skins during their upward climb. Neutral and charged (electron)
particle beams offer other possible kill mechanisms. These would damage
Soviet missiles by heating them or disrupting their electronics with beams of
atomic particles. Two varieties of "kinetic energy" weapons have also been
conceived: space-borne "rail guns" that would fire pellets at extremely high
velocities at Soviet missiles; and space-based rocket launchers that would
destroy rising Soviet missiles with very fast rockets. Finally, another scheme
would bathe Soviet missile fields with microwave energy, hopefully destroying
the missiles as they flew upwards.[7]

SDI program designers suggest that these weaponries might be based in
space; or they might be rocketed into space on warning of attack, in a "pop
up" arrangement; or, in some cases, they might be based on the ground,
perhaps in the United States, with their energy relayed to their targets by a
space-based network of mirrors. Most systems now under discussion would
be based in space, or popped up into space. (Hence the origins of the "Star
Wars" label that has stuck to the program.) However, a ground-based system
with mirror relays is envisioned for systems that would employ excimer or
free electron laser weapons, chiefly because these lasers need bulky power
generators that would be very costly to base in space.

SDI critics note that countermeasures that could defeat boost-phase interception are already envisioned. For instance, the Soviets might develop fast-
burn rocket boosters that burn out before the missile rises past the atmosphere,
which would defeat most beam and kinetic energy systems. Alternately, the
Soviets might spin their boosters to prevent damage by the American beams;
or they might give their boosters protective coating; or they might proliferate
boosters to overwhelm the American defense with great numbers; or they
might interfere with American warning systems (which must issue very rapid
warning for any boost-phase interception to succeed); or they could attack
the American defensive systems before or during their own attack; or they

---

7. Detailed discussions of these technologies are found in the contributions in this volume by Ashton Carter,
Sidney D. Drell, Philip J. Farley, and David Holloway.

could do some combination of the above. The feasibility of boost-phase interception depends heavily on its vulnerability to countermeasures of this kind.

Post-boost phase interception is less lucrative than boost-phase intercept, because the missile "bus" has by then begun dispensing its warheads and decoys, so that killing the missile no longer kills the whole cargo. Post-boost intercept is also more difficult than boost-phase intercept because the warhead-carrying "bus" has no easily-located rocket plume, and it can be shielded more easily against directed energy weapons.[8] United States government scientists have therefore focused on schemes for mid-course and terminal interception to destroy those Soviet warheads that survive boost-phase attacks.

To perform midcourse interception the United States could use the same directed-energy or kinetic-energy weapons that would be used to attack Soviet missiles in their boost phase. In addition, American planners envision using clouds of pellets or aerosols to destroy Soviet decoys, and conventional warheads carried on missiles launched from the United States to kill incoming warheads.[9] During the brief two-minute terminal phase American planners would destroy incoming warheads with two varieties of rocket interceptors—one above and one within the atmosphere.[10] Other kill mechanisms, such as rapid-fire guns, also might be used to defend local targets. The main elements of these mid-course and terminal interception schemes resemble the abandoned American ABM schemes of the 1960s, although SDI advocates note that some of the technologies they incorporate, such as radars, computers, and interceptor missiles, have since become much more capable.[11]

## Questions at Issue in the SDI Debate

Two questions lie at the heart of the SDI controversy: Are ballistic missile defenses feasible? Are they desirable? Answers to these questions often depend in turn on how a third question is answered: What are defenses supposed to do? Analysts often dispute the feasibility and desirability of defenses because they differ on the purpose that defenses should serve.

The most severe criticisms of the SDI concern its technical feasibility. SDI supporters have maintained that recent technological advances—especially in

8. Carnesale, "Strategic Defense Initiative," p. 190. "Directed energy" weapons include both laser and particle beam weapons.
9. Charles Mohr, "Antimissile Plan Seeks Thousands of Space Weapons," *New York Times*, November 11, 1985, pp. 1, 18; and Carnesale, "Strategic Defense Initiative," p. 190.
10. Mohr, "Antimissile Plan," p. 18.
11. Carnesale, "Strategic Defense Initiative," pp. 190-91.

laser, particle beam and kinetic-energy target destruction mechanisms—finally make boost-phase interception feasible; that technologies for midcourse and terminal phase interception, such as radar and computing technologies, have also improved (as just noted); and that these improvements now make a whole defensive scheme newly possible.[12] Hence President Reagan declared that "current technology has attained a level of sophistication where it's reasonable to us to begin this effort," concluding that "I believe we can do it."[13] His Fletcher Panel more tentatively, but hopefully, noted that it finished its work "with a sense of optimism," and concluded that the technological challenges of strategic defenses were "great but not insurmountable."[14]

However, many other experts have concluded that comprehensive defenses are infeasible, because current defensive technologies are too underdeveloped; because the task of providing a near-perfect area defense is inherently too demanding; because the Soviets can take effective countermeasures against American defenses; and because even a perfect ballistic missile defense would still leave the United States open to attack by Soviet bombers and cruise missiles, torpedoes, or even smuggled suitcase bombs, which could not be defeated unless the United States embarked on a defensive program similar in scale and cost to the SDI, if then. Many experts grant that some kind of point defense—that is, a defense of a small geographic location, such as a missile silo—may be possible,[15] but most experts view area defenses covering the entire country as a wholly different and vastly more difficult enterprise that will remain infeasible for the foreseeable future. Thus Ashton Carter concludes that the prospect of a perfect or near-perfect area defense system "is so remote that it should not serve as the basis of public expectations or national policy about ballistic missile defense."[16] The Union of Concerned Scientists similarly suggests that "total ballistic missile defense . . . is unattainable if the Soviet Union exploits the many vulnerabilities intrinsic to all the schemes that have been proposed thus far."[17] Former Secretary of Defense Harold Brown likewise warns that "technology does not offer even a reasonable prospect of a successful population defense," and suggests that the cost-ex-

12. See the "Fletcher Report," below, p. 305.
13. March 23 speech, below, p. 258.
14. "Fletcher Report," below, p. 325. An optimistic note is also sounded by physicist Robert Jastrow, who declares that American plans for defenses are "very promising" and asserts that America could deploy a defense in the early 1990s that could destroy at least 90 percent of incoming Soviet warheads. Robert Jastrow, *How to Make Nuclear Weapons Obsolete* (Boston: Little, Brown, 1985), pp. 48, 100. It should be noted, however, that others would argue that defenses must far surpass 90 percent efficiency to be effective.
15. See, for instance, the UCS, *Fallacy of Star Wars*, p. 12, granting the possibility of preferential defense of hard targets, such as missile silos.
16. Carter, *Directed Energy Missile Defense*, below, p. 253.
17. UCS, *Fallacy of Star Wars*, p. 151.

change ratio (the ratio between the cost of maintaining defenses and the cost of deploying countermeasures against them) for the active defense of urban areas would favor the offense by five or ten to one even if the defender only sought 50 percent survival of its urban building structures.[18] Even the "Hoffman Report" sounded a hesitant note: "In the long term [BMD] systems might provide a nearly leakproof defense against large ballistic missile attacks. However, their components vary substantially in technical risk, development lead time, and cost, and in the policy issues they raise."[19]

This criticism persuaded many SDI proponents to stop advocating comprehensive national population defenses, and to recommend instead more modest defensive schemes that would give the American population only marginal protection, or would protect only selected military targets, leaving the population completely unprotected. Such limited defenses are now the only variety that are seriously discussed by most military scientists and defense experts.

Nevertheless, President Reagan still endorses his original aim of building comprehensive population defenses that would make nuclear weapons "impotent and obsolete,"[20] and the public debate largely proceeds as if an impermeable "Astrodome" defense remains the official aim of the SDI. Thus two very different "Star Wars" proposals have emerged, dubbed "Star Wars I" and "Star Wars II" by defense analysts. Star Wars I embodies the President's plan for a comprehensive "Astrodome" population defense; Star Wars II encompasses the range of proposals for more limited defensive schemes. The existence of these two proposals has bifurcated the Star Wars debate into a surreal public dialogue focused on the merits and drawbacks of a Star Wars I "Astrodome" defense that defense professionals dismissed long ago, while these professionals debate limited Star Wars II proposals that are largely unknown to the public.

The desirability of both comprehensive Star Wars I and limited Star Wars II defenses is less controversial than is their feasibility: most analysts assume that defenses would be desirable if they were feasible. Nevertheless, some critics suggest that one or both varieties of defense are unnecessary, or are

18. Harold Brown, "The Strategic Defense Initiative: Defensive Systems and the Strategic Debate," *Survival*, Vol. 27, No. 2 (March/April 1985), pp. 56, 58. See also former Defense Secretary James Schlesinger in this volume (p. 15), who suggests that the cost-exchange ratio favors the offense by three to one, and concludes that "there is no serious likelihood of removing the nuclear threat from our cities in our lifetime—or in the lifetime of our children." "Rhetoric and Realities in the Star Wars Debate," pp. 20, 18. Sidney D. Drell and Wolfgang K.H. Panofsky have similarly concluded: "We see no practical prospect whatsoever of constructing a strategic defense that can—lacking prior *drastic* arms control restraints and reductions—enhance deterrence, much less render nuclear weapons impotent and obsolete." "The Case Against Strategic Defense: Technical and Strategic Realities," *Issues in Science and Technology* (Fall 1984), p. 64.
19. "Hoffman Report," below, p. 280.
20. March 23, 1983 speech, below, p. 258.

positively harmful, or both, and argue that defenses therefore should not be built even if they are judged to be feasible.

SDI advocates recommend defenses on both moral and strategic grounds. A comprehensive Star Wars I defense, advocates say, will defend the United States against Soviet nuclear blackmail, protect American society from destruction should war break out, and provide a more moral basis for American defense, by removing American dependence upon threats to destroy others' civilian populations.

Advocates recommend more limited Star Wars II defenses for a wide range of essentially different reasons. Such defenses, advocates suggest, can enhance deterrence by better protecting American strategic forces from attack; can protect American conventional forces from nuclear or perhaps even conventional missile attack, thus deterring both conventional war and nuclear escalation during conventional war; can reduce American casualties in an all-out nuclear war; can demonstrate American resolve, thus inducing the Soviets to bargain more generously in arms control and other negotiations; can diminish the value of Soviet ICBMs and SLBMs by diminishing their effectiveness, thus inducing the Soviets to bargain away these weapons more readily in arms control negotiations; can create uncertainties about the results of a nuclear war that may help to deter a Soviet attack; can deter a limited Soviet nuclear attack by forcing the Soviets to use a larger number of nuclear weapons, which the Soviets may shrink from doing for fear of further escalation; and can protect the United States from accidental or unauthorized Soviet missile launches, or from ballistic missile attacks by third countries. Finally, expressing a little-discussed but important argument for limited defenses, some advocates suggest that a thin population defense, if married to American first-strike counterforce weapons (such as the MX ICBM and the Trident D5 SLBM), can give the United States a meaningful first-strike counterforce capability against the Soviet Union. Defenses would thereby restore American nuclear superiority, re-establish the credibility of American nuclear threats, and extend American deterrence over Europe and other areas. This, these advocates suggest, would help to prevent war by deterring Soviet aggression.

Opponents of SDI answer by charging that some of the alleged benefits of defenses are simply illusory (such as the boost that the SDI would allegedly provide to arms control), while other alleged benefits are real but trivial (such as the protection that SDI would provide against accidental launches or third country missile attacks). SDI opponents also suggest that problems solved by defenses could be solved more easily in other ways—claiming, for instance, that American ICBMs can be better protected by making them mobile or by burying them in mountains than by building a BMD to protect them. Such opponents further note that American defenses will conjure up Soviet defenses, and that these Soviet defenses will complicate American military requirements in ways that may offset the benefits of American defenses. These critics argue, in essence, that the United States is just as well off militarily if neither superpower has defenses than if both possess them, as both eventually will if the SDI proceeds.

Some SDI opponents go on to warn that effective American defenses will do positive harm. Some suggest that defenses will fuel the arms race by forcing the Soviet Union to expand its offensive nuclear arsenal; in this view the SDI will promote arms racing, not arms control. These analysts also note that by forcing the amendment or abrogation of the 1972 ABM Treaty the SDI will destroy a cornerstone of the East-West arms control edifice, thus damaging the arms control process and, in so doing, further promoting arms racing.

Other analysts warn that defenses will directly raise the risk of war. They note that a transition from the current world of mutual assured destruction (MAD) to a defensive world (sometimes called "BAD": "both are defended")[21] could open "windows" of vulnerability and opportunity that one side might decide to close by launching a preventive war. For instance, they argue that the Soviets may feel compelled to attack American space-based systems before these systems become operational, if the Soviets fear that otherwise they will lose the race to deploy effective defenses. These critics note that such a war then could escalate.

Critics also warn that a BAD world might present situations in which the side striking first would gain the upper hand, thereby sparking a pre-emptive war.[22] This risk could arise, for instance, if both sides deployed defenses that could destroy the other side's missiles *and* its missile defenses. If so, only the side striking first would have an effective defense, creating a hair-trigger situation.

Such critics further warn that a BAD world probably would decay back into a MAD world, in the process opening more "windows" that would invite preventive war. Lastly, they argue that even limited American defenses, if married to effective American first-strike counterforce weapons, would raise the risks associated with unduly pre-emptive and offensive weapons, and thus

21. The term is from Robert J. Art, "The Role of Military Power in International Relations," in B. Thomas Trout and James E. Harf, eds., *National Security Affairs: Theoretical Perspectives and Contemporary Issues* (New Brunswick: Transaction, 1982), pp. 23ff, who uses "BAD" to abbreviate "both assured of defense."
22. On the risks of pre-emptive war see Thomas C. Schelling, *Arms and Influence* (New Haven: Yale University Press, 1966), pp. 221-59; and idem, *Strategy of Conflict* (Cambridge, Mass.: Harvard University Press, 1960), pp. 207-54.

would raise, rather than lower, the overall risk of war.[23] These critics note that ballistic missile defenses, arranged in such a fashion, would not be "defensive" in a strategic sense, but would rather be highly offensive: armed in this way, the United States could disable Soviet nuclear forces with a first strike and could then compel a Soviet surrender, if American defenses could provide an effective shield against surviving Soviet nuclear forces. This prospect could impel the Soviets to act boldly, perhaps dangerously, to restore their military strength. For instance, the Soviets might run high risks to gain new basing rights if these could restore their deterrent, as they did in 1962 by secretly moving missiles to Cuba, sparking the Cuban Missile Crisis. Alternately, the Soviets might even attack American defenses. Thus instead of deterring the Soviets, American defenses could elicit more Soviet aggressiveness.

These arguments are treated at length in the articles presented below, all of which originally appeared in *International Security*.[24] Fred S. Hoffman, who also chaired the Hoffman Panel, advances the case for the SDI in his essay on "The SDI in U.S. Nuclear Strategy." He notes that American interests are endangered in a MAD world by the incredibility of American nuclear threats, and by the nuclear anxiety that MAD engenders among the American public. He also argues that critics of the feasibility of defenses are measuring them against unreasonably high standards of performance. Even imperfect defenses could lower the Soviet Union's confidence in its attack, he suggests, or could induce the Soviets to avoid attacking targets that are located near large populations, which could save many lives by lowering the collateral damage that would ensue if a counterforce nuclear war were fought. He further argues that defenses will not significantly raise the risk of pre-emptive or preventive war, and that defenses would more probably serve to limit than to fuel the arms race.

The essays by James R. Schlesinger, Charles L. Glaser, Sidney Drell, Philip Farley, and David Holloway then present critical views of the SDI from various perspectives. Former Defense Secretary Schlesinger provides a skeptical assessment of the feasibility of defenses in his essay on "Rhetoric and Realities in the Star Wars Debate." In his view the SDI can succeed only if the Soviet Union cooperates by agreeing to restrain its offensive capabilities. Otherwise

---

23. On the dangers of unduly offensive military postures see Robert Jervis, "Cooperation Under the Security Dilemma," *World Politics*, Vol. 30, No. 2 (January 1978), pp. 167-214. On the origins of such postures and their consequences in 1914, see Steven E. Miller, ed., *Military Strategy and the Origins of the First World War* (Princeton, N.J.: Princeton University Press, 1984).

24. A useful volume that also addresses many of these issues is Ashton B. Carter and David N. Schwartz, eds., *Ballistic Missile Defense* (Washington, D.C.: Brookings, 1984).

defenses will fail because the Soviets can put effective penetration aids on their missiles, or can attack the American defense directly, or can circumvent the American defense with bombers and cruise missiles. The cost-exchange ratio between offense and defense favors the offense by about three to one, Schlesinger notes, which will allow the Soviets to neutralize American defenses at much lower cost to themselves. This will allow the Soviets to drive up the cost of American defenses, which could lead future American leaders to finance missile defenses by imposing unwise cuts on American conventional forces. He concludes that "there is no realistic hope that we shall ever again be able to protect American cities," and argues that the American people should adjust to living with the kind of vulnerability that other societies have long endured.

In his essay on "Why Good Defenses May Be Bad," Charles L. Glaser explores the desirability of perfect "Star Wars I" defenses. Such defenses, he concludes, are not unambiguously bad, but neither are they unambiguously good. On the positive side, effective defenses could reduce the damage that would follow if war broke out. However, good defenses are likely to degrade eventually, leading the world back from BAD toward MAD, and this transition could entail the risk of preventive war. The risk of pre-emptive war ("crisis instability") would also appear in a world of good defenses, if forces on either side were at all vulnerable to surprise attack—as they probably would be. Lastly, Glaser argues that good defenses might raise the risk of conventional wars, by removing the risk of nuclear punishment for conventional aggression.

Glaser then explores the desirability of more limited "Star Wars II" defenses in his essay asking "Do We Want the Missile Defenses We Can Build?" in which he outlines and evaluates nine principal arguments that have been advanced for partial or limited "Star Wars II" defenses. While conceding these arguments some validity, he finds none persuasive. In each case, he suggests, the benefits advertised would not in fact be provided by defenses, or they would not have much value, or they could be better provided by other means than missile defenses. Glaser also notes that the Soviets would probably deploy defenses once the Americans abrogated or amended the ABM Treaty, and these Soviet defenses could weaken the effectiveness of the British, French, and Chinese nuclear forces, and also might diminish American offensive counterforce capability.

The essay by Drell, Farley, and Holloway, "Preserving the ABM Treaty: A Critique of the Reagan Strategic Defense Initiative," is a condensation of a widely-noted monographic overview and assessment of the SDI that doubted both the feasibility and the desirability of defenses. The authors conclude their detailed survey of possible boost-phase intercept arrangements by arguing that none show technical promise, and all could probably be defeated by Soviet countermeasures. They further argue that the pursuit of defenses will probably fuel the East-West arms race, and the deployment of effective defenses may raise the risk of pre-emptive war.

Finally, David S. Yost explores European views on BMD in his essay on "Ballistic Missile Defense and the Atlantic Alliance." He concludes that America's NATO allies will oppose American defensive programs that require the amendment or abrogation of the 1972 ABM Treaty, because many Europeans believe that this Treaty has stabilized both the arms race and East-West relations generally. Limited American defenses aimed at protecting American ICBMs will be less controversial in Europe than defenses protecting American population, Yost suggests, but he concludes that even limited defenses may draw European opposition. Europeans will oppose American population defenses more strongly, he suggests, on grounds that these would undermine mutual deterrence, which many Europeans believe is peace-enhancing; would conjure up Soviet defenses that would weaken the independent French and British deterrents and NATO's other theater nuclear forces; would fuel the East-West arms race; and would foster the self-deterrence of the United States by constraining America's ability to execute limited number options.

In this volume we also present five government documents that figure prominently in the SDI debate. As noted above, the SDI was announced by the President in a speech to the nation on March 23, 1983, and we reprint the relevant sections of that speech (found below as Appendix A to Ashton Carter's study of *Directed Energy Missile Defense*).

In addition, we reprint the official unclassified summaries of the "Hoffman Report" and the "Fletcher Report" which the President commissioned after his March 23 speech to explore the strategic desirability and feasibility of pursuing strategic defenses. These two documents present the official case for the SDI, and they remain the most authoritative explication of the official reasoning behind it. Though difficult to read, they bear close study by those seeking the Reagan administration rationale for the SDI, and administration plans for its future development.

The debate over the technical feasibility of ballistic missile defenses was joined in April 1984 with the release of Ashton Carter's study of *Directed Energy Missile Defense*, which we also reprint below. This study, written under the auspices of the Congressional Office of Technology Assessment on the basis of full access to classified data, stirred a major controversy by presenting a deeply skeptical view of the feasibility of ballistic missile defenses. It remains one of three prominent unclassified technical criticisms of the feasibility of the SDI. (The others are the Drell, Farley and Holloway study condensed in this volume, and a study by the Union of Concerned Scientists, *Space-Based Missile Defense*.)[25]

Finally, we present the text of the 1972 ABM Treaty (found below as Appendix B of Ashton Carter's *Directed Energy Missile Defense* study), which stands today as a major legal obstacle to the SDI, and whose abrogation the SDI will ultimately require.

The meaning of the 1972 Treaty already lies at the heart of several disputes in the SDI debate. One controversy involves the legality of the American SDI testing and development program: the Reagan administration claims that Agreed Statement D contains language that allows the testing of BMD components, while most other American analysts (and the Soviet Union) argue that Article 5, Section 1 flatly forbids the development of BMD components, and that nothing in Agreed Statement D overrides this proscription.[26] Another controversy surrounds the legality of the Soviet radar near Krasnoyarsk, which Americans claim is an early warning radar and is therefore proscribed by language in Article 6 forbidding the deployment of early warning radars. The Soviets counter-claim that their radar has space tracking capability, and is thus permitted by language in Agreed Statement F permitting the deployment of space tracking radars.[27] This answer has failed to satisfy most American analysts—including most SDI opponents. As the SDI program collides more directly with this Treaty in the future, further controversies of Treaty interpretation can be expected to emerge, and the fate of the SDI will depend in part on the way these disputes are resolved.

25. The UCS study was released in March 1984, and is reprinted in revised form in: Union of Concerned Scientists, *The Fallacy of Star Wars*, pp. 39-176. Another important study of Star Wars technology is Office of Technology Assessment, *Strategic Defenses: Ballistic Missile Defense Technologies and Anti-Satellite Weapons, Countermeasures, and Arms Control* (Princeton, N.J.: Princeton University Press, 1986).

26. A critical assessment of the new Reagan administration position is Alan B. Sherr, *A Legal Analysis of President Reagan's "New Interpretation" of the Anti-Ballistic Missile Treaty of 1972* (Boston: Lawyers Alliance for Nuclear Arms Control, 1986, a special monograph available through LAFNAC, 43 Charles St., Boston). See also Charles Mohr, " 'Stars Wars' Dispute," *New York Times*, October 17, 1985, p. A4; and Gerard C. Smith, letter to the editor, *New York Times*, October 23, 1985, p. A22.

27. On this controversy see James Schear, "Arms Control Treaty Compliance: Buildup to a Breakdown?" *International Security*, Vol. 10, No. 2 (Fall 1985), pp. 154-62.

*The Star Wars Controversy*

# The SDI in U.S. Nuclear Strategy

*Fred S. Hoffman*

## Senate Testimony

**A**s we approach the second anniversary of President Reagan's speech announcing the SDI, it is useful to review the development of the issue. Critics and supporters alike now recognize that the central question concerns the kind of R&D program we should be conducting. Virtually no one on either side of the issue, here or among our allies, contests the need for research on the technologies that might contribute to a defense against ballistic missiles, and it is clear that the Administration does not propose an immediate decision on full-scale engineering development, let alone deployment of ballistic missile defenses.

Nevertheless, the issue continues to occupy a dominant place in discussions of national security issues and arms negotiations, far out of proportion to its immediate financial impact (significant as this is), to its immediate implications for existing agreements (current guidance limits the R&D to conformity with them), and to its near-term impact on the military balance. Reactions by the public and media in this country and among our allies, as well as the public response by Soviet leaders, suggest that the President's speech touched a nerve. Such extreme reactions to a program that has such modest immediate effect suggests that the President's initiative raises basic questions about some deep and essential troubles with the drift of NATO declaratory and operational strategy for the last 20 years, and about the direction in which we need to move during the next 20 years. The debate has only ostensibly been about the pros and cons of spending next year's funds on research and development. That the basic issues have been largely implicit is unfortunate. Entrenched Western opinion resists rethinking a declaratory strategy that has stressed a supposed virtue in U.S. vulnerability. And the Soviets have been campaigning furiously to aid a natural Western resistance to change. The Soviet campaign is also natural since in the 20-year

This statement was made by Fred S. Hoffman before the Subcommittee on Strategic and Theater Nuclear Forces of the U.S. Senate Armed Services Committee on March 1, 1985. It is a result of collaboration with Albert Wohlstetter and other colleagues at Pan Heuristics. Fred Hoffman is solely responsible for the statement in its present form.

*Fred Hoffman is Director of Pan Heuristics, a Los Angeles-based policy research group. He was director of the study group that prepared the report "Ballistic Missile Defenses and U.S. National Security" in October 1983 for the Future Security Strategy Study (generally known as the "Hoffman Report").*

*International Security,* Summer 1985 (Vol. 10, No. 1) 0162-2889/85/013-12 $02.50/1

period in which the West has relied on threats of Mutual Assured Destruction, the Soviets have altered what they call the "correlation of forces" in their favor.

The orthodoxy reflected in the SALT process and in much of the public discussion of the SDI is that of Mutual Assured Destruction (MAD)—a doctrine that holds that the only proper role of nuclear weapons on both sides is to deter their use by the other side, and that they must perform this role through the threat of massive and indiscriminate attacks on cities, designed to inflict the maximum destruction on the adversary's civilian population. On this view, any use of nuclear weapons is and *should* be clearly suicidal. Anything that interferes in any measure with the other side's ability to inflict "assured destruction" is "destabilizing"—in crises it is supposed to induce preemptive attack and, in the long-term military competition, a "spiralling nuclear arms race" with unlimited increases in the potential for indiscriminate destruction on both sides. MAD was the Western, though not the Soviet, strategic foundation for the ABM Treaty and the SALT offense agreements. It is largely unconscious dogma dominating the media discussions of nuclear strategy, SDI, and arms agreements.

Some who advocate this policy like to think of it as not a policy, but a "fact." A supposedly unalterable fact of nature. There is a grain of truth and a mountain of confusion in this assertion. The grain is the unquestioned ability of nuclear weapons to inflict massive, indiscriminate, and possibly global destruction. The mountain is the conclusion that this is the way we *should* design and plan the use of nuclear forces, and even more important, the assumption that this is the way the Soviet Union *does* design and plan the use of its nuclear forces. The prescription for our own strategy and the assumption about Soviet strategy are not unalterable facts of nature but matters of policy choices in each country. The contrasting U.S. and Soviet choices brought about the relative worsening of the U.S. position.

This is not the place for a detailed critique of MAD, but a summary of its principal deficiencies is essential to assess the potential role for defenses in our strategy. A central point on which most critics and supporters of SDI agree is that the assessment of defenses depends critically on what you want them to do. And what we want them to do depends on our underlying strategy.

MAD as a strategy might have something to recommend it (not nearly enough in my view) if the tensions between the Soviet Union and the U.S. were restricted to the threat posed by nuclear weapons. Relations between the United States and the Soviet Union have not been dominated by the possibility of border conflicts between the two countries or the fear of invasion by the other. Rather the post-World War II military competition arose from the desire of the Soviet Union to dominate the countries on the periphery of its Empire and the desire of the United States to preserve the independence of those countries. No nuclear strategy can long ignore the role of nuclear weapons in managing this underlying conflict of interests, nor can it ignore the asymmetry in the geostrategic situations of the two countries. The U.S. guarantees a coalition of independent countries against nuclear attack by the Soviet Union. We have also affirmed in NATO strategy that we would respond to overwhelming nonnuclear attack with whatever means proved to be necessary to defeat such an attack. Do we now mean to exclude a U.S. nuclear response in both these cases? What if the Soviets launch a nuclear attack, but one directed solely at our allies and which avoids any damage to the U.S.? How long can an explicitly suicidal nuclear response remain a credible threat in the eyes of our allies or the Soviet Union?

On the Soviet side, there is abundant evidence that they have never accepted MAD as a strategic basis for their military programs (in contrast to their rhetoric designed to influence Western opinion). They continue to maintain and improve, at massive cost, air defense forces, ballistic missile defenses, and protective measures for their leadership and elements of their bureaucracy intended to ensure the continuity of the Soviet state. Their military strategy has increasingly focused on qualitative improvements to their massive forces intended to give them the ability to win a quick and decisive military victory in Europe using their nonnuclear forces to attack our theater nuclear forces as well as our conventional forces while deterring the use of our nuclear forces based outside the theater. Deterring a suicidal use of nuclear force is not very difficult. They have steadily improved the flexibility of their own nuclear forces in what Lt. Gen. William Odom, a leading professional student of Soviet military thought, has called their "strategic architecture." They design that architecture for the pursuit of Soviet political goals as well as military operations.

They clearly wish to dominate on their periphery and to extend their influence over time. By creating conditions that weaken ties between the United States and other independent countries they serve both ends. They clearly prefer to use latent threats based on their military power, but have shown themselves willing to use force either directly or indirectly and in a degree suited to their political goals. They regard wars, especially long and large wars, as posing great uncertainties for them. Because they cannot rule out the occurrence of such wars, they attempt to hedge against the uncertainties in their preparations. There is no reason to suppose that their plans for the use of nuclear weapons are inconsistent with their general approach to military planning.

From the Soviet point of view, Western public espousal of MAD is ideal. Western movement away from such a strategy based on indiscriminate and suicidal threats would increase the difficulty of Soviet political and strategic tasks. The consequences of Western reliance on threats to end civilization can clearly be seen in the increasing level of Western public anxiety about a nuclear cataclysm. While the incumbent governments among our allies have successfully resisted coercion, trends in public opinion and in the positions of opposition parties give us little reason for comfort. In the U.S. as well, public attitudes reflected in the freeze movement will make it increasingly difficult to compete with the Soviets in maintaining parity in nuclear offensive forces. The Soviet leaders have reason to believe that the West will flag in its efforts to make up for the ground it lost in the quantitative offense competition. Proponents of MAD have also impeded and delayed qualitative improvements in the name of "stability." Finally, a broad and increasing segment of the public is questioning the morality and prudence of threats of unlimited destruction as a basis for our strategy.

The specific relevance of MAD to the assessment of SDI is best illustrated in the assertion by critics of the hopelessness of the SDI's task. They observe that if even one percent of an attack by 10,000 warheads gets through the defenses, this means 100 nuclear weapons on cities and that for more likely levels of defense effectiveness, the ballistic missile defenses would be almost totally ineffective in protecting cities. They generally leave implicit the remarkable assumption that the Soviets would devote their entire (and in this example, presumably undamaged) missile force to attacks on cities, ignoring military targets in general and not even making any attempt to reduce our retaliatory blow by attacking our nuclear offensive forces. If the Soviet attack, for example, devoted ⅔ of their forces to attacking military targets, then only ⅓ of the warheads surviving a defense like a boost phase intercept system would be aimed at cities. In one particularly remarkable exercise of this sort, the authors concluded that defenses would cause the Soviets to concentrate their forces on our cities, *even if their attack were to result in nuclear winter.*

Such a bizarre assumption suggests the absence of serious thought about the objectives that might motivate Soviet leaders and military planners if they ever seriously contemplated the use of nuclear weapons. Whatever we may think of the heirs of Karl Marx, the followers of Lenin, and the survivors of Stalin, nothing in their background suggests suicidal tendencies. Certainly, their strictest ideological precepts call for the preservation of Soviet power and control. Neglect of the actual motivation of our adversaries is particularly strange in a strategic doctrine that professes to be concerned with deterrence. Despite the fact that deterrence is in the mind of the deterred, those who espouse MAD rarely go beyond the assumption that the attacker's purpose is to strike preemptively before he is attacked.

MAD doctrine takes it as axiomatic that to deter such a Soviet attack we must threaten "assured destruction" of Soviet society. A consequence of this view is that only offensive forces can directly contribute to deterrence. Defensive forces can contribute only if they are useful in protecting our missile silos and the "assured destruction" capability of the missiles in them. Beyond this ancillary role in deterrence, MAD relegates defenses along with offensive counterforce capability and civil defenses to the role of "damage limiting" if deterrence fails. But since our damage-limiting capability diminishes Soviet assured destruction capability, eliciting unlimited Soviet efforts to restore their deterrent, MAD dismisses damage limiting (and with it defenses) as pointless and destabilizing.

To recapitulate, acceptance of MAD doctrine implies for SDI:

• Defenses must be essentially leakproof to be useful;
• Defenses can at best serve an ancillary role in deterring attack;
• Defenses that reduce civilian damage are inherently destabilizing.

Even a leakproof defense would not satisfy the last condition. Together these three conditions implied by MAD are an impenetrable barrier—a leakproof defense against SDI. Since I have indicated above reasons for rejecting MAD as a doctrine, I believe we should reexamine each of these in turn.

Most important, if defenses must be leakproof to be useful, then the odds of success for the SDI R&D program are much lower than if lesser levels of effectiveness can contribute to our security objectives. The record is replete with instances of faulty predictions about the impossibility of technological accomplishments by those with the highest scientific credentials, and we should view current predictions about the impossibility of effective ballistic missile defenses in the perspective of that record. Nevertheless, if everything in a complex and diverse R&D program must work well to derive any benefit, the odds of success will be low and the time required very long.

The critics compound the problem further by demanding that the SDI research program prove and guarantee at its outset that the defenses that might ultimately be developed and deployed will be able to deal with a wide variety of ingenious, but poorly specified and, in some cases, extremely farfetched countermeasures. Critics can produce countermeasures on paper far more easily than the Soviets could produce them in the field. In fact, the critics seldom specify such "Soviet " countermeasures in ways that seriously consider their costs to the Soviet Union in resources, in the sacrifice of other military potential, or the time that it would take for the Soviets to develop them and incorporate them into their forces. The countermeasures suggested frequently are mutually incompatible.

If, instead, we replace MAD with a view of deterrence based on a more realistic assessment of Soviet strategic objectives, we arrive at a radically different assessment of the effectiveness required for useful defenses and of the appropriate objectives of the SDI R&D program. The point of departure ought to be reflection on the motives that might induce Soviet leaders and military planners to contemplate actually using nuclear weapons. The test of deterrence would come if we and the Soviet Union found ourselves in a major confrontation or nonnuclear conflict.

In such circumstances, Soviet leaders might find themselves facing a set of alternatives all of which looked unpleasant or risky. If, for example, they lacked confidence in their ability to bring a nonnuclear conflict to a swift and favorable conclusion, they might consider ensuring the futility of opposing them by a militarily decisive use of nuclear weapons. A decisive nuclear attack in this sense might or might not have to be "massive," in the sense of "very large." Its primary motivation would be the destruction of a set of general purpose force targets sufficient to terminate nonnuclear resistance. If Soviet leaders decided that the gains warranted the risks, they would further have to decide whether to attack our nuclear forces or to rely on deterring their use in retaliation. The extent and weight of such an attack would be a matter the Soviet leaders would decide within the context of a particular contingency, based on their assessment of our probable responses.

The alternative risks they would face would be the prospect of nuclear retaliation to an early nuclear attack on one hand; on the other hand, those of gradual escalation of a nonnuclear conflict in scope and violence with the ultimate possibility of nuclear conflict in any case. In either case their primary concern would be to achieve military victory while minimizing the extent of damage to the Soviet Union and the risk of loss of Soviet political control.

Their targets would be selected to contribute to these goals. Wholesale and widespread attacks on civilians would not contribute but would only serve to ensure a similar response by the large nuclear forces remaining to us even after a relatively successful Soviet counterforce attack. And this does not even take account of the possibility that, should they launch a massive attack on cities, that might trigger nuclear winter, making our retaliation irrelevant.

The magnitude of collateral damage to Western civilians from a Soviet attack with military objectives would depend on the extent of Soviet attack objectives and the weight of attack required to achieve those objectives. Like us, they have been improving the accuracy of their weapons and reducing their explosive yield. As this trend continues, motivated by the desire for military effectiveness and flexibility in achieving strategic objectives, they will become increasingly capable of conducting effective attacks on military targets while limiting the damage to collocated civilians and while remaining below the threshold of uncertainty of global effects that would do serious harm to themselves. At present, a Soviet attack on a widespread set of general purpose force and nuclear targets would undoubtedly cause very great collateral damage but could be conducted so as to leave the bulk of Western civil society undamaged and to remain safely under the threshold for a major climatic change affecting the Soviet Union.

We should judge the utility of ballistic missile defenses in the light of their contribution to deterring such attacks and their ability to reduce the collateral damage from such attacks if they occur. The relevant question for the foreseeable future is not whether defenses should replace offensive weapons but whether we should rely exclusively on offensive weapons or whether a combination of militarily effective and discriminating offense and defenses will better meet our strategic requirements for deterrence and limiting damage.

This change in the criterion by which we judge defenses from the one imposed by MAD has profound consequences for the level of effectiveness required of defenses, for the treatment of uncertainty about defense effectiveness and for the terms of the competition between offense and defense. Instead of confining the assessment to the ability of *defense* to attain nearly leakproof effectiveness, a realistic consideration of the role of defense in deterrence recognizes that an *attacker* will want high confidence of achieving decisive results before deciding on so dangerous a course as the use of nuclear weapons against a nuclear-armed opponent. Analysis will show that defenses with far less than leakproof effectiveness can so raise the offensive force requirements for attacks on military target systems that attacks on limited sets of critical targets will appear unattractive and full-scale attacks on military targets will require enormous increases in force levels and relative expense to achieve pre-defense levels of attack effectiveness and confidence in the results. Because of an attacker's desire for high confidence in a successful outcome, he must bear the burden of uncertainty about defense effectiveness and is likely to bias his assumptions toward overestimating it. This is particularly important for his willingness to rely on sophisticated countermeasures such as those liberally assumed by critics of the SDI.

In addition, the technical characteristics of the defenses that are contemplated in the SDI would pose particularly difficult problems for a Soviet attack planner. A particularly prevalent and misguided stereotype in current discussion contrasts "an impenetrable umbrella defense over cities" with a hard-point defense of silos as though these were the only choices. Reality offers more types of targets and defenses than are dreamt of in this "city-silo" world. The preceding discussion has attempted to show the importance of general purpose force targets in motivating a possible nuclear attack. With respect to the characteristics of future defenses, the technologies pursued under the SDI have the potential for a multi-layered defense that begins with boost phase intercept, continues in the exoatmospheric mid-course phase, and terminates with systems for intercept after reentry into the atmosphere. Each successive layer is more specific in terms of the target coverage it provides, but none is effectively so circumscribed that it is properly described as a point defense.

This means that defenses can simultaneously protect several military targets and can simultaneously protect military targets and collocated population. The problem this poses for the attacker is that he cannot, as he could against point defenses, economize in his use of force by predicting which defenses protect which targets and planning his attack precisely to exhaust the defense inventory (even assuming that he can afford to forgo attacks on some military targets). Moreover, to the extent that there is redundancy in military target systems (or in their possible unknown locations), and the defense can identify the targets of particular enemy warheads in the mid-course, or terminal, phase, the defense can defend targets "preferentially." To have an expectation of destroying the desired fraction of a preferentially defended target system in the absence of information about the defense allocation of its resources, the attacker would have to treat each target as defended by a disproportionate share of the defense resources. This greatly enhances the competitive advantage of the defense.

Another implication of the foregoing discussion is that defenses do not come in neat packages labelled "protection of military targets" and "protection of civilians." Warheads aimed at military targets will, in general, kill many collocated civilians and defenses that protect against such attacks will reduce civilian casualties. Again, in contrast to the kind of nightmare attack assumed by MAD theorists, when we consider more realistic Soviet attacks, effective but far from leakproof defenses can protect many civilians against collateral damage. If, moreover, a Soviet attack planner knows that we will protect collocated military targets more heavily and he must choose between attacking similar targets some of which are collocated and others of which are isolated, he will opt for the isolated targets if he wishes to maximize his military effectiveness (the reverse of what is generally assumed by critics of defenses). When we understand that the problem of protecting civilians is primarily the problem of dealing with collateral damage, it becomes clear that we do not need leakproof defenses to achieve useful results. The more effective the defenses, the greater the protection, but there is no reason to expect a threshold of required effectiveness.

Another charge levied against defenses is that they are "destabilizing." The prospect of leakproof defenses is allegedly destabilizing because they present an adversary with a "use it or lose it" choice with respect to his nuclear offensive capability. Defenses with intermediate levels of effectiveness are also held to be destabilizing because they work much better if an adversary's force has previously been damaged in a counterforce strike, intensifying incentives for preemption in a crisis. The first charge hardly needs response. Leakproof defenses, if they ever become a reality, are unlikely to appear on short notice or all at once. The Soviets know that they can live under conditions of U.S. nuclear superiority without any serious fear of U.S. aggression because they have done so in the past. In fact, they survived for years under conditions of U.S. monopoly. They can also and are pursuing defense themselves, and undoubtedly will continue. The notion that they would have no choice for responding to U.S. defenses other than to launch a preventive war is not a serious one.

The crisis stability argument is also a weak one. The analysis generally advanced to support it is incomplete and inadequate to determine the strength of the alleged effect because it is unable to compare meaningfully the importance of the difference between striking "first" and striking "second" with the difference between either and "not striking at all." Such analyses ignore, therefore, one of the most important elements of the theory of crisis stability contained in the original second-strike theory of deterrence. Moreover, since defenses would contribute to deterrence by denying achievement of Soviet attack objectives, it would at least be necessary to determine the *net* effect of strengthening deterrence with the effect of intensifying incentives to preempt and this the analysis cannot do. Finally, the argument focuses on the wrong culprit. The grain of relevance in the argument is its identification of the problems presented by vulnerable offensive forces. It then superimposes partially effective defenses on the vulnerable offensive forces and concludes that the defenses are destabilizing. But it would be a virtuoso feat to design SDI-type, multi-layered defenses that would not, willy-nilly, reduce the vulnerability of the offensive nuclear forces, and it would certainly be possible by proper design to reduce that vulnerability far enough to eliminate the so-called destabilizing effect while realizing the other benefits of defenses.

Turning next to the effect of introducing defenses on the long-term military competition, we once again encounter the charge that defenses are destabilizing. A common assertion is that the offense will always add force to overwhelm the defense with the net result of larger offensive forces and no effective protection. This stereotyped "law of action and reaction" that flourished in the 1960s and early 1970s was also supposed to imply that if we reduce defenses, the Soviets will inevitably reduce their offenses. It has no basis in theory, and it has been refuted by reality. The United States drastically cut its expenditures on strategic defense in the 1960s and 1970s while the Soviets tripled their expenditures on strategic offense. After we abandoned any active defense against ballistic missile attacks even on our silos, the Soviets deployed MIRVs for the first time and increased them at an accelerating rate. The action-reaction theory of the arms race led to some of our worst intelligence failures in the 1960s and early 1970s.

The effects of U.S. defenses on the incentives governing Soviet offensive forces are likely to depend on the terms of the competition as they are perceived by each side. The incremental increase in effort or force size by the offense required to offset an increment of effort or force in the defense (the "offense-defense leverage") is particularly important in determining the character of the long-term response by the offense to the introduction of defenses. The leverage in turn as suggested by the foregoing discussion is extremely sensitive to the strategic criterion we adopt, the specific targets being protected, and the characteristics of the defenses. When we assess the role of defense within a strategic framework like the one outlined above and take account of the defense characteristics that could result from the technologies pursued under the SDI, the leverage is radically shifted in favor of the defense compared with the results suggested by evaluations within the MAD doctrine and under the misleading stereotype of defense characteristics prevalent in public discussion.

More fundamentally, ballistic missiles now offer an attack planner a degree of simplicity and predictability associated with no other weapon system. Planning a ballistic missile attack is much more like building a bridge than it is like fighting a war. The distinguishing characteristic of warfare, an active and unpredictable opponent, is missing. Introduction of defenses will change that radically and the change will reduce the strategic utility of ballistic missiles, now the keystone of U.S. and Soviet military forces. President Reagan called for defenses to make ballistic missiles "impotent and obsolete." Defenses of relatively moderate capability can make them obsolete to a military planner long before they are impotent in terms of their indiscriminate destructive potential.

If this point is reached or foreseen, the incentives governing negotiations over arms agreements will be fundamentally changed in a direction offering much more hope of agreement on substantial reductions in forces on both sides. Moreover, the growing problem of verification of limitations on nuclear offensive systems makes it increasingly difficult to foresee the possibility of agreeing to sizable reductions in the absence of defenses. One of the contributions of defenses can be to increase the ability to tolerate imprecision in the verifiability of arms limitations.

The point of view advanced here has major implications for the conduct of the SDI R&D program as well as for the criteria we should apply to evaluating its results when we approach the decision for full-scale engineering development and deployment. If we adopt the MAD view of the role and utility of defenses, and require essentially leakproof defenses or nothing, then we will conduct the SDI on what has been called the "long pole" approach. We will seek first to erect the "long pole in the tent," that is, we will devote our resources to working on those technical problems that are hardest, riskiest, and that will take longest, and we will delay working on those things that are closest to availability. The objective of this approach will be to produce a "fully effective" multi-layered system or nothing. Unfortunately such an approach increases the likelihood that we will in fact produce nothing, and it is certain that it delays the date of useful results into the distant future.

If instead, as argued here, we believe that defenses of moderate levels of capability can be useful, then we will conduct SDI in a fashion that seeks to identify what Secretary Weinberger has called "transitional" deployment options. These may be relatively near-term technological opportunities, perhaps based on single layers of defenses or on relatively early versions of technologies that can be the basis for later growth in system capability. Or if they are effective and cheap enough, they might serve for a limited lifetime against early versions of the Soviet threat while the SDI technology program continues to work on staying abreast of qualitative changes in the threat. Such an approach would incorporate a process for evaluating the transitional deployment options in terms of their effectiveness, their robustness against realistic countermeasures, their ability to survive direct attack on themselves, their cost, and their compatibility with our long-term strategic goals. Such an approach represents the best prospect for moving toward the vital goals enunciated by President Reagan two years ago.

# Rhetoric and Realities in the Star Wars Debate

**D**uring the mid-1960s when I was at Rand, the initial deployment of the Soviet ABM system caused a good deal of concern. The perplexing question of how to assure penetration of that system was argued and re-argued. The final judgment—the canonical solution of Secretary McNamara—was that the United States would counter the Soviet ABM by greatly expanding the number of warheads that we could throw against the Soviet Union. Indeed, by the time I left Rand, we were already talking about some 50,000 warheads to overcome Soviet defenses. In other words, we were going to expand our offensive capabilities geometrically to deal with Soviet defense. That was the initial American reaction to the problem of ballistic missile defense. Therein also, more than coincidentally, lay the birth of the MIRV. The way we were going to add large numbers of warheads was to fractionate the payload of our missiles. Several years later we proceeded to do precisely that—for entirely different and perhaps more dubious reasons. In all this there is a moral to be learned, which I shall attempt to develop later on.

In the late 1960s Secretary McNamara was informed by his President, Lyndon Johnson, that contrary to the Secretary's own advice the United States was going ahead with its own ABM system—then known as the Sentinel. The Sentinel would provide a thin-area defense designed to stop a limited number of warheads coming into the United States. It was, I think you will all recall, the period when the Red Chinese (more recently known as the People's Republic of China) were supposedly on the march under the malevolent guidance of Lin Piao. Supposedly the Chinese were preparing to encircle the cities from the rural areas, which we interpreted to mean they were going to destroy the industrial nations through guerilla warfare—as in Vietnam. A good deal of apprehension was expressed at the time by Secretary Rusk and by the President about the Chinese threat. What would happen when this billion people were armed with nuclear weapons? That small

This paper was presented as a speech at the National Security Issues Symposium at the MITRE Corporation on October 25, 1984, and appears in the Symposium Proceedings.

_James Schlesinger was U.S. Secretary of Defense 1973–75 and Secretary of the Department of Energy 1977–79. He is presently Counselor at the Center for Strategic and International Studies at Georgetown University._

_International Security_, Summer 1985 (Vol. 10, No. 1) 0162-2889/85/03-10 $02.50/1

prospective Chinese capability turned out to be the principal argument for thin-area defense.

When President Nixon came into office, our ABM potential was carefully reexamined—and the Sentinel system was transformed into the Safeguard system. That transformation occurred as a result of a study led by David Packard which concluded that the thin-area defense did not really serve our purposes and that the appropriate objective was to defend our missile fields. That was what the Safeguard would do. Unfortunately, it was exactly the same hardware that had previously been intended for a substantially different mission, and it was not particularly suitable for the new mission.

At that time I was at the old Bureau of the Budget, where one of my duties was to supervise the flow of water projects, dams, and post offices that would lubricate the creation of the Safeguard system. I was also in charge of reviewing the '71 Army budget. The Safeguard system turned out to be the only weapon system development in my experience in which staggering overruns were already revealed *prior* to the inception of work. Those experiences provided much of my background in ballistic missile defense. Later, as Secretary of Defense, even after the signing of the Moscow ABM Treaty in 1972, I strongly supported steady research and development activities in BMD, despite Congressional opposition. I want to stress these credentials: I have no objection, and have had no objection, to a vigorous R&D program— which at the moment is the principal activity of the Strategic Defense Initiative. I want especially to stress that I have no objection to such an R&D program in light of the cautionary comments that will follow.

My only other association with ABM was being briefed as part of the Scowcroft Commission on the Fletcher and Hoffman reports. I believe I should confess that I may be the source of that rather rough-and-ready estimate of one trillion dollars for the complete SDI. That was simply an extrapolation based upon the old days of the Safeguard system and the cost overruns I observed at that time.

Let me turn now from these historical reminiscences to what is hopefully called the "real world of today." I go back to the President's speech of March 1983—in particular, the suggestion in that speech that some day nuclear weapons would be rendered impotent and obsolete, and that American cities might be safe from nuclear attack. The speech did not provide a complicated assessment of the role defense might play in strengthening deterrence. It held out to American citizens the unqualified hope that they need not forever live with the nuclear threat over their heads. Through the vigor of American technology, some day—even if not until the 21st century—our cities would once again be safe from nuclear attack, as they have been for most of the nation's history. In that lay the political appeal of the speech (and we should understand that its appeal is fundamentally political), for invulnerability to nuclear assault is what the American public believes is going to be achieved.

Since that time there has been a dramatic change in the nature of the argument. In the follow-up to the President's speech, a rather loose rhetoric developed within the Administration in which the most fervent supporters of the SDI began to speak of *the immorality of deterrence*. Let me make this admonition clear. Within the Air Force, within the Administration, and within the society as a whole, the justification for strategic defense should never be based on assertions regarding the "immorality" of deterrence. For the balance of our days, the security of the Western world will continue to rest on deterrence. Those were—and are—reckless words.

There is no realistic hope that we shall ever again be able to protect American cities. There is no leak-proof defense. Any defense is going to suffer some erosion at best. An effective opponent will develop defense suppression techniques and will punch a hole through any space-based defense that is deployed. (I cannot go into more detail on that.) Moreover, even if we were discussing a hypothetically leak-proof defense, we would need to bear in mind that there are means of nuclear weapons delivery other than by ballistic missile. For a nation that has very limited air defense capabilities, compared for example to the Soviets, we should recognize the relative ease with which our defenses can be penetrated by air-breathing vehicles. If we were ever to deploy ballistic missile defense, it would impose upon us the corresponding costs of developing or attempting to develop comparably effective air defenses.

I point this out here at MITRE, because the United States Air Force has long argued that air defense systems are penetrable and will always be penetrable. If that were not the case, we would not be expending the re- sources we presently are on the B-1. The United States has long seacoasts. In contrast to the Soviet Union, the bulk of our population lies along the coast. We are also the very nation that has led the way in the development of the sea-launched cruise missile (SLCM). There is no foreseeable way that we can preclude such missiles' impacting on our cities—even if we had a perfect ballistic missile defense. The fact that we have moved ahead with SLCMs is perhaps analogous to our early movement into MIRVs. Since we are more vulnerable to SLCMs, it suggests there may be an absence of coordination between the development of the ballistic missile defense and of the offensive weapons.

There is no serious likelihood of removing the nuclear threat from our cities in our lifetime—or in the lifetime of our children. If those cities are going to be protected, they will be protected either through effective deterrence or through the forbearance of those on the other side. And it is for that reason that cries of the immorality of deterrence are both premature and pernicious.

Only the United States and the Soviet Union have in their historic roles been so powerful that defense against all threats might appear to be within their grasp. This is in sharp contrast to the European experience. No European power has been in a position to believe that by its own unaided efforts it could unilaterally provide perfect defense. The historic experiences of the U.S. and Russia have been different. These two societies might reasonably hope to achieve defense or deterrence unilaterally. Traditional attitudes in both the United States and the Soviet Union have stressed this unilateralism—in contrast to the presuppositions of our allies.

We in the United States have been even more inclined than the Soviets to believe in the unilateral capacity to achieve *perfect* defense. Russia, both Soviet and Imperial, has been repeatedly invaded, has suffered grievous damage, and has survived largely through its own efforts. But the United States throughout its history has been secure here in the Western hemisphere. The American psyche believes that perfect defense *should* be attainable. In that we differ from all other nations. It is this unique belief that underlies the current hope for the SDI. What I have said, I believe, indicates that this hope is illusory.

Where does the Strategic Defense Initiative stand today? As I have indicated, it has undergone a remarkable transformation. The argument is no longer that somehow we can protect American cities perfectly. Instead the argument has become that maybe, not definitely but maybe, strategic defense would permit us to improve deterrence—and that the mix of offense and defense would lead to a more stable world. This is a plausible argument, but one should be keenly aware of the dramatic change that has occurred.

It is certainly not impossible that the introduction of defensive capabilities might improve deterrence. Indeed, that had been the general aspiration, if not the conviction, prior to the signing of the Moscow ABM Treaty in 1972. Particularly in light of the impressive growth of Soviet counterforce capabilities since 1972, such a possibility deserves careful examination, or more precisely, reexamination.

A number of studies have been done that have, I believe, effectively demonstrated—within the assumptions of the study—that there are certain cases in which a mixture of defense and offense would improve the position of the United States, improve the position of the Soviet Union, improve world stability, and provide a strategic relationship in which, if nuclear war did nonetheless come, there would be less damage. *All of these studies rest upon the assumption that the offense is constrained.*

Remember the 1960s, and Secretary McNamara's reaction to the Soviet ABM system. Are the Soviets likely to be any less "offensive-conservative" than we were then? Given the Soviet Union's political ambitions, or its neurosis, or its quest for world domination, or its Marxist–Leninist creed (depending upon whose eyes one is looking through), how likely is it that in the event of an American deployment of substantial strategic defense, the Soviets would agree to a constraint on offensive capabilities?

I think the actual history of the American reaction in the past is one thing we should bear in mind as we look to a hypothetical Soviet reaction in the future. If the Soviets were to accept a constraint on their offense, it would require a minimum of trust. It would require a mutual approach to arms control, and that mutuality would almost certainly be reduced by our own efforts unilaterally to achieve general strategic defense capabilities of the type involved in the SDI. Moreover, we should not forget that we have more or less been putting aside the air-breathing threat—in a period in which the Soviets could have many more submarines equipped with SLCMs at sea. Indeed, 15, 20, or 30 years from now, they might just have developed the stealth technology that we ourselves are developing today—thus making the penetration of an air defense system relatively more easy than at the present time.

Let us go beyond the nagging question of the *likelihood* of a constrained offense, which is the only way in which greater stability is achieved in these models. Let us now look at some other issues. First is the issue of cost ratios. The historic judgment (or really, intuition) in the mid-60s was that the cost ratio between defense and offense was on the order of five to one. In other words, one's opponent could, by an investment of 20 percent of one's own investment in defense, create the offensive forces that would neutralize that investment in defense. Conversely, it would require an expenditure of five times as much on defense to neutralize the effect of the opponent's creation of additional offensive capabilities.

It is now hypothesized that these cost ratios have modestly improved since the 1960s, although that argument is somewhat flimsy. It is suggested that the ratio is now on the order of three to one. But that judgment rests primarily upon a single change: the belief that we can intercept Soviet missiles during the boost phase prior to separation of the re-entry vehicles. Nonetheless, it is clear that the ratio is still strongly weighted against defense and will remain so. If one is to put up a defense, it will require the opponent to constrain his offense. Otherwise he will be able to force you to misallocate resources to the point that you may no longer be able to protect yourself. And this may be true, even aside from the air-breathing threat.

Given constrained budgets, the adverse cost ratio means quite simply that, if one starts down the full-defense track, one is inevitably facing the drawdown of conventional and other forces. Here, amongst our Air Force friends, I must confess that if I were an officer in TAC, I would be very much concerned about marching along that trail beyond deployment. Indeed, I might even be concerned if I were a SAC officer. The problem may be even more painful for the Army. To illustrate the point, the fiscal year 1971 budget, as first submitted by the Pentagon, proposed a reduction of seven Army divisions and a substantial Europeanization of European defense. Of course, the Army did have the honor of developing the Safeguard system. But it also seemed embarked on the path to eating itself out of house and home—at the cost of its conventional capabilities. That problem has not gone away; it is likely to recur.

All that I have discussed to this point involves conditions *before* the Soviet Union begins to take serious countermeasures. One of the reasons the Scowcroft Commission stressed the need for a larger missile like the MX was to provide the throw-weight that might be needed to carry penetration aids. If one looks at the two strategic force structures at this time, and if one asks which of the two sides has the throw-weight to move a substantial array of PENAIDS, it is clearly the Soviet Union. That does not constitute an advantage for the United States.

On another point, I mentioned earlier that the great improvement visualized for the defense-offense cost ratio rests on the belief that we can now intercept Soviet missiles during the boost phase. But this could remain quite hypothetical. Some of the proposed kill mechanisms cannot reach down into the atmosphere. Thus the Soviets could shorten the boost phase, separate the warheads at an earlier point, and thereby preclude the gains in cost leverage that we now think we see in defense.

As yet, we have not solved the decoy problem, although we believe that we may. Nor have we solved the problem of assuring communications with those satellite systems that protect the United States. That will continue to depend upon ground-based facilities. I do not believe our targeting people in Omaha would have much difficulty in designing an attack with SLCMs that would take out those communications points.

So much for some of the technical problems in developing a workable and cost-effective system. However, let me reiterate: I strongly support a vigorous R&D program for strategic defense. But all of these matters must be soberly and responsibly faced—*before* we seriously consider deployment.

Let me turn now to the special problem of our Western European allies. During the first burst of enthusiasm here for the SDI, there was a good deal of protest from Western Europe—indeed horrified protest. There was a kind of ironic quality to that. If we were really able to do what the original speech suggested we might do—provide total protection for American cities—there might be a full restoration of the American strategic dominance of the 1950s and 1960s. Western Europe would again be fully protected—*if* one believes that strategic dominance can be restored. However, it is not believed, largely for the reasons I have already laid out. Because of the worries of our Western European allies, it has now been stated that the SDI is not just for North America, that instead we are prepared to provide it to everybody. Our European allies can have this defense; the Japanese can have it; indeed, in the latest variant, the Soviets can have it.

Assuming the Europeans wanted to deploy, who would pay for such a deployment? Would it be the American taxpayer? I rather doubt it. Those do not seem to be the noises that I hear coming from Capitol Hill. Would it be the Europeans? If so, what would be the consequences for their conventional capabilities—if they were actually to move down this line? And how effective would such a defense be in Western Europe? Far less effective than it would be in the United States. There is, first of all, much less warning time. Flight times to Western Europe are shorter, so there would be less opportunity to intercept in the boost phase and less opportunity to intercept after the boost phase.

In brief, rather than relying upon satellite systems in Western Europe, we would quickly discover the primary dependency on terminal defense. I think that there would be great political difficulties in deploying such terminal defenses in Western Europe. Indeed, I fear that deployment of such terminal defenses would make the recent start on the deployment of the Pershing II seem a relative political picnic. Moreover, given short flight times, the Europeans are subject to attack by air-breathing vehicles, even more than we are in the United States. And finally, if the Europeans were to proceed down this line, the inevitable consequence would be a drastic weakening of the direct defense embodied in their conventional forces. And the conventional deterrent remains perhaps the weakest link in the entire Alliance structure. So it would seem ill-advised for us to urge our friends to direct their resources away.

I turn from the special problems of Western Europe back to the bilateral relationship. What would be the impact of the introduction of space-based capabilities, not assuming that they had suddenly and successfully been deployed, but during the actual process of deployment. The likely outcome would be to create instabilities during the entire period of deployment. The process would be rendered particularly unstable because the system would be space-based. And the advantage of striking first, for either side, would be far greater than is the case for terrestrial capabilities. We must not assume that the Soviets will allow us any unilateral advantage. And in the instabilities of the unavoidable superpower competition lies the potential for disaster.

Finally, we should be aware that, even if the strategic defense system were to work reasonably well, and even if it were to enhance stability, it is still not certain that so large an investment—a trillion dollars is probably a good number—would be cost-effective in light of the other capabilities, particularly the conventional forces, that would need to be sacrificed under prospective budget constraints. A heavy additional burden on the defense budget is scarcely what is required, if we are to maintain a balanced force.

As you no doubt have observed, I have not been entirely positive in my assessment of the SDI. So let me conclude on the issue of what to do. Enough has already been said to suggest that, aside from proceeding with appropriate R&D activities, we should proceed *very cautiously*. Yet, there is a good deal of rhetoric floating around that we are now going to replace "mutual assured destruction" with "mutual assured survival." I find that rhetoric interesting. I do not know that anyone has attempted to define, as yet, precisely what mutual assured survival is. But even were it better defined, I rather doubt that it is achievable. We must be careful not to be swept away by rhetoric. It would be irresponsible for us to base our defense posture on rhetoric that may sell well on the political scene but bears little relationship to the under-lying technical, budgetary, and strategic realities.

Whether or not the President should have said what he did in March 1983 has now been overtaken by events. One cannot excise the words that the President spoke. It is an illusion of the critics of the SDI that somehow or other all this can be rolled back. It cannot be. We shall now have to deal with all of the consequences—creating alarm amongst our allies, reinforcing the Soviet belief that the United States is now attempting to restore strategic dominance, and the impact of that belief on negotiations and on arms control. We are obliged to look at what strategic defense might give us, not only in terms of force deployment, but at the bargaining table. It is, of course, incumbent upon us to think through the strategic consequences—before we proceed to deploy such forces. In its final report to the President, the Scow-croft Commission offered the following guidance:

The Commission was requested to review the Administration's proposals for research on strategic defense. In the Commission's view, research per-mitted by the ABM Treaty is important in order to ascertain the realistic possibility which technologies might offer as well as to guard against the possibility of an ABM breakout by the other side. But the strategic implica-tions of ballistic defense and the criticality of the ABM Treaty to further arms control agreements dictate extreme caution in proceeding to engineering development in this sensitive area.[1]

The Commission's comments were deliberately understated. It is no longer reasonable to pretend that we have not entered a new era touched off by the President's remarks. On the other hand, it is incumbent upon us not to push ahead willy-nilly, while neglecting the impact upon our relations with the Soviets and the consequences for arms control. I can think of no better guidance in the period ahead than this most sensible admonition in the Commission's final report.

Finally, what purpose might these strategic defense possibilities serve? The Soviet Union has historically shown an immense, perhaps exaggerated, re-spect for American technological capabilities. And, in the course of the last 18 months, we have certainly gotten their attention with respect to strategic defense. As I have indicated, that has its unfavorable side, but it also has the potential for being immensely useful. The President has repeatedly spo-

---

1. Brent Scowcroft, "President's Commission on Strategic Forces," March 21, 1984, p. 8.

ken of his desire to proceed with arms control. It is apparent that we should now have delineated a grand design for an arms control agreement with the Soviet Union. The new grand design is remarkably similar to the old grand design—the one of 1972. You may recall that the Soviets were keenly aware of the inadequacies of their ABM system (whose capabilities we had much exaggerated). When the United States began actually to deploy the Safeguard system, the Soviets were deeply alarmed about the immense advantages of American technology. They therefore proceeded in the negotiations to insist on a limitation on ABM systems.

That ultimately resulted in the 1972 Treaty. Throughout the entire period President Nixon took the position—I believe correctly and certainly courageously—that there would be no ABM treaty unless the Soviets agreed to limitations on offensive forces. Although the Soviets wanted no agreement at all on offensive forces, their eagerness for the ABM treaty forced them, in effect, to accept the 1972 agreement on offensive forces.

That grand design—of limits on Soviet offensive forces in exchange for constraint on American defense technologies—lies before us again, beckoning. If, through Soviet fears of American space technology, we were able to achieve a breakthrough in arms control negotiations (in a rather unpromising era), the President's launching of his new initiative would have fulfilled its most laudable purpose. In short, perhaps the best use of the Strategic Defense Initiative lies in that much maligned role of bargaining chip. Indeed, one might say, the Strategic Defense Initiative is the quintessential bargaining chip.

# Why Even Good Defenses May Be Bad

*Charles L. Glaser*

$O$nce again, the United States is in the midst of a debate over whether to deploy defenses designed to protect U.S. cities and population from Soviet missile attack. This debate is, most immediately, the result of President Reagan's "star wars" speech, in which he asked the rhetorical question: "wouldn't it be better to save lives than to avenge them?" He offered a future vision of "truly lasting stability" based upon the "ability to counter the awesome Soviet missile threat with measures that are defensive."[1] Just six months later a senior interagency group recommended to the President that the "U.S. embark on early demonstrations of credible ballistic missile defense technologies to its allies and the Soviet Union."[2]

There is, in addition to this most recent catalyst, a deep-seated, enduring reason why the possibility of defending the United States from Soviet nuclear attack is a recurrent issue. Put most simply, it is quite natural for the United States to want to remove itself from a situation in which the Soviet Union has the capability to virtually destroy it. The United States cannot, today, physically prevent the Soviet Union from wreaking such destruction. U.S. security therefore depends upon its ability to deter Soviet nuclear attack. If deterrence works, then the United States will be able to avoid nuclear war with the Soviet Union. Unfortunately, the possibility that deterrence could fail cannot be easily dismissed. Deterrence will have to work for decades and centuries—that is, unless the current situation, in which the United States is vulnerable to Soviet nuclear attack, is dramatically altered. While one cannot

---

The author would like to thank Robert Art, Albert Carnesale, Lynn Eden, Michael Nacht, Thomas Schelling, Stephen Van Evera, Stephen Walt, and the members of the Avoiding Nuclear War working group for their helpful comments on earlier drafts of this article.

---

*Charles L. Glaser is a Postdoctoral Fellow on the Avoiding Nuclear War Project at the John F. Kennedy School of Government, Harvard University, and a Research Fellow at the School's Center for Science and International Affairs.*

---

1. *The New York Times*, March 24, 1983, p. 20.
2. Clarence A. Robinson, Jr., "Panel Urges Defense Technology Advances," *Aviation Week and Space Technology*, October 17, 1983, p. 16.

---

*International Security*, Fall 1984 (Vol. 9, No. 2) 0162-2889/84/020092-32 $02.50/1

specify with confidence the way in which the superpowers' nuclear arsenals might come to be used, knowing that deterrence could fail in a variety of ways is sufficient to create a feeling that, given enough time, deterrence will fail. Consequently, as long as the United States remains vulnerable to Soviet nuclear attack, the possibility of nuclear attack will create an interest in defense against it.

The current debate over the deployment of ballistic missile defense (BMD), like the one in the late 1960s, is highly polarized. Defense, according to its opponents, is undesirable on all scores. They argue that defense will not work effectively, will increase the probability of war, and will cause arms races. Proponents, on the other hand, see few, if any, disadvantages with defense. They argue that defense will reduce the damage the Soviet Union could inflict on the United States, will not increase the probability of war and might decrease it, and might even improve the prospects for achieving arms control agreements which limit offensive nuclear forces.[3]

The vast majority of the debate has pivoted on the technological feasibility of effective BMD. The implicit assumption is that if effective BMD could be developed and deployed, then the United States should pursue the BMD route and the associated change in its nuclear strategy.[4] The principal argu-

---

3. A ballistic missile defense is a system capable of destroying Soviet missiles (or warheads) in flight. The terms "ballistic missile defense" (BMD) and "anti-ballistic missile" (ABM) are usually used interchangeably. BMD programs which might contribute to the goals described in President Reagan's so-called "star wars" speech are also referred to as the Strategic Defense Initiative (SDI).

Influential arguments made in opposition to BMD in the earlier debate are found in: Abram Chayes and Jerome Wiesner, eds., *ABM: An Evaluation of the Decision to Deploy Anti-Ballistic Missile Systems* (New York: Harper and Row, 1969); and Richard Garwin and Hans Bethe, "Anti-Ballistic Missile Systems," *Scientific American*, March 1968, pp. 164–174. Arguments in favor of BMD were presented in Johan J. Holst and William Schneider, Jr., eds., *Why ABM? Policy Issues in the Missile Defense Controversy* (New York: Pergamon Press, 1969). A recent book on the subject which does not promote any specific BMD policy is Ashton B. Carter and David N. Schwartz, eds., *Ballistic Missile Defense* (Washington, D.C.: Brookings Institution, 1984). Recent arguments against BMD are found in: *Space-Based Missile Defense*, A Report by the Union of Concerned Scientists (Cambridge, Mass.: Union of Concerned Scientists, 1984); and William E. Burrows, "Ballistic Missile Defense: The Illusion of Security," *Foreign Affairs*, Vol. 62, No. 4 (Spring 1984), pp. 843–856. For current arguments in favor of BMD, see Keith B. Payne and Colin S. Gray, "Nuclear Policy and the Defensive Transition," *Foreign Affairs*, Vol. 62, No. 4 (Spring 1984), pp. 820–842.

4. For an explicit statement of this belief from strong opponents of BMD, see *Space-Based Missile Defense*, in which the Union of Concerned Scientists states, "if it were possible to put in place overnight a fully effective, invulnerable defense against nuclear weapons, there could hardly be serious objections to doing this" (p. 71).

ment against defenses is that they will not work. Opponents of defense, presumably because they believe that effective defense is infeasible, tend not to examine carefully either the advantages or the disadvantages of effective defense. As a result, examination of a world in which the superpowers have deployed effective defense has been left to the advocates of defense, and a question of fundamental importance continues to be overlooked by the debate:

Could the deployment of effective defenses by both superpowers create a nuclear situation preferable to our current one, in which both countries maintain redundant assured destruction capabilities?

I am using the term "defense" to refer only to area defense, i.e., systems designed to protect cities and other value targets. BMD that would protect the United States by reducing the Soviet Union's ability to inflict damage is an area defense. By contrast, a point defense is designed principally to protect nuclear force capabilities.[5]

By "effective defenses," I have in mind systems that are capable of denying one's adversary an assured destruction capability. Defenses which cannot eliminate assured destruction capabilities are far less interesting because they would not significantly reduce the damage the United States would suffer in an all-out nuclear war.[6] Another way, then, of stating the above question is:

---

5. This distinction is important because these two types of defense have fundamentally different strategic implications: a country's area defense, if sufficiently effective, could *reduce* the size of the *adversary's* deterrent threat; a country's point defense, by increasing the size of its offensive force that would survive a counterforce attack, could *increase* the size of the *country's* deterrent threat.

6. An assured destruction capability is generally understood to be the capability, following a full scale counterforce attack against one's forces, to inflict an extremely high level of damage upon one's adversary. The levels of potential damage which analysts believe assured destruction requires are usually similar to those prescribed by Robert McNamara, U.S. Secretary of Defense from 1960 to 1968. McNamara's criteria for assured destruction, which were influenced by the diminishing marginal damage potential of increasing the size of the U.S. force, required that the United States be able to destroy, in a retaliatory attack, approximately 25 percent of the Soviet population and 50 percent of Soviet industry. He judged that such a level of destruction would be intolerable to the Soviet Union and, therefore, that the capability to inflict this level of damage would be sufficient to deter deliberate Soviet nuclear attacks on the United States. See Alain C. Enthoven and K. Wayne Smith, *How Much Is Enough? Shaping the Defense Program 1961–1969* (New York: Harper and Row, 1971), pp. 172–184, 207–210.

A related, but conceptually distinct, interpretation of assured destruction focuses on the relationship between the costs a decision-maker associates with the nuclear attack and the damage that would result from such an attack. Assured destruction in this interpretation requires that the potential damage in one's retaliatory capability should be sufficiently high that increasing the potential damage would not result in significantly higher costs to the adversary. In this article, assured destruction is intended to have this second meaning. Clearly, any evaluation of

Could the United States be more secure than it is today if, as a result of mutual deployment of defenses, neither the United States nor the Soviet Union had assured destruction capabilities?

My objective in this essay is to analyze this question.

A country with an assured destruction capability can inflict extremely high levels of damage. Nuclear situations in which the Soviet Union lacks an assured destruction capability, therefore, range from those in which the United States is invulnerable to attack to those in which the Soviet Union could destroy a sizable fraction of the U.S. population. The most interesting alternatives to mutual assured destruction situations are, of course, those in which the U.S. defense reduces the potential damage the Soviet Union could inflict far below the level required by assured destruction. Analytically, however, there are important similarities that cross the full range of nuclear situations in which assured destruction capabilities do not exist. As a result, this analysis applies equally well to all such situations. In fact, a distinguishing feature of this analysis is that it examines the requirements of strategic nuclear deterrence in situations in which defenses have eliminated assured destruction capabilities. In contrast, most analyses of strategic nuclear deterrence require that the United States possess an assured destruction capability, almost as if this were a prerequisite for deterrence.

There is general agreement that defenses capable of eliminating assured destruction capabilities do not exist today, and are extremely unlikely to be developed in the foreseeable future.[7] However, to facilitate examination of

---

the costs associated with such unprecedented damage is highly subjective. Many people believe that the United States would have to be able to reduce damage to itself far below the levels specified by McNamara before it could significantly improve the outcome of an all-out war; others believe that any reduction in damage, even if damage remained well above these levels, would be significant. The two different understandings of assured destruction are often not distinguished because McNamara said that an assured destruction capability would be sufficiently large to annihilate one's adversary in retaliation and because analysts tend to assume that costs to one's adversary could not be increased if the adversary could already be annihilated.

The arguments in this article do not depend upon a specific assessment of the level of retaliatory damage required for assured destruction. Instead, the arguments view the level of damage required for assured destruction as an imprecise boundary, above which additional damage does not significantly increase the costs of an attack, and below which reductions in damage would significantly reduce the costs. People disagree on the location of this boundary, but the arguments apply in all cases.

7. For an authoritative analysis of the technical feasibility of BMD, see Ashton B. Carter, *Directed Energy Missile Defense in Space* (Washington, D.C.: U.S. Government Printing Office, 1984). Carter judges as "extremely remote the prospect that directed-energy BMD (in concert with other layers if necessary) will succeed in reducing the vulnerability of U.S. population and society to the neighborhood of 100 megatons or less" (p. 68). See also: *Space-based Missile Defense;*

the issues that lie beyond the technical feasibility of BMD, I hypothesize in this article that highly effective defenses are available. Although effective defenses are, at best, a distant prospect, their presumed advantages have a significant influence on the BMD debate. Assuming, for the sake of analysis, that effective defenses are available makes possible a closer examination of the desirability of defensive situations. (I will use the term "defensive situation" to refer to nuclear situations in which defenses have eliminated assured destruction capabilities.)

The article focuses on situations in which *both* the United States and the Soviet Union deploy defenses. This case is important because it is the most probable outcome of U.S. deployment of defense. The Soviet Union is extremely likely to deploy defenses in response to a U.S. deployment. There is little reason to assume that in the long run the United States could maintain a technological advantage that enabled only the United States to have effective defense. Furthermore, the case of symmetric deployment is especially interesting due to the intuitive appeal of reducing U.S. vulnerability to attack without creating an advantage that threatens Soviet security.[8]

This analysis of how mutual deployment of effective defense would affect U.S. security proceeds through a number of stages. I identify three features of the nuclear situation that affect the United States' ability to avoid nuclear war with the Soviet Union: 1) the United States' ability to deter premeditated Soviet attack; 2) the crisis stability of the nuclear situation; and 3) the robustness of the U.S. deterrent to changes in Soviet forces. Next, I compare the probability of nuclear war in defensive and assured destruction situations by examining these three features for both types of nuclear situations. The final stage of the analysis compares U.S. security in defensive and assured destruction situations based upon expected costs. This requires considering

---

Spurgeon M. Keeny and Wolfgang K.H. Panofsky, "MAD Versus NUTS," *Foreign Affairs*, Vol. 60, No. 2 (Winter 1981–82), pp. 297–303; and Carter and Schwartz, *Ballistic Missile Defense*, in which Carter concludes: "the prospect that BMD will thwart the mutual hostage relationship— if this is taken literally to mean the ability of each superpower to do socially mortal damage to the other with nuclear weapons—is so remote as to be of no practical interest" (p. 11).

8. It should not go unmentioned that many of the advocates of BMD favor asymmetric deployment—that is, situations in which the United States can gain a strategic advantage by deploying BMD which is superior to Soviet BMD. See, for example, Colin S. Gray, "Nuclear Strategy: The Case for a Theory of Victory," *International Security*, Vol. 4, No. 1 (Summer 1979), pp. 54–87; and Colin S. Gray and Keith Payne, "Victory Is Possible," *Foreign Policy*, No. 39 (Summer 1980), pp. 14–27. In contrast to these earlier articles, in the recent "Nuclear Policy and the Defensive Transition," Gray and Payne argue as though the defensive situation they advocate would be symmetric. They do not explain the origins of this apparent inconsistency.

the damage that would result if nuclear war occurred as well as the probability of its occurrence.

The conclusion of this analysis is that defensive situations, even those in which defenses were perfect, are *not* clearly preferable to assured destruction situations. This conclusion is indeterminate because in defensive situations the probability of certain types of wars would increase, but the damage of other types of wars would decrease.

The indeterminacy of this conclusion should not obscure its policy significance. The assumptions used in the analysis have made a best case for defense: effective defense is assumed to be technically achievable; the enormous economic costs required to deploy any effective defense are overlooked; and the deployment of defenses is assumed to avoid the creation of asymmetries in the superpowers' capabilities that could create incentives for preventive attack and could encourage adventurous, crisis-provoking behavior. Even in this best case, defensive situations are not clearly preferable to the current assured destruction situation. In addition, many of the nuclear situations that could result from starting down the BMD route are far less desirable than our current mutual assured destruction situation. Without the possibility of a best outcome that is clearly preferable to our current situation, there is now no good reason to invest enormous resources in strategic defense and to risk creating a more dangerous world. The arguments for not dramatically altering the nuclear status quo are much stronger than those that call for U.S. deployment of an area defense.

*Perfect Defense*

It is important to begin with an examination of the strategic implications of perfect defenses, however distant they may seem, because that is the goal towards which many advocates of strategic defense, including President Reagan, wish to move. Despite the widespread presumption that perfect defenses are desirable if feasible, there are two major shortcomings of a world of perfect defenses that draw into question whether it would be safer than our current nuclear situation.[9]

---

9. The following discussion assumes that both countries would know the effectiveness of both their own defense and the adversary's defense. This is admittedly unrealistic, since there would always be uncertainties about the effectiveness of the defenses, and because the implications of these uncertainties could be significant. The reason for assuming that the effectiveness of the defenses would be known, however, is to focus the examination of perfect defenses on other

First, there could be no guarantee that perfect defenses would remain perfect. The technical challenge of developing and deploying a defense that would make the U.S. invulnerable to nuclear attack is enormous. Such a defense is commonly referred to as "perfect." The difficulty of *maintaining* a perfect defense indefinitely is likely to be far greater than developing it in the first place. Consequently, so-called perfect defenses should not be envisioned as a permanent technological solution to the dangers posed by nuclear weapons. The far more likely course of events is that a world of perfect defenses would decay into a world of imperfect defenses.[10]

A nuclear situation in which both superpowers were invulnerable to nuclear attack would be extremely sensitive to even small improvements in the ability of one country's offense to penetrate the adversary's defense. For example, the ability to penetrate the adversary's defense with ten warheads would provide the potential for enormous destruction when compared to no destruction. The country that first acquired even a small capability to penetrate the adversary's defense would have attained an important coercive advantage: nuclear attack could be threatened with impunity since effective retaliation would be impossible given the adversary's inability to penetrate one's own defense. Recognizing that the adversary is likely to acquire a similar capability—that is, that one's defense will not remain impenetrable— could create pressure to reap the benefits of the strategic advantage quickly. This time pressure would be especially strong if one's advantage could be used to prevent the adversary from acquiring the capability to penetrate one's defense.

By contrast, when both superpowers possess redundant assured destruction capabilities, as is the situation today, the addition of tens or hundreds or even thousands of warheads would not significantly change the nuclear

---

issues. This assumption of certain information strengthens the arguments for perfect defense and therefore reinforces the best case assumptions used in this analysis. Some of the complications that would likely result from uncertainties about effectiveness are discussed later in this article.

10. Many advocates of pursuing highly effective defense argue that even if the prospects for effective defense do not look extremely promising today, history suggests that major technological changes should be expected. For example, Payne and Gray observe in "Nuclear Policy and the Defensive Transition" that: "All of recorded history has shown swings in the pendulum of technical advantage between offense and defense. For the strategic defense to achieve a very marked superiority . . . would be an extraordinary trend in the light of the last 30 years, but not of the last hundred or thousand years. Military history is replete with examples of defensive technology and tactics dominating the offense" (p. 826). This argument would, however, apply at least as well to the maintenance of the defensive world they advocate and points to the major problems that would exist in defensive situations.

situation. As a result, the probability of gaining a strategic advantage is extremely low, especially when both superpowers are aware of and react to changes in the other's nuclear force.

The dangers, in a world of impenetrable defenses, that result from this sensitivity to small offensive improvements would be increased by the strong incentives the superpowers would have to defeat each other's defense. Each country could be expected to make the acquisition of a strategic advantage a priority. Moreover, because there would be no guarantee that perfect defenses would remain perfect, even a country that did not want to acquire an advantage would feel compelled to acquire additional strategic capabilities. Such a country would want to improve its defense to offset anticipated improvements in the adversary's offense. In addition, there would probably be a strong instinct to improve one's offense as well as a hedge against the possibility of not being able to offset, with improvements in one's defense, the adversary's enhanced offense. One's adversary, however, would not be able to know with confidence that these strategic programs were intended only to maintain a situation of equal capability. Consequently, even if both countries preferred to remain in a world of perfect defense, an interactive competition which threatened to reduce the effectiveness of the defenses would be likely to ensue. (Nuclear situations would continue to be sensitive to relatively small changes when the defenses were imperfect. This lack of "robustness" to changes is examined in detail below.)

The second problem with perfect defenses is that they could increase the probability of superpower conventional wars. Today's nuclear forces greatly increase the potential costs of any direct U.S.–Soviet military confrontation. As a result, nuclear weapons increase the risk of starting a conventional war, and therefore contribute to the deterrence of conventional war. Impenetrable defenses would eliminate this contribution. There is disagreement among strategic analysts about which features of the superpowers' extensive survivable strategic arsenals are most critical for deterrence of conventional war. Few, if any, commentators however believe that the existing arsenals do not contribute at all to the deterrence of conventional war.[11]

---

11. For an insightful discussion of why large nuclear arsenals reduce the probability of superpower conventional wars, even when neither superpower has an advantage in purely military terms, see Robert Jervis, "Why Nuclear Superiority Doesn't Matter," *Political Science Quarterly*, Vol. 94, No. 4 (Winter 1979–80), pp. 617–633. At the other end of the spectrum, Gray and Payne in "Victory Is Possible" find the U.S. strategic force inadequate to meet its extended deterrence commitments, but admit that U.S. strategic nuclear forces do contribute to deterrence of Soviet conventional attack in Europe (p. 16).

Perfect defenses might be in the U.S. security interest despite the increased probability of conventional war. That conventional war would be more likely does mean, however, that there is an important trade-off to consider. As World Wars I and II demonstrated, global conventional wars can be extremely destructive. The net effect of increasing the probability of major conventional war, while eliminating the possibility of more destructive but extremely unlikely nuclear war, might not be positive. The evaluation of this trade-off would involve many factors, including estimates of the probability of nuclear and conventional wars with and without perfect defenses, estimates of the size and costs of these wars, and the availability of options for reducing the probability and costs of conventional war. The objective of this short discussion is to call attention to this trade-off, not to resolve it.

In short, then, what are commonly called perfect defenses would have two shortcomings. First, they would probably not be truly perfect, but instead only temporarily impenetrable. The undermining of one country's defense would create a situation in which the incentives to initiate a nuclear war would be greater than today. Second, even if the temporary nature of impenetrable defenses is ignored, the net effect of both superpowers' deploying impenetrable defenses remains unclear because major conventional wars could become more likely.

*Imperfect Defense and the Probability of Nuclear War*

Understanding security in a world of perfect defense is relatively easy because as long as the defenses remain impenetrable, there is no possibility of a strategic nuclear war.[12] Assessing security in a nuclear situation in which imperfect defenses have been deployed is more difficult. Since, in this case, the United States would be vulnerable to Soviet strategic nuclear attack, we need to evaluate the United States' ability to reduce the probability of these attacks.

The following analysis considers nuclear situations in which both countries have imperfect defenses, but each is capable of denying the other an assured destruction capability. Implicit in this formulation is a relationship between one country's offensive force and the adversary's defensive force. When

---

12. This assertion depends on the assumption made above that both countries know that the defenses are perfect. If defenses were not known to be perfect, although in fact they were, then nuclear attack might be carried out (but would not result in damage) and nuclear threats might be used coercively.

defenses are imperfect there will always be, at least in theory, an offense which is sufficiently large to have an assured destruction capability. Therefore, for one country's imperfect defense to deny the adversary an assured destruction capability, either the size of the adversary's offense must be limited or the defense must be able to expand and improve to offset increases in the size of the offense. This analysis does not examine the feasibility of achieving these conditions. It assumes the establishment of a nuclear situation in which neither the United States nor the Soviet Union has assured destruction capabilities.

The probability that the United States will avoid war with the Soviet Union depends upon the following three features of the nuclear situation:

1) *The United States' ability to deter Soviet nuclear attack during periods when war does not appear to be imminent, that is, when there is not a severe crisis.* Deterrence of this type of attack requires that the Soviet Union believe that the net effect of starting a nuclear war would be negative, that is, that the Soviet Union would be worse off after the war than before it. I will term these "premeditated attacks." Surprise attacks, including the infamous "bolt from the blue," fall within this category.

2) *The crisis stability of the nuclear situation.* In a crisis, one or both superpowers might fear a nuclear attack by the other. If striking first is believed to be preferable to being struck first, and if a country believes the probability that the adversary will strike first is sufficiently high, then launching a first strike would be preferable to taking a chance on being struck first. This type of first strike is commonly termed a "preemptive attack." Unlike the case of premeditated attack, the country launching a preemptive attack would expect to be less well off after the war than before it. The crisis stability of the nuclear situation is a measure of how severe a crisis must be (or how high one's estimate that the adversary will strike first must be) before striking first becomes one's best option.[13]

---

13. The probability of preemptive nuclear war depends on the probability of crises, as well as the crisis stability of the nuclear situation. For example, a change in the nuclear situation which increases crisis stability but also increases the probability and severity of crises could increase the probability of preemptive nuclear war. The following comparison of the probability of nuclear war in defensive and assured destruction situations does not take into consideration the relative probability of crises. Because defensive situations are likely to increase tensions between the superpowers and because superpower cooperation will be more difficult than in assured destruction situations, the effect of not including the probability of crises in this analysis probably favors defensive situations. Therefore, this simplification tends to reinforce the best case which this analysis makes for defensive situations.

3) *The robustness of the nuclear situation.* The adequacy of U.S. forces depends not only on their ability to reduce the probability of preemptive and premeditated attacks, but also on how sensitive this ability is to potential changes in the Soviet forces. The more easily the Soviet Union could build forces that either would make a premeditated attack attractive or would significantly increase the incentives for preemptive attack, the greater the probability of a nuclear war. The robustness of the U.S. nuclear force is a measure of the difficulty the Soviet Union would encounter in trying to reduce U.S. security.

These three measures of the quality of the nuclear situation (the United States' ability to deter premeditated attacks, the degree of crisis stability, and the robustness of U.S. forces to change) are frequently used to assess the adequacy of U.S. nuclear forces. What distinguishes the following analysis from standard analyses of the nuclear situation is the assumption that assured destruction capabilities do not exist. Past analyses have asked the question: what capabilities are required to minimize the probability of war? The answers all include the need for an assured destruction capability (or at least a large retaliatory capability). This analysis, by examining the effect on these three measures of the nuclear situation, explores how the elimination of assured destruction capabilities by mutual deployment of defenses would affect the probability of nuclear war.

PREMEDITATED ATTACKS: IS ASSURED DESTRUCTION NECESSARY FOR DETERRENCE?

Consider a nuclear situation in which Soviet defenses could deny the United States an assured destruction capability. In this situation, the most basic and generally accepted U.S. deterrent requirement (that is, possession of an assured destruction capability) would not be satisfied. A natural conclusion is that the U.S. deterrent would be inadequate. This belief fueled opposition to strategic defense during the earlier BMD debate.[14] But closer examination of nuclear situations in which *both* superpowers deploy defenses shows that U.S. deterrent requirements could be satisfied without U.S. possession of an assured destruction capability.

---

14. That deterrence requires assured destruction capabilities was rarely made as a separate argument. It was, however, an integral part of the argument that BMD would necessarily result in an arm race. The inevitability of this arms race was based in part on the assertion that each superpower, to maintain an effective deterrent, would have to possess an enormous retaliatory capability. See, for example, Chayes and Wiesner, eds., *ABM*, pp. 49–54.

The requirement that the United States have an assured destruction capability implicitly assumes that the Soviet Union can annihilate the United States: the standard argument is that to deter an annihilating attack, the United States should be able to threaten credibly to annihilate the Soviet Union in retaliation. But if the United States could, by deploying defenses, eliminate the Soviet Union's annihilation capability, then deterrence of this attack would not be necessary. Furthermore, it is difficult to imagine any other Soviet actions the deterrence of which requires the United States to threaten the annihilation of the Soviet Union. So, if the Soviet Union did not have the ability to annihilate the United States, then the United States would not need to be able to annihilate the Soviet Union in retaliation. Consequently, a mutual deployment of defenses that eliminated both U.S. and Soviet annihilation capabilities need not result in an inadequate U.S. deterrent. The United States would, of course, still need a nuclear retaliatory capability to deter other Soviet nuclear attacks.

What capability would the United States need to deter attacks against its homeland when defenses had denied the Soviet Union an annihilation capability? Deterrence requires that the United States have the ability following any Soviet attack to inflict costs greater than the benefits the Soviet Union would achieve by attacking. To determine the U.S. retaliatory requirement, we must estimate the value the Soviet leaders would place on attacking the United States. We need to consider why the Soviet Union might attack the United States and what it would hope to gain by doing so. In the most general terms, the Soviet Union could use its nuclear force to damage or weaken the United States and to coerce the United States. The U.S. forces required to deter these actions are examined briefly below.

For all of the concern about attacks against U.S. cities, it is not clear why the Soviet Union would ever launch an all-out countervalue attack. Still such an attack is not impossible, so we need to estimate the value the Soviet Union might place on attacking U.S. cities. One possible reason for attacking U.S. cities would be to weaken the United States, thereby reducing the U.S. ability to oppose the Soviet Union's pursuit of its foreign policy objectives. Presumably people believe the Soviet Union is interested in annihilating the United States because this would make it the dominant world power. The analogy, if U.S. defenses had eliminated the Soviet ability to annihilate the United States, would be a countervalue attack designed to weaken the United States.

To deter this type of attack, the United States would need a retaliatory capability that could weaken the Soviet Union as much as the Soviet countervalue attack could weaken the United States. A countervalue capability roughly equivalent to the Soviet countervalue capability should be sufficiently large to satisfy this requirement. In fact, this is a very conservative requirement because U.S. retaliation would not only deny the Soviet Union the desired increase in relative world power, but would also inflict direct costs by destroying Soviet value targets. Because the Soviet Union could first attack U.S. forces, and then attack U.S. cities, the United States should have forces that provide a countervalue capability essentially equal to the Soviets' both before and after a Soviet counterforce attack.[15] I will call this an "equal countervalue capability."

The second way in which the Soviet Union might use its nuclear capability is to coerce the United States. While the benefits to the Soviet Union of attacking U.S. cities can be questioned, the potential benefits of coercing the United States are far more obvious. If the Soviet Union could inflict enormous damage on the United States and the United States lacked the ability to deter these attacks, then the Soviet Union might be able to compel the United States to compromise its security and vital interests.

As in other cases, deterrence would require that the United States be able to threaten the Soviet Union with expected costs greater than expected benefits. In the case of coercion, however, the United States could deny the Soviet Union any benefit simply by refusing to perform the action the Soviet Union demanded. The costs threatened by the United States need not be greater than the benefits the Soviet Union hopes to gain through its coercive demand because any U.S. attack combined with refusal of the Soviet demand would result in a net Soviet loss. If faced with a coercive threat, the United States could refuse the Soviet demand and tell the Soviet Union that attacks against value targets would be reciprocated. To adopt this strategy, the

---

15. A similar argument is made by Donald Brennan in "The Case for Population Defense," in Holst and Schneider, *Why ABM?*, pp. 100–106. An earlier version of this argument appeared in Donald Brennan and Johan Holst, *Ballistic Missile Defense: Two Views*, Adelphi Paper No. 43 (London: International Institute for Strategic Studies, 1967), pp. 9–11.

Including in this analysis uncertainty and imperfect information about the level of vulnerability to countervalue attack would weaken this argument. Redundant assured destruction capabilities are extremely large by any reasonable evaluation. There is little opportunity to misjudge this destructive potential, and assessments of damage are therefore not sensitive to relatively small differences in force size. In contrast, in a defensive situation in which each country's ability to inflict damage has been greatly reduced, relative force capabilities would be harder to evaluate, and uncertainties, misevaluations, and misperceptions would be more likely to result in a perceived advantage that could result in a failure of deterrence.

United States would have to be confident that it could deter the Soviet Union. This would require that the United States believe that the Soviet Union finds the U.S. retaliatory threats credible.

A large disparity in U.S. and Soviet countervalue capabilities could undermine U.S. credibility. So, a reasonable force requirement for denying the Soviet Union the coercive use of its nuclear forces is that the Soviet Union not have an advantage in countervalue capabilities: an advantage should not exist in the deployed forces, nor should the Soviet Union be able to gain a countervalue advantage in surviving forces by launching a counterforce attack. Therefore, U.S. forces which satisfy the equal countervalue requirement should be sufficient to deny the Soviet Union a capability which enables it to coerce the United States.[16]

In summary, a reasonable requirement for deterrence of Soviet attacks on the United States is possession of an equal countervalue capability.[17] Requiring that the United States possess an equal countervalue capability is significantly different from requiring an assured destruction capability. The equal countervalue requirement explicitly couples U.S. and Soviet capabilities to inflict countervalue damage. The equal countervalue requirement could be

---

16. This does not mean, however, that the Soviet Union would necessarily be unable to coerce the United States. As in a situation of mutual assured destruction capabilities, if the Soviet Union were able to convince the United States that it would carry out a threat to attack U.S. cities, then the Soviet Union might be able to coerce the United States. The U.S. possession of an equal countervalue capability, by making possible a highly credible retaliatory threat comparable to the Soviet threat, would make it difficult for the Soviet Union to make its coercive threat convincing. If the Soviet Union were able to coerce the United States, the key to its success would be greater resolve and willingness to take risks than the United States, and not an advantage in nuclear forces.
17. There is a third reason, not discussed in the text, why the Soviet Union might attack the United States: to eliminate the U.S. ability to deter the Soviet Union from pursuing its foreign policy objectives. For example, consider a hypothetical case in which the United States deters Soviet attack on Western Europe entirely with threats of strategic nuclear retaliation. If a Soviet counterforce attack could sufficiently reduce the potential cost of U.S. retaliation, then the Soviet Union could judge that attacking the United States, incurring U.S. retaliation, and invading and acquiring Western Europe could result in a net benefit. This type of Soviet nuclear attack on the U.S. homeland, unlike the two discussed in the text, does not depend upon the Soviet Union's ability to inflict countervalue damage on the United States. In contrast, it is a purely military attack, motivated entirely by the desire to reduce the damage the United States could inflict on the Soviet Union. The United States could eliminate the Soviet incentive for launching this type of attack by making its forces invulnerable. However, even with invulnerable forces, and with the equal countervalue requirement satisfied, there might be nuclear situations in which the U.S. countervalue threat would be insufficiently large to deter Soviet attack on Europe. This would be true for the same reason that perfect defenses could increase the probability of conventional war: U.S. escalation to the nuclear level would no longer be sufficiently costly to deter Soviet attack.

satisfied by both the United States and the Soviet Union at all levels of vulnerability to attack. In contrast, the assured destruction requirement demands that the United States have a retaliatory force capable of inflicting a specific level of countervalue damage independent of the size of the Soviet ability to inflict damage. According to the equal countervalue requirement, if the United States can reduce the Soviet Union's ability to inflict countervalue damage, then the United States can afford to have its ability to inflict countervalue damage in retaliation reduced. Moreover, improvements in Soviet defenses which reduce the damage the United States could inflict on the Soviet Union could be compensated for by improvements in U.S. defenses. The assured destruction requirement, on the other hand, demands that improvements in Soviet defenses be offset either by an increase in the size of the U.S. offense or by an increase in the ability of the offense to penetrate the Soviet defense.

CRISIS STABILITY: WHAT WOULD BE THE EFFECT OF DEFENSES?
There is a common belief that defenses capable of eliminating an adversary's assured destruction capability would decrease crisis stability: a country that can protect itself (that is, a country that can deny its adversary a second strike annihilation capability) is more likely to strike preemptively in a crisis.[18] The following analysis explores this proposition and identifies the conditions under which it is correct.

Crisis stability depends upon the decision-maker's incentives to strike preemptively in a crisis, that is, during times when there is reason to believe one's adversary is likely to launch a first strike. The decision to preempt in a crisis would depend upon how the costs of being struck first compare to the costs of being struck second.[19] If the adversary has an assured destruction capability, then there would be little if any incentive for a rational decision-maker to preempt: a preemptive attack could not deny the adversary an annihilating retaliatory capability, so there would be little difference between

---

18. See, for example, *Space-Based Missile Defense*, pp. 79–80; William C. Foster, "Strategic Weapons: Prospects for Arms Control," *Foreign Affairs*, Vol. 47, No. 3 (April 1969), pp. 414–415; and Robert L. Rothstein, "The ABM, Proliferation and International Stability," *Foreign Affairs*, Vol. 46, No. 3 (April 1968), pp. 498–499.
19. Extensive discussions of crisis stability include: Thomas C. Schelling, *The Strategy of Conflict* (Cambridge, Mass.: Harvard University Press, 1960), pp. 207–254, and *Arms and Influence* (New Haven: Yale University Press, 1966), pp. 221–248; and Glenn H. Snyder, *Deterrence and Defense* (Princeton: Princeton University Press, 1961), pp. 97–114.

the costs of being struck first and second. In an assured destruction situation, the vulnerability of the adversary's forces does not create a preemptive incentive. The adversary's force is sufficiently large and survivable that the fraction of the force that would survive a counterforce attack would still be able to inflict the damage required for annihilation.

For the same reason, the adversary would have little incentive to preempt if one's own surviving force would be sufficiently large to annihilate the adversary. Since a leader's decision to preempt would be fueled by anticipation of the adversary's preemption, possession of an assured destruction capability by either country should be sufficient to create a highly crisis-stable nuclear situation.

If one's own defense eliminates the adversary's assured destruction capability *and* if the adversary's retaliatory capability is partially vulnerable, then preemption would reduce the damage from an all-out countervalue attack. As a result, if the decision-maker anticipates a countervalue first strike, then there would be an incentive to preempt.[20] Since without defenses there would be virtually no incentive to preempt (because the adversary could maintain his assured destruction capability), deploying defenses that eliminate assured destruction capabilities would decrease crisis stability.

But there is another important case, the one in which the adversary's retaliatory capability is invulnerable to a first strike. In this case, there is nothing to be gained by striking first because the magnitude of the adversary's retaliatory strike would be no less than if he had struck first—that is, the costs to one's own country of suffering an all-out countervalue first strike or second strike would be equal. So, because the adversary's forces were invulnerable, there would be no incentive to preempt. This would be true when defense had not been deployed, even if the adversary's ability to inflict countervalue damage were far below the annihilation level. Deploying defenses would reduce the adversary's ability to inflict damage, but would not

---

20. The assumption that decision-makers would anticipate a countervalue strike is implicit in many discussions of crisis stability. It underlies the logic that says if a counterforce attack could reduce the adversary's countervalue potential, then there will be an incentive to strike first. A crisis, however, should provoke fears of a counterforce attack. If we assume the adversary's first strike would be counterforce, then the nuclear situation is far more crisis-stable than if we assume the attack would be countervalue. For a good discussion of this argument see Snyder, *Deterrence and Defense*, pp. 104–109. If we assume that both countries anticipate counterforce first strikes, then the effect of defenses on crisis stability is likely to be minimal. Given this assumption, the incentives to preempt would be small, or nonexistent, with or without defenses.

create an incentive to preempt. This example shows, at least in principle, that defenses that reduce the adversary's retaliatory capability below the annihilation level would not always decrease crisis stability.

The practical significance of this observation should not be overestimated. An invulnerable retaliatory capability requires not only that the forces be invulnerable, but also that attacks against the command and control system would not reduce the size of the possible retaliatory attack. These conditions might not be achievable. Submarines in port are vulnerable to attack and much of the command system is now highly vulnerable. The combination of one's effective defenses with the adversary's offensive force vulnerabilities would result in an incentive to preempt. Consequently, while in theory defenses that eliminated assured destruction capabilities need not decrease crisis stability, in practice they probably would.

The fundamental insight we can draw from this discussion is that defenses do not by themselves create incentives for preemption. The source of preemptive incentives is offensive force vulnerabilities. Therefore, the effect of defenses on crisis stability should not be evaluated without considering the vulnerabilities of the offensive force to a counterforce attack. By reducing retaliatory capabilities, defenses can increase the significance of offensive vulnerabilities.

Because reducing the degree of offensive force vulnerability would enhance crisis stability, one way to offset the decrease in crisis stability that would result from deploying effective defenses would be to accompany the deployment with programs to reduce offensive force vulnerabilities. One approach for reducing force vulnerability is to protect offensive forces with active defenses. Area defenses, although not designed specifically for this mission, could increase force survivability. In addition, there are many other ways to increase force survivability, including deploying point defenses. If effective area defense were feasible, then defenses that could provide a high degree of force survivability, including survivability of command and control, would also be feasible. In this case, the reduction in crisis stability that would result from deploying effective defenses could be small.

In summary, effective defenses would be likely to decrease crisis stability. It would probably be possible, however, to keep this negative effect of defenses quite small. The source of preemptive incentive is offensive force vulnerability. Therefore, if offensive forces could be made highly survivable, then the effect on crisis stability of defenses that eliminate assured destruction capabilities would be small.

ROBUSTNESS: THE PRIMARY INADEQUACY OF DEFENSIVE SITUATIONS

We do not live in a static world. Consequently, in addition to evaluating U.S. security as if U.S. and Soviet forces could be held constant, we must also examine the effect of possible changes in Soviet forces on U.S. security and the probability of these changes. More specifically, we must evaluate not only the United States' ability to deter premeditated Soviet attack and the degree of crisis stability, but also the probability of changes in Soviet forces that could reduce the United States' ability to deter premeditated attacks or that could reduce crisis stability. The robustness of U.S. forces is a measure of the difficulty the Soviet Union would encounter in trying to reduce U.S. security.[21]

All other things being equal, the more easily U.S. security could be jeopardized by changes in Soviet forces, the less desirable the nuclear situation. A nuclear situation which would be highly desirable when the two countries' forces could be held fixed, but which lacks robustness, might not be preferable to one which is less desirable when the forces are held fixed, but which is more robust.

I have already discussed the lack of robustness of nuclear situations in which perfect defenses have been deployed. This section extends that analysis by considering cases in which imperfect defenses have been deployed. The conclusion remains the same: nuclear situations in which defenses significantly reduce the vulnerability of value targets would lack robustness.

The following discussion compares the difficulty the Soviet Union would have undermining U.S. deterrence of premeditated attacks in defensive and

---

21. Arms race stability is the standard measure of this characteristic of the nuclear situation. I have chosen to use the term "robustness" to avoid the confusion that surrounds the term "arms race stability." Arms race stability brings to mind at least two issues which are related to robustness, but which are conceptually distinct. First, arms race stability is often considered an indicator of the likelihood and/or intensity of arms races that will occur in a specific nuclear situation. Arms races, however, can occur for a variety of reasons which are only peripherally related to the effect of building nuclear forces on the adversary's security. Consequently, arms races can occur in highly robust nuclear situations, as has occurred in our current highly redundant and diversified assured destruction situation.

Second, use of the term "arms race stability" can connote a belief that arms races cause wars. Whether arms races actually cause wars is a theoretical issue on which there is substantial disagreement. But one can assert that the probability of war depends upon the robustness of the nuclear situation without believing that, in general, arms races cause wars. Robustness is a measure of how sensitive a country's security would be to the adversary's buildup of forces. It does not imply that the process of competitive armament itself leads to war. Rather, assuming a force buildup takes place either competitively or unilaterally, a war is more likely when the initial nuclear situation is less robust.

in assured destruction situations. It assumes that the requirement for deterrence of premeditated attacks, that is, the equal countervalue requirement, is satisfied in the initial nuclear situation. The United States' deterrence of premeditated attacks could be undermined by two types of changes: improvements in Soviet defenses that reduce the United States' ability to retaliate, and improvements in the penetration capability of Soviet offenses that increase the vulnerability of U.S. value targets to attack.

The robustness of U.S. nuclear forces to these changes depends upon two interdependent factors. The first is the magnitude of the change in potential countervalue damage required so that the Soviet Union would no longer be deterred from launching a premeditated attack. Specifically, how much must the Soviet Union reduce the U.S. countervalue threat to gain a strategic advantage? or how large an increase in Soviet countervalue capability is required to provide a significant advantage? The second factor is the technical difficulty of changing the threat to value targets by this amount. For example, assuming that in a specific nuclear situation the Soviet Union, to gain an advantage, must increase its countervalue capability by 50 warheads, how difficult would it be for the Soviet Union to achieve this change? The combination of these two factors determines the overall difficulty of acquiring a strategic advantage.

The discussion of perfect defenses focused on the first factor, the magnitude of the change, and argued that even small changes could have strategic significance. Situations in which imperfect defenses had been deployed would suffer, although less severely, from the same sensitivity. The following example illustrates this observation. Imagine three nuclear situations, one in which both superpowers have impenetrable defenses, one in which each superpower can penetrate the other's defense with ten warheads, and one in which both superpowers have assured destruction capabilities. Now consider how a change in one country's nuclear force that enabled it to penetrate the adversary's defense with ten additional warheads would affect the adversary's security in each situation. The addition of ten warheads of countervalue capability to one country's force would be less significant when added to the nuclear situation in which both countries started with ten penetrating warheads than when added to a situation in which both countries had perfect defenses. The advantage in countervalue capability would be harder to use coercively when the adversary would be able to threaten retaliation against one's own value targets.

In contrast, the addition of ten penetrating warheads to one force when both countries had assured destruction capabilities comprised of thousands of warheads would be far less significant than when added to the nuclear situation in which both countries had ten penetrating warheads. The addition to the mutual assured destruction situation might not even change the country's ability to inflict damage; the addition in the ten warhead situation, while it might be difficult to use coercively, could result in a significant difference in the two countries' ability to inflict damage.

The general conclusion to be drawn from this specific example is that the lower the vulnerability of value targets in a given nuclear situation, the smaller the change in their vulnerability required to gain an advantage. This conclusion can be restated specifically in terms of defenses: the smaller the number of warheads that could penetrate a country's defense, the more sensitive that country's security would be to offensive changes that reduce the effectiveness of its defense.

The second factor affecting robustness, the technical difficulty of changing countervalue capability to gain an advantage, depends upon the type, size, and number of changes required to achieve a strategic advantage. The type of change is determined by whether the status quo is an assured destruction situation or a defensive situation. In assured destruction situations, it is the difficulty of reducing the adversary's offensive threat that affects robustness. In defensive situations, on the other hand, both the difficulty of further reducing the adversary's offensive threat and the difficulty of penetrating the adversary's defense would affect robustness.

Assessing the relative difficulty of penetrating a specific defensive system with an offensive system or of defeating a specific offensive system with a defensive system is beyond the scope of this paper. Moreover, such an assessment would necessarily be highly speculative because effective defensive systems have not yet been developed. Consequently, it is impossible to compare the difficulty of defeating those defensive systems with the difficulty of developing defensive systems to defeat today's offenses or the offenses of the future. One fact that bears upon this issue should, however, be mentioned. Even if defenses were developed that were perfect against currently deployed offenses, experts believe that the task of developing offensive countermeasures to defeat those defenses would be relatively easy.[22] The defensive system would be understood by its adversary, enabling the development of countermeasures designed specifically with the defense in mind. The defense, by contrast, to remain effective, would have to be able to

---

22. On the existence of countermeasures, see Carter, *Directed Energy Missile Defense in Space*, pp. 69–70.

overcome the full range of possible countermeasures. This asymmetry means that defenses may always be at a disadvantage, that is, the development of effective defenses against a competitive threat may always be more difficult than developing offenses that can penetrate defenses.

The size of the change required to gain a strategic advantage affects the technical difficulty of achieving the change. (This is why the two factors affecting robustness are interdependent.) A defense which must reduce the offensive threat by a large amount is harder to build than one that must reduce the same offensive threat by a small amount. Similarly, a new offensive system which must be able to penetrate the adversary's defense with many weapons would be harder to build than one that had to penetrate the same defense with only a few weapons. Even taking into account the likely asymmetry between offense and defense mentioned above, it is not possible to say with certainty whether the changes required to gain an advantage in an assured destruction situation would be easier or harder to achieve than in a defensive situation. As discussed above, however, the size of the requisite change in assured destruction situations is larger than in defensive situations. Due to this difference, gaining an advantage will tend to be more difficult in assured destruction situations than in defensive situations.

The larger the number of changes in a country's forces required to gain an advantage, all else being equal, the harder the advantage will be to obtain. The number of force changes required to achieve an advantage depends upon the diversity of the adversary's forces. In assured destruction situations, ensuring one's ability to destroy large numbers of the adversary's value targets is the strategic requirement. Diversification of one's offensive force helps to ensure the continuing achievement of this objective by increasing the number of defensive changes that are required before the adversary could eliminate one's assured destruction capability. For example, an offensive force which could annihilate the adversary with either an air-breathing threat or a ballistic missile threat requires that the adversary develop two types of highly effective defense. Obviously, this is a harder task than developing an effective defense against a single threat.

This article has discussed defenses in general, not defenses against specific types of offensive threats. But when we think about the feasibility of defense, it is crucial to keep in mind the potential diversity of offensive threats. If BMD were technologically feasible, but defense against advanced technology bombers or cruise missiles were impossible, then the strategic significance of the BMD would be greatly reduced. The technological feasibility of defenses that would reduce vulnerability to attack is determined by the difficulty of defending against all offensive threats.

By contrast, in a nuclear situation in which one's own defenses have significantly reduced the vulnerability of value targets, the strategic requirement is the maintenance of a low level of vulnerability. In this case the adversary's ability to diversify offensive forces makes maintaining low vulnerability more difficult. Each of these offensive threats must be defended against, and the adversary's ability to defeat any of the defenses would be sufficient to make maintenance of low vulnerability impossible.

To make defensive situations more resistant to the adversary's offensive improvements, a country could diversify its defenses against each offensive threat. It would be more difficult, however, to diversify one's defenses to require the same number of force changes as in an assured destruction situation. Because the number of required offensive changes would be determined by the offense against which the defense was the least diversified, it would be necessary to diversify one's defenses against each of the adversary's offenses. So, for example, if the adversary had two offenses with different penetration modes, then to force him to have to make two changes would require a total of four defenses. By comparison, in an assured destruction situation, two offenses are sufficient to require the adversary to make two changes, that is, to deploy two effective defenses. Consequently, the relative difficulty of diversification would tend to make it more difficult to make a defensive situation resistant to the adversary's offensive changes than to make an assured destruction situation resistant to the adversary's defensive changes. For what could be termed "structural" reasons, defensive situations could not be made as resistant to change as assured destruction situations. In addition, a defensive situation which is not diversified is susceptible to catastrophic failure, because one offensive breakthrough by the adversary could render one's country vulnerable to large attacks.

To summarize, nuclear situations in which defense significantly reduces the vulnerability of value targets would lack robustness for two reasons: relatively small changes in vulnerability could threaten one's security; and the difficulty of achieving these changes would be relatively low due to the small change required and to the greater difficulty of making defensive situations resistant to offensive changes.

A lack of robustness would not be so dangerous if the United States and the Soviet Union would not have incentives to try to change the equal countervalue condition of the nuclear situation. If a political environment could be created in which the superpowers chose not to attempt to gain a strategic advantage, then the need to make the nuclear situation resistant to change would be reduced. Moreover, superpower cooperation in structuring the nuclear situation could contribute to the situation's robustness.[23] But, in a world of imperfect defenses, as described for the case of perfect defense, countries would feel tremendous pressure to pursue, or at least to prepare to pursue, capabilities for defeating the adversary's defense. Even in the unlikely event that a highly robust situation could be achieved (which would require making one's defense highly resistant to the adversary's offensive innovations), it would be hard to have high confidence in this robustness: a country could not know with certainty that offensive threats that would undermine the defense could not be developed. This uncertainty could not be overlooked because the change required to gain an advantage would still be small and the adversary's incentive to try to alter the nuclear situation would be obvious.

These conditions would make establishing a political environment in which cooperation was possible far harder under conditions of reduced vulnerability than under assured destruction.[24] And, given our limited success in negotiating strategic arms control treaties when both countries have redundant assured destruction capabilities, there is little reason to be optimistic about the prospects for cooperation.

In conclusion, situations in which defenses have eliminated countries' large retaliatory capabilities would suffer a lack of robustness. The danger of this condition would be increased by the incentives and pressures that both the United States and the Soviet Union would feel not to cooperate and to increase the other's vulnerability.

*Could Defense Create a Preferable Nuclear Situation?*

The preceding discussion of imperfect defenses capable of eliminating both superpowers' assured destruction capabilities compared the probability of nuclear war in defensive and assured destruction situations. Specifically, it compared the United States' ability to deter Soviet premeditated attack, the

---

23. For a discussion of how arms control might help to increase the robustness of defensive situations see Charles L. Glaser, "The Implication of Reduced Vulnerability for Security in the Nuclear Age," Ph.D. dissertation, Harvard University, 1983, pp. 217–228.
24. For a discussion of the factors which affect the probability that countries will be able to cooperate, see Robert Jervis, "Cooperation Under the Security Dilemma," *World Politics*, Vol. 30, No. 2 (January 1978), pp. 167–214.

crisis stability, and the robustness of defensive and assured destruction situations. This analysis of the probability of nuclear war is, however, not by itself sufficient to determine in which type of nuclear situation the United States would be more secure. This is because U.S. security depends upon the cost if war were to occur, as well as the probability of war.

Comparison of nuclear situations requires a measure that combines these U.S. security objectives, that is, to minimize the probability of war and to minimize the costs if war occurs.[25] These objectives should be evaluated simultaneously. For example, if a change in the nuclear situation would reduce the damage of a nuclear attack but would also increase the probability of the attack, then the change might not increase U.S. security. Examining only the probability of war or the costs if war were to occur is insufficient to understand the net effect of the change. The correct measure of security is the probability of the war multiplied by the costs if there were a war, which is the expected cost.

It is crucial to keep these two aspects of U.S. security in mind when analyzing defenses. Much of the debate over BMD tends to ignore the need for simultaneous evaluation. Proponents of defenses emphasize the reductions in damage that defenses could provide. Opponents of BMD argued during the 1960s that deployment would increase the probability of nuclear war by undermining deterrence and decreasing crisis stability. Neither of these arguments is sufficient to draw a conclusion about area defenses: each looks at only one aspect of U.S. security.

A complete analysis of the expected costs in a specific nuclear situation would describe the full spectrum of wars in which the United States might become involved, would evaluate the probability that they would occur, and would estimate the damage that would result if they did occur. This spectrum of wars would include conventional wars and limited nuclear wars in addition to all-out premeditated and preemptive wars. The preceding sections of this article do not focus on conventional and limited nuclear wars.[26] However, the analysis is sufficient to draw the fundamental conclusion about nuclear

---

25. This formulation of objectives assumes that the United States is a strictly status quo power. This may not be entirely accurate, but is a reasonable assumption for this discussion of nuclear weapons policy. A further concern about this formulation is that it does not include the objective of minimizing losses that could result from nuclear coercion. While this is clearly an objective of U.S. policy, the nuclear capabilities required to achieve it closely resemble those required to minimize the probability of nuclear war. Consequently, not explicitly including this objective does not bias the analysis.
26. For a more complete analysis see Glaser, "The Implications of Reduced Vulnerability for Security in the Nuclear Age."

situations in which effective defenses have been deployed: the probability of nuclear war would be higher in these situations than assured destruction situations, but the costs of certain types of nuclear wars would be lower. Therefore deciding whether these defensive situations are preferable to assured destruction situations requires making a trade-off. This trade-off is highly subjective. Analysts are likely to disagree about the level at which reducing the size of the adversary's potential attack begins to make a significant difference, about the relative probability of different types of nuclear war, and about how to balance a believed increase in the probability of war against a decrease in its potential costs. Because individuals will make these judgments differently, defensive situations cannot be said, in general, to be inferior or superior to assured destruction situations.

Before clarifying how the preceding analysis leads to this conclusion, I want to pause for a moment to stress its significance. Saying that defenses are neither clearly desirable nor clearly undesirable might appear to provide little policy insight. However, although indeterminate, this conclusion differs markedly from the conventional wisdom that effective defenses would be desirable, and weighs heavily against deploying defense to limit U.S. vulnerability to attack. Recall that the analysis has considered a best case for defense. Even making these optimistic assumptions, a defensive situation might not be preferable to our current assured destruction situation. More realistic, less optimistic assumptions about defenses result in nuclear situations which would be more dangerous than today's. Advocates of strategic defense are driven by the promise of a world far safer than the current one. If defenses could create such a world, then taking a chance on ending up in one of the more dangerous, and more likely, defensive situations might be justified. But to run great risks and to spend enormous resources in the hope of reaching a nuclear situation that might not be preferable to our current assured destruction situation, and might be worse, make little sense.

The remainder of this section explains how this conclusion follows from the preceding analysis. The analysis first identified three factors that influence the probability that the U.S. will be able to avoid nuclear war with the Soviet Union: 1) the U.S. ability to deter premeditated attacks; 2) the crisis stability of the nuclear situation; and 3) the robustness of the nuclear situation. I then evaluated how the deployment by both superpowers of effective defenses, that is, defenses sufficiently capable to deny the adversary an assured destruction capability, would affect these factors. The findings of this evaluation are:

1) Effective defenses need not undermine deterrence of premeditated attacks. Assured destruction is not necessary for deterrence; an equal countervalue capability (i.e., the possession of a countervalue capability equal to the Soviets' both before and after a Soviet counterforce attack) is sufficient for deterrence of premeditated attacks. The equal countervalue requirement could be satisfied when defenses of any level of effectiveness had been deployed and at all levels of vulnerability of value targets to attack.

2) Crisis stability would be likely to decrease if assured destruction capabilities were eliminated by defense. It might, however, be possible to keep this negative effect of defense quite small. At least in principle, a defensive situation could be made as crisis-stable as an assured destruction situation by deploying invulnerable retaliatory capabilities. While making one's retaliatory capability entirely invulnerable might not be possible, achieving a high level of invulnerability might be possible.

3) The Achilles heel of defensive situations is the tremendous difficulty of maintaining them. Defensive situations, unlike assured destruction situations, would be likely to lack robustness. This means that changes in the adversary's forces that could undermine deterrence of premeditated attack and create incentives for preemptive and preventive attack would be far more likely in defensive situations. The lack of robustness would likely result in tense superpower relations, making security cooperation extremely difficult. Due to the difficulty of creating robust defensive situations, the probability of nuclear war would be higher than in assured destruction situations.

Overall, then, this evaluation concludes that the probability of nuclear war in defensive situations would be higher than in assured destruction situations. This increase in the probability would not be due primarily to a decrease in the United States' ability to deter premeditated attacks or to maintain crisis stability: if the superpowers could not change their forces, then there might be defensive nuclear situations in which these wars would not be more likely than in assured destruction situations. This constraint, however, is unrealistic. The superpowers would be able to alter the status quo, that is, to change the offensive and defensive forces which are deployed at a given time. Due to the lack of robustness in defensive situations, this competitive armament would be more likely to result in a nuclear war in defensive situations than in assured destruction situations.

So, if security depended only on the probability of nuclear war, then defenses would unambiguously decrease security. But security depends also on the damage that would result if there were a war. Because defenses would decrease the damage of certain wars, they would thus have some positive effects as well as negative ones.

To illustrate this in greater detail, I will examine briefly each path to nuclear war. Consider first a nuclear situation in which effective defenses have been deployed, the equal countervalue requirement is satisfied, and assume this situation characterized by offensive and defensive forces of a given size and capability is maintained, that is, neither country builds forces that increase its countervalue capability. Deterrence could fail even when the equal countervalue requirement was satisfied: this is true today when the assured destruction requirement is satisfied, and it would be true in a defensive situation in which the equal countervalue requirement was satisfied. The damage in an all-out countervalue war, however, would be lower in the nuclear situation with defenses. Since the probability of premeditated nuclear war might not be greater in the defensive situation, the expected costs along this path to nuclear war would be lower than in an assured destruction situation.

In this defensive situation, the damage from a preemptive attack would be lower than in an assured destruction situation. The probability of preemptive war, and therefore the expected costs of a preemptive attack, would depend upon the situation's crisis stability. If the defensive situation were less crisis-stable than an assured destruction situation, then the probability of preemptive attack would be higher. In this case, the defensive situation would have a higher probability of preemptive attack and lower costs if the preemptive attack were to occur. Therefore the comparison of expected costs from preemptive attack in defensive and assured destruction situations would be indeterminate. If, on the other hand, the defensive situation were as crisis-stable as the assured destruction situation (which is possible if the retaliatory capability is invulnerable), then the expected cost along the preemptive path would be lower than in an assured destruction situation.

Now consider the case in which we drop the constraint that the status quo defensive situation would be maintained, that is, we include the possibility that countries might deploy forces that increase the adversary's vulnerability or decrease their own. Unlike the preceding case, in this case the damage in defensive situations might not be lower than in assured destruction situations. The changes in the countries' forces might return the defensive situation to an assured destruction situation. The potential damage would not be determined by the status quo forces. If a country could build its way out of the defensive situation (i.e., regain its assured destruction capability), then the damage to the adversary could be the same as in a mutual assured destruction situation. The expected costs of nuclear wars resulting from changes in the status quo could be greater in defensive situations than in assured destruction situations: the damage in defensive situations is not constrained by the status quo forces and might not be lower than in assured destruction situations; and, due to the lack of robustness in defensive situations, the probability of nuclear war along this path is higher.

In summary, defenses could reduce the damage that could occur in certain types of wars. This positive effect would, however, tend to be offset by the increased probability of wars resulting from the lack of robustness of defensive situations. The net result, therefore, of both superpowers' deploying effective defenses might not be to increase U.S. security.

As I stressed above, this indeterminancy is extremely significant. It undermines the commonly held belief that defensive situations, and so-called defense-dominance, would be preferable to assured destruction situations. Even after making the best case for defenses (that is, ignoring questions of technical and economic feasibility, the effect on the probability of superpower conventional wars, and a number of other issues discussed briefly at the end of this article), assured destruction nuclear situations might be preferable to those in which defenses drastically reduce the superpowers' vulnerability to nuclear attack.[27]

The implications of this conclusion for U.S. policy are obvious, and profound. Whether the United States could be more secure in a world of highly effective defenses should no longer be viewed primarily as a technological issue. The United States should examine more completely the strategic and political issues associated with defense against nuclear attack before making a decision to pursue such a fundamental change in nuclear strategy. Given the reservations about deploying defenses raised by the preceding analysis, the recent enthusiasm for BMD must be judged imprudent.

*Additional Problems with Strategic Defense*

This examination of strategic defense has analyzed a best case for defense. Effective defense was hypothesized to be technologically and economically

---

27. The transition to a defensive situation deserves extensive analysis that is beyond the scope of this paper. The preceding analysis can, however, be used to analyze U.S. security during a symmetric transition. A symmetric transition would likely be the safest possible transition.

feasible. Even with these highly controversial assumptions, the decision to pursue the deployment of defense and to make the associated fundamental shift in nuclear strategy is found to have serious shortcomings. The case for effective defense and for starting to deploy defenses in the foreseeable future is further weakened by a number of factors:

1) *Uncertainty.* The effectiveness of U.S. defenses would be uncertain and small uncertainties would be highly significant. In addition to the uncertainties inherent in the operations of complex systems, the effectiveness of defenses would be uncertain due to the severe limits on testing. The defense could not be tested against a full scale attack or against Soviet offenses. And while estimates could be made of effectiveness against deployed Soviet offenses, there would always be reasonable questions about Soviet penetration aids that could be quickly added to their offensive force.

Small uncertainties would be significant because, with the large offensive forces which are currently deployed, a small difference in the percentage of penetrating weapons would translate into a large difference in destructive potential. The uncertainties involved with a defense which was in fact perfect would be likely to be large enough to leave the United States unsure about whether it was vulnerable to an annihilating attack by the Soviet Union.

The effect of uncertainty would affect U.S. policy in a number of ways. The United States would never feel adequately defended. (Nor would the Soviet Union.) Even without uncertainties, there would always be arguments that the United States needed additional defense to improve its protection against Soviet attacks and as a hedge against Soviet offensive breakthroughs. The strength of these arguments would be greater than those made about

---

Therefore, because a transition would probably not be symmetric, the significance of the symmetric transition should not be overestimated. However, because the transition to a defense situation is generally believed to be very dangerous, it is interesting to briefly consider this best case.

Security in transition states would be determined by the same factors as in the defensive situations already studied. Applying the conclusions of the analysis of defensive situations to the transition, we find: 1) Premeditated attacks could be deterred equally well throughout the transition; 2) Crisis stability might not decrease during the transition; and 3) Robustness would decrease during the transition. It follows that the probability of nuclear war in the final state of the transition is likely to be greater than in any of the transition states.

The analysis of defensive situations presented in this article turns the conventional wisdom about strategic defenses on its head. It finds that the desirability of effective defenses, which is usually taken for granted, is at best highly questionable; and that the probability of war during a symmetric transition, which is usually believed to be high, could be lower than in the final defensive state.

the inadequacy of today's offensive forces since defensive capability would start to become redundant only once the defenses were perfect. The existence of uncertainties would be likely to result in unrelenting requests for additional defenses, yet fulfilling these requests would yield little satisfaction and add little to the public's sense of security.

A second effect of uncertainty would be the creation of fears that the Soviet Union had a superior defensive capability. Prudent military analysis could require assessing uncertainties in favor of Soviet defense and against U.S. defense. As a result, if the United States and the Soviet Union had comparable defensive capabilities, U.S. defenses would not provide confidence that the United States was maintaining a strategic nuclear balance and would likely be judged inferior. This conclusion would contribute to the demands for improving defensive capabilities.

2) *Allies.* Any comprehensive analysis of defensive situations must consider the reaction of U.S. allies and the implications for their security.[28] One issue of great importance to them has already been raised, that is, the effect of defenses on the probability of conventional war. If strategic defense were believed to increase the probability of conventional war, then tremendous resistance from the European allies should be anticipated. Conventional wars in Europe are expected to be so costly that they are barely less unacceptable than are nuclear wars to many Europeans. A second concern would focus on the vulnerability of allies to nuclear attack. A policy that drastically reduces the United States' vulnerability to nuclear attack while leaving its European and other allies highly vulnerable cannot look good from their perspective. A third concern would be the effect of defenses on the independent deterrent capabilities of the French and British. A highly effective but imperfect Soviet defense would leave the United States with a modest retaliatory capability, but would probably eliminate the value of these independent European deterrents.

3) *Suitcase bombs.* The ability to defend effectively against ballistic missiles, cruise missiles, and bombers could greatly increase the importance of clandestinely delivered nuclear weapons. Nuclear bombs could be placed on Soviet ships and commercial airplanes, or could be carried into the United States by Soviet agents. These alternative types of delivery are possible today,

---

28. For a discussion of the likely alliance reaction to extensive U.S. homeland defense, see David S. Yost, "Ballistic Missile Defense and the Atlantic Alliance," *International Security*, Vol. 7, No. 2 (Fall 1982), pp. 154–158.

but are not of great importance due to the Soviets' large ballistic missile and air-breathing threats.

These alternative forms of delivery would not necessarily render defense useless: the Soviet ability to deliver clandestinely weapons in a crisis might be severely limited; hiding weapons before a crisis would be risky unless early detection would be impossible; and the damage from clandestine attacks might be less extensive than is currently possible without defenses. Still, the observation that defense against the delivery systems which are most important today would not eliminate vulnerability to nuclear attack raises basic issues about strategic defense: What threats must the United States be able to defend against? How would a "partial defense," that is, a defense against standard delivery systems, affect the nuclear threat? How would highly effective or perfect defense against standard delivery systems affect the political and military uses of nuclear weapons?

*Conclusion*

Strategic defense and the prospect of being invulnerable to nuclear attack have undeniable appeal. But there is no excuse for being romantic or unrealistic about the nature of a world in which the superpowers have built tens of thousands of nuclear weapons and sophisticated delivery systems, and in which the knowledge about these technologies cannot be destroyed. Strategic defense cannot return us to a pre-nuclear world. Defensive situations have not been studied as carefully or extensively as assured destruction situations. There is, however, no reason to believe defensive situations would be either less complex or easier to manage than assured destruction situations.

The best of worlds in which both superpowers have effective defense would not be so good and might not be preferable to today's redundant assured destruction situation: in all but the case of perfect defense, the U.S. would still depend upon deterrence for its security; the lack of robustness in defensive situations would make them sensitive to small changes in forces and would create strong incentives to pursue threatening improvements in offensive forces; the acquisition of these forces would increase the probability of nuclear war; the probability of large conventional wars between the superpowers and their allies might well increase; and, the threat posed by clandestinely delivered nuclear weapons would be much more significant than today.

Any serious policy for deploying defenses must address the dangers that would result from the difficulty of maintaining the defensive situation. This article has argued that no defensive situation could be highly robust. The most robust defensive situations will require superpower cooperation. This brings to the forefront the issue of U.S.–Soviet relations in a defensive world. Recent statements by the President have suggested that effective defenses would eliminate the need for offensive weapons.[29] This outcome is not impossible, but is extremely unlikely. A more realistic assessment is that deploying defenses would lead to an intense offensive and defensive nuclear weapons competition between the superpowers and to tense, strained relations. We should expect that arms control agreements to limit or reduce offensive nuclear forces would be difficult, if not impossible, to negotiate. Careful thought should be given to whether, in a defensive situation, a cooperative relationship between the superpowers would be possible, and to whether the pressures for confrontation could be kept low. If these would not be possible, and I believe they would not be, then the prospects for improving security by shifting to a world of effective defenses must be judged to be especially gloomy.

The United States appears to be at the beginning of a major shift in nuclear weapons policy. There is no evidence that the decision to pursue highly effective defense was based upon a complete analysis of defensive situations. Unfortunately, a world in which both superpowers deployed effective defense is far less attractive than its proponents suggest: even after making the most optimistic assumptions, defensive situations might not be more secure than assured destruction situations; and the more likely outcomes of deploying BMD would place the U.S. in a situation far less secure than today's. Until a convincing argument is presented for this fundamental change in U.S. nuclear weapons policy, the United States should cut back severely on its enthusiasm and funding for strategic defense, attempt to repair the damage that is likely to have occurred in Soviet understanding of U.S. nuclear weapons policy, and pursue with renewed determination a prudent policy of offensive weapons acquisition and arms control.

---

29. *The New York Times*, March 30, 1983, p. 14.

# Preserving the ABM Treaty

## A Critique of the Reagan Strategic Defense Initiative

*Sidney D. Drell,
Philip J. Farley, and
David Holloway*

In his speech on March 23, 1983, President Reagan offered a vision of escape from grim reliance on the threat of retaliation to deter aggression and prevent nuclear war: "What if free people could live secure in the knowledge that their security did not rest upon the threat of instant U.S. retaliation to deter a Soviet attack, that we could intercept and destroy strategic ballistic missiles before they reached our soil or that of our allies?" The way to realize this vision, he said, was to "embark on a program to counter the awesome Soviet missile threat with measures that are defensive. . . . I call upon the scientific community in our country, those who gave us nuclear weapons, to turn their great talents now to the cause of mankind and world peace, to give us the means of rendering these nuclear weapons impotent and obsolete."[1]

This vision appeals to powerful moral sentiments. The impulse to look to our weapons and armed forces to defend ourselves rather than threaten others is a natural one, and is not new. At a press conference in London on February 9, 1967, the Soviet Premier A.N. Kosygin said: "I think that a defensive system which prevents attack is not a cause of the arms race. . . . its purpose is not to kill people but to save human lives."[2] Two years later President Richard Nixon, in explaining his reluctant decision to forgo a nationwide anti-ballistic missile (ABM) defense in favor of the limited Safe-

---

*Sidney Drell is Deputy Director of the Linear Accelerator Center and Co-Director of the Center for International Security and Arms Control, both at Stanford University. Philip Farley is Senior Fellow at the Center for International Security and Arms Control at Stanford and former Deputy Director of the Arms Control and Disarmament Agency. David Holloway is Senior Research Associate at the Center for International Security and Arms Control at Stanford and Reader in Politics at the University of Edinburgh.*

---

This is a much abridged version of *The Reagan Strategic Defense Initiative: A Technical, Political and Arms Control Assessment*, by Sidney D. Drell, Philip J. Farley, and David Holloway, published by the Center for International Security and Arms Control, Stanford University, in 1984. Many people helped in preparing that Report, and the authors would like to express here, in abridged form, the gratitude which is acknowledged at greater length in the Report.

---

1. *The New York Times*, March 24, 1983, p. 20.
2. *Pravda*, February 11, 1967. For a discussion of Kosygin's statement, see Raymond L. Garthoff, "BMD and East–West Relations," in Ashton B. Carter and David N. Schwartz, eds., *Ballistic Missile Defense* (Washington, D.C.: Brookings, 1984), pp. 295–296.

---

*International Security*, Fall 1984 (Vol. 9, No. 2) 0162-2889/84/020051-40 $02.50/1

guard system, said that "although every instinct motivates me to provide the American people with complete protection against a major nuclear attack, it is not now within our power to do so."[3]

The ABM Treaty of 1972 did not reflect any lack of awareness on the part of the United States or the Soviet Union of the arguments for strategic defense. But however desirable nationwide ABM defenses might be in the abstract, they were judged in the early 1970s to be futile, destabilizing, and costly:

*futile* because in a competition between offensive missiles and defensive systems, the offense would win, especially against urban areas;
*destabilizing*, first, because they would speed up the arms race, as both sides developed and deployed not only defensive systems, but also offensive systems to overpower, evade, or attack and disable the opposing ABM defense; second, because each side would fear the purpose or the capability of the other's ABM (especially against a weakened retaliatory strike), and in a crisis these fears might create pressure to strike first;
*costly*, because the offensive countermeasures to maintain the deterrent threat of intolerable retaliatory damage appeared not only capable of overwhelming the defense, but also less costly.

Have these earlier judgments ceased to apply? Has it now become possible to build a leak-proof defense which renders nuclear weapons impotent and obsolete?

Early in 1984, the Department of Defense (DoD) submitted a Strategic Defense Initiative (SDI) program to the Congress. Although that program addresses the President's broad goal of *escaping from* reliance on deterrence, it lays more stress on the alternative or interim goal of *enhancing deterrence.* Dr. Richard DeLauer, Under Secretary of Defense for Research and Engineering, told the House Committee on Armed Services that DoD studies to date have

"concluded that advanced defensive technologies could offer the potential to enhance deterrence and help prevent nuclear war by reducing significantly the military utility of Soviet preemptive attacks and by undermining an aggressor's confidence of a successful attack against the United States or our allies."[4]

---

3. U.S. Arms Control and Disarmament Agency, *Documents on Disarmament 1969* (Washington, D.C.: U.S. Government Printing Office, 1970), p. 103.
4. *Statement on the President's Strategic Defense Initiative* by the Hon. Richard D. DeLauer, Under Secretary of Defense for Research and Engineering before the Subcommittee on Research and Development of the House Committee on Armed Services, March 1, 1984, mimeo, p. 1.

The Fiscal Year 1985–89 SDI program (projected to cost $26 billion) is designed, he said, to obtain the technical information and "experimental evidence" on which to base "an informed decision on whether and how to proceed."[5] Other administration spokesmen have argued that a partially effective defense would have the further advantage of providing some protection to urban-industrial areas and thus limiting damage and saving lives if deterrence should fail.

No specific systems deployment has been proposed. The SDI has been presented as a long-term research and development (R&D) program leading to options from which a President may choose in the 1990s. But it should be understood that the new program differs significantly from a "hedge" against Soviet technical advances or unilateral ABM deployment. It is directed towards a Presidentially stated goal that is at odds with the premises on which Soviet–American strategic relations have been conducted, and arms control negotiations based, for the last fifteen years. The change may eventually prove salutary, the new direction sound. But just now it is unilateral change, to which the Soviet Union will have to respond in accordance with its view of how its own security may be affected. The Soviet Union is not likely to defer adjustments in its own strategic programs and arms control policy until the United States makes its formal deployment decisions five to ten years hence. Thus the SDI is not merely a theoretical enterprise, for it may have a serious impact on the Soviet–American strategic relationship even before any deployment decisions are taken.

Whether the SDI is seen as exploring a leak-proof defense *transcending* deterrence, or as an effective but partial defense *enhancing* deterrence, President Reagan's initiative has raised basic issues that go to the heart of the superpower strategic relationship.

*The ABM Treaty and the U.S.–Soviet Strategic Relationship*

Ever since the atomic bomb was built in 1945, governments have sought to bring nuclear weapons under control. The Baruch Plan proposed international control leading to the eventual elimination of nuclear weapons. When this proved unnegotiable, elaborate schemes were devised and presented in the 1950s for general and complete disarmament, with elimination of nuclear

---

5. Ibid., p. 2.

weapons at the final stages. Also unnegotiable, such approaches were put aside at the end of the 1950s as quixotic in the foreseeable international setting. The goal of eliminating nuclear weapons remains with us, but it is not entertained by governments as feasible in the near term.

The possibility of defense against nuclear attack was examined in the 1950s and 1960s in the United States and the Soviet Union. National leaders, military planners, and defense scientists were naturally reluctant to accept that they could not provide the means of shielding their nations against nuclear attack. Air defenses and civil defenses were adapted to the new conditions of warfare, especially in the Soviet Union. ABM development and initial deployments took on momentum in the 1960s, until parallel reassessments in the Soviet Union and the United States led to their virtual abandonment in the ABM Treaty of 1972.

As a result of this assessment, mutual deterrence became the bedrock of the strategic relationship. In this context the ABM Treaty, along with the agreements on offensive systems and on reducing the risk of nuclear war, has a broader purpose than specific limitations on strategic arms, valuable though these may be in themselves. It defines the common premises of the U.S.–Soviet strategic relationship and lays the basis for the pursuit of political measures to avoid nuclear war and its "devastating consequences for all mankind" (to quote the Preamble to the Treaty).

The basic purpose of severely limiting ABM was not to save money, and surely not to achieve "mutually assured destruction." No government has ever had a policy of pursuing mutual destruction. But during SALT I the two superpowers recognized that mutual destruction could not be escaped (in that sense, was "assured") if they were drawn into nuclear war. If destruction was to be avoided, the driving purpose of national policy had to be the prevention of nuclear war.

In the circumstances, the continued existence of deterrent forces had to be accepted. These forces carried risks, however, which the United States and the Soviet Union tried to reduce by means of various agreements. The 1971 agreement to update the "hot line" ensures that national leaders can communicate with each other if crisis impends, or even during crises. The 1971 agreement on measures to reduce the risk of war commits the two sides to safeguard against accidental or unauthorized use of nuclear weapons, and to provide information in order to lessen suspicions and misinterpretations and to prevent overreaction, in the event of accidents or ambiguous actions. The 1973 agreement on the prevention of nuclear war recognizes that war might grow out of actions by third nuclear powers, or out of military confrontations or other crises, and commits both superpowers to act to avoid nuclear war in such situations.[6]

At the 1972 Moscow Summit, it was envisaged that the two sides would continue to try to improve their relations and mutual understanding. In the *Basic Principles of Relations between the U.S. and the U.S.S.R.*, agreement is set forth on these points:

First. They will proceed from the common determination that in the nuclear age there is no alternative to conducting their mutual relations on the basis of peaceful coexistence. Differences in ideology and in the social systems of the USA and USSR are not obstacles to the bilateral development of normal relations. . . .

Second. The USA and the USSR attach major importance to preventing the development of situations capable of causing a dangerous exacerbation of their relations. Therefore, they will do their utmost to avoid military confrontations and to prevent the outbreak of nuclear war.[7]

Such general statements often raise excessive hopes, to be followed by disappointment; this one has been no exception. The quoted principles, nevertheless, are not discredited by the inability of the rival nations and their changing leaders to apply them consistently. They remain the only basis on which arms control can progress or have more than a precarious *ad hoc* effect. And these principles did have one concrete effect on the strategic relationship through the ABM Treaty.

The ABM Treaty is a milestone in the political approach to preventing nuclear war.[8] It is a *realpolitik* approach, not an idealistic one. It accepts deterrence as a present necessity and objective condition—not as an active threat, which would be intolerable. In the phrase reiterated by the Soviets, it is a recognition that under current conditions it would be tantamount to suicide for a nation to initiate nuclear war—a sobering and restraining ("deterring") realization.[9] In that sense deterrence is prudent reciprocal *self-deterrence* of initiation of nuclear war, not only a "threat" against an opponent.

---

6. The texts of these agreements and of the ABM Treaty may be found in *Arms Control and Disarmament Agreements: Texts and Histories of Negotiations* (Washington, D.C.: U.S. Arms Control and Disarmament Agency, 1980).

7. The text of this agreement may be found in Coit D. Blacker and Gloria Duffy, eds., *International Arms Control: Issues and Agreements*, 2nd ed. (Stanford, Calif.: Stanford University Press, 1984), pp. 429–430.

8. This approach is set out particularly in the Preambles to the ABM Treaty and the "Accident Measures" Agreement, and in the Agreement on the Prevention of Nuclear War.

9. See below, notes 15 and 19.

ABM deployments were severely limited, not from any antipathy to defending one's nation and population, but because, in a world of nuclear weapons and ballistic missiles, they were judged to be not only futile and costly, but destabilizing and dangerous as well. Only by actively preventing nuclear war, not by attempting defense against missile attack in the course of such a war, could nations and people be protected from its catastrophic effects.

The approach by both sides in SALT I was thus to give priority to prevention of nuclear war in a context of common vulnerability to the devastation that such a war would bring. Taken together, the measures adopted in the early 1970s constitute a strategy for managing the nuclear peril which should be assessed both in terms of its rationale and its continuing promise as a means of reducing the risk of nuclear war.

Although President Reagan said in his March 23 address that the first step of his initiative would be consistent with the ABM Treaty, his speech took direct issue with the rationale of the Treaty, because he promised escape from the relationship of mutual deterrence that the Treaty is designed to help manage. Even though the SDI is conditionally framed (to examine *whether* there are technologies that *might emerge* in the distant future and that hold *some promise* of defense against missile attack), it will sooner or later run up against specific provisions of the ABM Treaty.

Article V.1 of the Treaty, for example, specifies a number of types of ABM components, or modes of deployment, which are categorically prohibited:

Each Party undertakes not to develop, test, or deploy ABM systems or components which are sea-based, air-based, space-based, or mobile land-based.

The inconsistency between Article V.1 and the ABM systems envisaged in the President's speech is self-evident. For the SDI program, the first issue in assuring adherence to the ABM Treaty relates to what should be understood by the terms "development" and "testing." This issue will arise as the DoD proceeds with technology demonstrations of air- or space-based ABM components.[10]

---

10. One broad area of ambiguity in applying the treaty is the scope of the term "development." Ambassador Gerard Smith provided the Nixon Administration's interpretation of the term on July 18, 1972, in his testimony before the Senate Armed Services Committee:

Article V of the ABM Treaty and an Agreed Interpretive Statement (E) obligate the U.S. and U.S.S.R. "not to develop, test, or deploy" mobile ABM systems, rapid reload devices, or ABM interceptor missiles for delivery of more than one independently guided warhead.

The Treaty was so drafted as to assist in coping with ambiguities, some identifiable in 1972 and some only vaguely foreseeable. Article II defines an ABM system as "a system to counter *strategic* [emphasis added] ballistic missiles or their elements in flight trajectory. . . ." Launchers, interceptors, and radars intended for use against aircraft and nonstrategic missiles might have, or be given, a significant capability against strategic ballistic missiles. Article VI specifies that the *capability* of such systems to serve in an ABM capacity is limited, and that testing of such components, regardless of their primary purpose, in an ABM mode is banned:

To enhance assurance of the effectiveness of the limitations on ABM systems and their components provided by the Treaty, each Party undertakes:
(a) not to give missiles, launchers, or radars, other than ABM interceptor missiles, ABM launchers, or ABM radars, capabilities to counter strategic ballistic missiles or their elements in flight trajectory, and not to test them in an ABM mode; . . .

The U.S. pushed vigorously for this provision during the negotiations, both to clarify the application of the treaty, and to facilitate verification.

---

The SALT negotiating history clearly supports the following interpretation. The obligation not to develop such systems, devices or warheads would be applicable only to the stage of development which follows laboratory development and testing. The prohibitions on development contained in the ABM Treaty would start at that part of the development process where field testing is initiated on either a prototype or breadboard model. It was understood by both sides that the prohibition on "development" applies to activities involved after a component moves from the laboratory development and testing stage to the field testing stage, wherever performed. The fact that early stages of the development process, such as laboratory testing, would pose problems for verification by national technical means is an important consideration in reaching this definition. Exchanges with the Soviet Delegation made clear that this definition is also the Soviet interpretation of the term "development."

Consequently, there is adequate basis for the interpretation that development as used in Article V of the ABM Treaty and as applied to the budget categories in the DoD RDT&E program places no constraints on research and on those aspects of exploratory and advanced development which precede field testing. Engineering development would clearly be prohibited. (*Military Implications of the Treaty on the Limitations of Anti-Ballistic Missile Systems and the Interim Agreement on Limitation of Strategic Offensive Arms,* Hearing before the Senate Committee on Armed Services, June–July 1972 [Washington, D.C.: U.S. Government Printing Office, 1972], p. 377.)

This statement, and particularly its final sentence, provides illuminating background for an observation of the Scowcroft Commission in its final report to the President on March 21, 1984: "the strategic implications of ballistic missile defense and the criticality of the ABM treaty to further arms control agreements dictate extreme caution in proceeding to engineering development in this sensitive area." (*President's Commission on Strategic Forces, Report of March 21, 1984,* mimeo, p. 8)

DeLauer, *Statement,* pp. 4, 5, lists a number of proposed hardware demonstrations for the late 1980s and early 1990s. One can foresee sharp disagreements between the parties, as each represents its own R&D activities as falling within the permitted early stages of the development process, but argues that detectable or suspected activities of the other raise serious questions of compliance.

In a Unilateral Statement, neither challenged nor endorsed by the Soviet side, the U.S. delegation listed a number of examples of what would be considered "tested in an ABM mode."[11] Since then, the U.S. has been vigilant in monitoring and challenging Soviet activities, mainly in Soviet air defense programs, which might be inconsistent with this provision. Currently the United States has programs underway to determine whether the U.S. Army's *Hawk* and *Patriot* surface-to-air missile systems for air defense could be given anti-tactical missile capabilities. If such developments go ahead, the question of inherent ABM capability and consistency with Article VI will arise.

Agreed Statement D attempts to lay a basis for dealing with ways in which technology might change the means used for ABM defense (which Article II describes as *currently* consisting of ABM interceptor missiles, ABM launchers, and ABM radars with defined characteristics). It reads as follows:

In order to insure fulfillment of the obligation not to deploy ABM systems and their components except as provided in Article III of the Treaty, the Parties agree that in the event ABM systems based on other physical principles and including components capable of substituting for ABM interceptor missiles, ABM launchers, or ABM radars are created in the future, specific limitations on such systems and their components would be subject to discussion in accordance with Article XIII and agreement in accordance with Article XIV of the Treaty.

Since the contingency referred to has not yet arisen—although it is imminent—the precise meaning of the statement has not been finally determined, and its application has not been established. The literal meaning is that such "exotic" ABM systems or components could only be developed, tested, or deployed *after* amendment of the Treaty (Article XIV). Taken in conjunction with Article V, with its categorical prohibitions, there seems to be no alternative in the case of sea-, air-, space-, and mobile land-based systems. Thus the effect of the Agreed Statement is to make clear that "exotic" systems or technologies are not exempt from the limitations of the Treaty because not mentioned specifically, but that they are not ruled out permanently *provided* the parties amend the Treaty first to permit them.

Such possibilities might be identified and explored in the Standing Consultative Commission (SCC), though that would not be the exclusive channel, of course. If the United States pursues a space-based ABM system development and testing program, the Soviet Union has a clear option to raise

---

11. *Arms Control and Disarmament Agreements,* p. 147.

questions in the SCC regarding its intent and character. The United States has an obligation under Article XIII.b to "provide on a voluntary basis" such information as will answer these questions if the Soviet Union should raise them.

Quite aside from the question of legal obligation, such action would appear prudent to avoid suspicions and reactions, particularly if we attach importance to the ABM Treaty until some better approach to strategic stability and arms control is jointly accepted by both nations.

SOVIET PERSPECTIVES ON THE ABM TREATY

Soviet ABM policy in the early and mid-1960s was rooted in both military strategy and political sentiment. ABM systems were designed, as Marshal V.D. Sokolovskii said in his *Military Strategy*, to limit the damage that an enemy nuclear strike would cause, and to enable the state and the armed forces to continue to function in time of war.[12] It was also hoped that they would make it possible for the Soviet Union to base its security on something other than the balance of terror.

But in the late 1960s, Soviet attitudes began to change. In 1967 and 1968, as the Soviet leaders were preparing for SALT, some military leaders expressed doubts about the effectiveness of ABM systems. The Commander in Chief of the Strategic Rocket Forces, Marshal N.I. Krylov, stated that offensive missiles had characteristics that in practice guaranteed the "invulnerability of ballistic missiles in flight, especially when employed in mass."[13] In 1967 and 1968, work on the ABM system around Moscow slowed down, and only four of the eight complexes were completed, suggesting that the Soviet authorities had little confidence in its effectiveness.

Alongside growing doubts about ABM systems came increasing confidence in the deterrent power of the Soviet Union's offensive strategic forces. The Soviet Union was now approaching strategic parity with the United States. In 1973 Major-General M. Cherednichenko, one of Sokolovskii's closest collaborators, wrote in the classified General Staff journal *Military Thought* that the Soviet Union had now

acquired the capability of delivering a devastating nuclear response to an aggressor in any and all circumstances, even under conditions of a sneak

---

12. Marshal V.D. Sokolovskii, *Voyennaya Strategiya* (Moscow: Voyenizdat, 1962), pp. 231, 271.
13. Marshal N.I. Krylov, "Raketnye voiska strategicheskogo naznacheniya," *Voyenno-istoricheskii zhurnal*, No. 7 (1967), p. 20.

nuclear attack, and of inflicting on the aggressor a critical level of damage. An unusual situation developed: an aggressor who would initiate a nuclear war would irrevocably be subjected to a devastating return nuclear strike by the other side. It proved unrealistic for an aggressor to count on victory in such a war, in view of the enormous risk for the aggressor's own continued existence.[14]

The Soviet Union acknowledged at SALT that each side was vulnerable to a devastating retaliatory strike if it attacked first. The opening Soviet statement (as summarized by Gerard Smith, the chief American negotiator) said that:

a situation of mutual deterrence existed. Even in the event that one of the sides was the first to be subjected to attack, it would undoubtedly retain the ability to inflict a retaliatory blow of destructive force. It would be tantamount to suicide for the ones who decided to start war.[15]

Soviet policy at SALT was based on the recognition that this nuclear balance could be upset by the deployment of either offensive or defensive systems. As Major-General V.M. Zemskov noted in *Military Thought* in May 1969, the balance could be disrupted either by a sharp increase in one side's offensive forces, or by "the creation by one of the sides of highly-effective means of protection from a nuclear attack of the enemy in conditions when the other side lags considerably in resolution of these missions."[16]

It seems likely that the fear of an unconstrained race in ABM systems played some part in convincing the Soviet leaders of the desirability of the ABM Treaty. But this was not the only factor in their decision to sign the Treaty. They were aware that the deployment of ABM systems would stimulate the further development of offensive forces. During the exploratory moves before SALT, the Soviet Union had asked that defensive and offensive systems be considered together. After the ABM Treaty was signed, Marshal A.A. Grechko, the Defense Minister, claimed that it would prevent "the development of a competition between offensive and defensive rocket-nuclear weapons."[17]

Although the Soviet ABM discussion was very different from the contemporary American debate, some of the arguments advanced against ABM

---

14. Maj. Gen. M. Cherednichenko, "Military Strategy and Military Technology," *Voyennaya Mysl'*, No. 4 (1973), FPD 0043, November 12, 1973, p. 53.
15. Gerard Smith, *Doubletalk: The Story of SALT I* (New York: Doubleday, 1980), p. 83.
16. Maj. Gen. V.M. Zemskov, "Wars of the Modern Era," *Voyennaya Mysl'*, No. 5 (1969), FPD 0117/69, p. 60.
17. *Pravda*, September 30, 1972.

systems were the same: they would not be effective against offensive missiles; they would spur the other side into increasing its offensive forces; they could upset the nuclear balance, and this might increase the danger of war (if the United States gained the upper hand).

The recognition of mutual vulnerability has remained central to the Soviet conception of the Soviet–American strategic relationship. Leonid Brezhnev, for example, told the 26th Party Congress in 1981 that

the military and strategic equilibrium prevailing between the USSR and the USA . . . is objectively a safeguard of world peace. We have not sought, and do not now seek, military superiority over the other side.[18]

He also declared that "to try to outstrip each other in the arms race or to expect to win a nuclear war is dangerous madness." In September 1983, Marshal N.V. Ogarkov, the Chief of the General Staff, wrote that

with the modern development and dispersion of nuclear arms in the world, the defending side will always retain such a quantity of nuclear means as will be capable of inflicting "unacceptable damage," as the former Defense Secretary of the USA R. McNamara characterized it in his time, on the aggressor in a retaliatory strike.[19]

And in words that echo the opening Soviet statement at SALT in November 1969, he added that "in contemporary conditions only suicides can wager on a first nuclear strike."

The Soviet reaction to President Reagan's March 23 speech shows a very clear awareness that this is the character of the present strategic relationship. Four days after the speech, Yuri Andropov stated that the defensive measures Reagan spoke of would seem defensive only to "someone not conversant with these matters."[20] The United States would continue to develop its strategic offensive forces, he said, and

under these conditions the intention to secure itself the possibility of destroying with the help of ABM defenses the corresponding strategic systems of the other side, that is of rendering it incapable of dealing a retaliatory strike, is a bid to disarm the Soviet Union in the face of the United States nuclear threat.

American weapons programs—notably the MX ICBM, the Trident D-5 SLBM, the Pershing II, and the cruise missile programs—have been portrayed by

---

18. *Pravda*, February 24, 1981.
19. *Krasnaya Zvezda*, September 23, 1983.
20. *Pravda*, March 27, 1983.

the Soviet leaders as part of a concerted American effort to achieve strategic superiority. Before the Reagan speech, they had expressed the fear that even if these programs did not enable the United States to escape from the threat of a retaliatory strike, they might give it a strategic advantage.[21] In this context, it is not surprising that the immediate Soviet reaction was to view President Reagan's speech as further evidence of the drive for superiority.

Andropov's criticism of Reagan's speech was rooted firmly in the ABM Treaty. At SALT I, he said, the Soviet Union and the United States "agreed that there is an inseverable interconnection between strategic offensive and defensive weapons." "Only mutual restraint in the field of ABM defenses" would make progress possible in limiting and reducing offensive weapons. Now, said Andropov, "the United States intends to sever this interconnection. Should this conception be translated into reality, it would in fact open the floodgates to a runaway race of all types of strategic arms, both offensive and defensive."

By the end of 1983, more detailed Soviet analyses of the issue had been made. In a report on the prospects for a space-based ABM system, a working group from the Academy of Sciences argued that the creation of a space-based ABM that could destroy 1,000 ICBMs in their boost phase was beyond current technological capabilities, and would require significant expansion and intensification of R&D work.[22] Even if it could be developed, such a system would be vulnerable to destruction by space mines, retroflectors of laser beams, antisatellite weapons, and powerful ground-based lasers.[23] Various countermeasures—such as shielding missile launchers with reflective and ablative materials—would make it more difficult for lasers to destroy the missiles. The report estimated that it would be much cheaper to develop effective means of destroying space-based systems than to develop the systems themselves.

---

21. See especially Defense Minister D.F. Ustinov's statement that "superiority is understood as synonymous with the attainment of a capability to strike a blow at the Soviet Union when and where Washington considers it expedient, calculating on the fact that the retaliatory blow at the USA will become less powerful than in other conditions." *Pravda*, July 12, 1982.

22. *Prospects for the Creation of a U.S. Space Ballistic Missile Defense System and the Likely Impact on the World Military Political Situation*, Report of the Committee of Soviet Scientists in Defence of Peace and Against the Threat of Nuclear War, Moscow, 1983, mimeo. The working group was headed by Academician R.Z. Sagdeev, Director of the Institute of Space Research, and Dr. A.A. Kokoshin of the Institute of the U.S.A. and Canada.

23. "The Soviet Union could deploy antisatellite lasers to several ground sites in the next ten years," according to *Soviet Military Power*, 3rd ed. (Washington, D.C.: U.S. Government Printing Office, 1984), p. 35.

The report's strategic conclusions follow from this analysis. If a space-based ABM is vulnerable to destruction, then it cannot provide effective defense against a first strike, since the attacking side will be able to destroy the system. But such a system might give rise to the illusion that it could provide a relatively effective defense against a ragged retaliatory strike that has already suffered some attrition. The deployment of such a system would therefore have to be seen by the other side as a very threatening move. Under these circumstances, each side—both the side that had ABM and the one that did not—would have an incentive to strike first. The net effect of deploying such a system would not be to provide an escape from mutual deterrence, but to make that relationship less stable.

The initial Soviet reaction to President Reagan's speech has been totally hostile. The SDI has been portrayed as potentially a very serious threat to Soviet security, and as a rejection of the ABM Treaty and its underlying premises. All the Soviet commentary has pointed to the inevitability of a Soviet response to counter an American effort to develop and build an effective nationwide ABM.

#### SOVIET ABM ACTIVITIES

Since the late 1960s the Soviet leaders have recognized mutual deterrence as an objective condition, but they have not embraced it with enthusiasm. They have shown concern about the possibility of war by accident or miscalculation. They have also thought it prudent to prepare for nuclear war, in case it should occur. According to the *Military Encyclopedic Dictionary*, published in 1983 and edited by Marshal Ogarkov,

in accordance with Soviet military doctrine, which has a profoundly defensive character, the main task of Soviet military strategy is to develop the means of repulsing the aggressor's attack and then defeating him utterly by means of conducting decisive operations.[24]

Even within the confines of mutual vulnerability, Soviet military strategy still focuses on how to wage a nuclear war and defeat the enemy, if such a war should occur. This has contributed to Western fears that the Soviet Union is planning to break out of the ABM Treaty.

Since 1972, the Soviet Union has maintained a steady R&D effort in ABM technologies. This effort has five main elements. First, the Soviet Union has

---

24. *Voyennyi Entsiklopedicheskii Slovar'* (Moscow: Voyenizdat, 1983), p. 712.

been upgrading the Moscow ABM system, which had become fully operational in 1970 or 1971. In 1980 it began to replace the *Galosh* interceptor missiles with SH-04 and SH-08 nuclear-armed missiles. The SH-04 is an exoatmospheric missile like *Galosh,* and the SH-08 is a hypersonic endoatmospheric missile like the American *Sprint.* Now the Moscow system can use atmospheric sorting to discriminate between real reentry vehicles and decoys. New phased-array radars (the Pushkino radar) are being built to perform the engagement function.[25]

These measures will make the Moscow system more effective against very limited or accidental strikes. But the radars are vulnerable to attack, and without them the system would be crippled. The number of interceptor missiles that the Soviet Union can deploy is still limited to 100 by the ABM Treaty, while the reentry vehicles on the United States' offensive missiles have increased greatly in number and sophistication since 1972. The upgrading of the Moscow system does not mark a significant shift in the Soviet–American strategic balance.

Second, it is possible that the SH-04 and SH-08 missiles might form part of a system (designated the ABM-X-3 defense system by the United States) that could be deployed rapidly to create a nationwide defense. New tracking and missile guidance radars have reportedly been developed. These radars are designed modularly (which is now standard production practice in the United States as well) and are said to be suitable for fairly rapid deployment. With its large radar and missile industries, the Soviet Union could produce the components of the ABM-X-3 system on an extensive scale. A recent report by the CIA has raised the specter of a Soviet breakout from the ABM Treaty.[26]

There is, however, no firm evidence that the Soviet Union is preparing to do this. Even with modularly designed radars it would take years rather than months to deploy the new system on a significant scale. Such a system would not be leakproof and could be overwhelmed, and the radars would be vulnerable to attack. If the Soviet Union attempted a breakout, the United States would have time to respond by improving the penetration of its offensive forces.

---

25. For an excellent analysis of Soviet ABM programs, see Sayre Stevens, "The Soviet BMD Program," in Carter and Schwartz, eds., *Ballistic Missile Defense,* pp. 182–220. On more recent developments, see Clarence A. Robinson, Jr., "Soviets Accelerate Missile Defense Efforts," *Aviation Week and Space Technology,* January 16, 1984, pp. 14–16.
26. The CIA report is discussed in Robinson, "Soviets Accelerate Missile Defense Efforts."

Third, near Krasnoyarsk the Soviet Union is constructing a radar that U.S. intelligence analysts believe is similar to the early warning radars at Pechora, Komsomol'sk-na-Amure, and Kiev. The Krasnoyarsk radar appears to be oriented outwards towards the northeast, thus filling a gap in the Soviet early warning system. The Reagan Administration claims that this radar "almost certainly" violates Article VI.b of the ABM Treaty, which limits the deployment of ballistic missile early warning radars to locations along the borders of the two countries and requires that they be oriented outwards. The Soviet Union has said that the radar is designed for space tracking and is thus consistent with the Treaty.[27]

The purpose of the Krasnoyarsk radar, which has not yet begun to radiate, remains unclear. It may indeed be an early warning radar. Some reports suggest that it has been built where it is because the permafrost makes construction nearer the northeast coast very difficult. But this would be a poor excuse for violating the ABM Treaty. The military significance of this radar is negligible while it stands in isolation from interceptor missiles, and it would be vulnerable to destruction or blackout in a nuclear attack. But the political significance of this violation would be considerable, for it would suggest that the Soviet Union did not take its Treaty obligations seriously.

Fourth, surface-to-air missiles (SAMs), which caused difficulties in the SALT I negotiations, have become even more problematic. Soviet air defenses have responded to the challenge of new aerodynamic systems, including cruise missiles, by improving their capability to deal with smaller radar cross sections and shorter reaction times. As a result, the capability of SAMs against ballistic missile reentry vehicles has been enhanced.

The problem is most dramatically illustrated by the development of the mobile SA-12, which is reported to have been tested not only against aerodynamic systems, but also against ballistic missile reentry vehicles.[28] These tests would violate Article VI.a of the ABM Treaty if those reentry vehicles were similar to the reentry vehicles on ICBMs or SLBMs. This suggests that it may be necessary to modify the Treaty to cope with the growing ambiguity created by advances in this area of technology.

---

27. *The President's Report to the Congress on Soviet Noncompliance with Arms Control Agreements* (Washington, D.C.: The White House, Office of the Press Secretary, January 23, 1984), mimeo, pp. 3–4. See also Robinson, "Soviets Accelerate Missile Defense Efforts"; and Philip J. Klass, "U.S. Scrutinizing New Soviet Radar," *Aviation Week and Space Technology*, August 22, 1983, pp. 19–20.
28. Stevens, "The Soviet BMD Program," pp. 214–216; Robinson, "Soviets Accelerate Missile Defense Efforts."

Fifth, the Soviet Union has been doing research into the use of directed energy for ABM purposes. Most estimates of the quality of this research put it on a level with that in the United States.[29] It would be surprising if the Soviet Union were not engaged in research of this kind, at least as a hedge against the possibility of a technological breakthrough. According to the Department of Defense, however, the Soviet Union lags in other technologies that are crucial for ABM: computers, optics, automated control, electro-optical sensors, propulsion, radar, software, telecommunications, and guidance systems. The only area in which the Soviet Union is said to enjoy a substantial lead is in large rockets that could lift heavy loads into space.[30] The level of Soviet technology suggests that if the Soviet Union tried to break out of the ABM Treaty by deploying an advanced ABM system, it could not develop an effective system so rapidly as to deprive the United States of its ability to deliver a massive retaliatory strike.

Several conclusions can be drawn about Soviet ABM R&D and deployment activities. First, although the Soviet Union may have violated some provisions of the ABM Treaty (and that is not proven), it has not broken out of the Treaty, in the sense of deploying an extensive ABM system. Second, the Soviet Union could not break out so rapidly as to endanger the deterrent power of the United States' offensive forces. Third, although the Krasnoyarsk radar may violate the Article VI.b of the Treaty (and that is not yet proven) its military significance would be negligible. Fourth, continuing Soviet SAM development against tactical and intermediate-range ballistic missiles, cruise missiles, and aircraft has complicated the problem of drawing the line between "ABM capabilities" and permitted SAM systems.

Soviet ABM R&D since 1972 has been steady and unfrenzied. It can be interpreted as an effort to develop defenses in those areas not covered by the Treaty, in order to cope with new offensive threats—cruise missiles and Pershing IIs, for example. It can be viewed as a hedge against an American breakout from the ABM Treaty—a hedge that may look sensible ten years from now. It may also be aimed at exploring new technologies that may one day permit a radically new balance between the offense and the defense. It

---

29. For example, in 1982 the DoD estimated that the Soviet Union and the United States were equal in the directed energy basic technology area. See *The FY 1983 Department of Defense Program for Research, Development, and Acquisition,* Statement by the Hon. Richard D. DeLauer, Under Secretary of Defense for Research and Engineering, to the 97th Congress, p. II-21.

30. Ibid. See also R. Jeffrey Smith, "The Search for a Nuclear Sanctuary (I)," *Science,* July 1, 1983, p. 32.

is impossible to be sure what has motivated the ABM R&D effort. The available evidence points to a Soviet reluctance to embrace the relationship of mutual deterrence with enthusiasm, but does not justify claims that the Soviet Union has broken out of the ABM Treaty, or is preparing to do so in the near future.

*SDI: A Technical Appraisal*

Up to the present, the dominance of offense over defense has been based on technical considerations. In recent years technology has advanced significantly, removing some of the shortcomings of previous defense concepts. Great strides have been made in the ability to produce, focus, and aim laser and particle beams of increasingly high power. These new "bullets" of directed energy, which travel at or near the velocity of light, have led to revolutionary new ideas for defense against ballistic missiles. There has also been a revolutionary expansion in our ability to gather, process, and transmit vast quantities of data efficiently and promptly. This makes it possible to provide high quality intelligence from distant parts of the earth and space in order to assess and discriminate the properties of attacks very promptly.

Does technology now offer a new promise of changing the conditions of offense dominance? Although we cannot build an effective defense based on what is known today,[31] can we now foresee the possibility of building an effective defensive system? And if so, under what conditions?

The major technical fact that has not changed is the destructive power of nuclear weapons. To speak of making nuclear weapons impotent and obsolete by defending one's people, industries, and cities against a massive nuclear attack still requires a defense that is almost perfect. Technical assessments of ABM concepts cannot escape this awesome systems requirement. If but 1 percent of the approximately 8,000 nuclear warheads on the current Soviet force of land-based and sea-based ballistic missiles succeeded

---

31. Dr. James C. Fletcher, Chairman of the Defensive Technologies Study Team (summer of 1983), whose report has been used by the Departments of Defense and Energy as the technical basis for the Strategic Defense Initiative, testified that "I would like to say at the outset that no one knows how effective defensive systems can be made, nor how much they might cost." The Fletcher report was concerned with defining a technology development program toward what is "conceivable" for an effective defense. *The Strategic Defense Initiative*, Statement by Dr. James C. Fletcher before the Subcommittee on Research and Development of the House Committee on Armed Services, March 1, 1984, p. 6.

in penetrating a defensive shield and landed on urban targets in the United States, it would be one of the greatest disasters in all history!

A defensive system against ballistic missiles is formed from many components, all of which are crucial to its effective operation. These include the sensors providing early warning of an attack; the communication links for conveying that information to analysis centers for interpretation, to the command centers with authority to make decisions as to the appropriate national response, and to the military forces to implement the decisions; the sensors of the ABM that acquire, discriminate, track, point, fire, and assess the effectiveness of the attack; and finally the interceptors or directed energy sources that make the kill. The systems for managing the battle and for delivering destructive energy concentrations with precision must be operational at the initiation of an attack and must remain effective throughout. This means being on station, yet being able to survive direct attack. The ability to satisfy these two requirements simultaneously is a major challenge.

Because no single technology can provide an impenetrable defensive shield, the concept of a layered defense has been developed. The first layer attacks the rising missiles during their boost phase while their engines are burning. Typically, this phase lasts three minutes for modern missiles with solid fuel motors, and up to five minutes for liquid-fuel boosters. During this time the missile rises above the atmosphere to heights of two to three hundred kilometers (km). The second (and perhaps third) layer of the defense attacks the post-boost vehicle as well as the reentry vehicles during their mid-course trajectories, which last about 20–25 minutes. The final or terminal layer of the defense attacks the reentry vehicles during the last minute or two of their flight as they reenter the atmosphere, which strips away the lighter decoys accompanying them in mid-flight. A three-layer system, each of whose layers is 90 percent effective, would allow only 8 out of an attacking force of 8,000 RVs to arrive on target and would, if achievable, provide a nearly perfect defense.

BOOST-PHASE INTERCEPT

The possibility of boost-phase intercept is the principal new element in considering ABM technologies. It also has the highest potential payoff for two reasons:

1. Whatever success is achieved in this initial layer of the defense reduces the size of the attacking force to be engaged by each subsequent layer.
2. If a missile is destroyed during boost, all of its warheads and decoys are destroyed with it.

Following the missile boost phase, the defense has more time to acquire and discriminate warheads from decoys, attack its targets, and confirm their destruction. But it must also cope with many more objects since a single large booster is capable of deploying tens of warheads and many hundreds of decoys. Thus the two defensive layers for boost-phase and for mid-course intercept face very different technological challenges. An effective boost-phase layer which greatly reduces the number of objects that subsequent layers must destroy is crucial to the overall effectiveness of a defensive system.

SPACE-BASED CHEMICAL LASERS

One of the most widely discussed systems for boost-phase intercept is a constellation of high energy lasers based on platforms orbiting the earth in space. Very well-focused laser beams have the attractive feature for ABM of traveling vast distances with the speed of light in space above the atmosphere. The disadvantages of space-based lasers are that they are complex and expensive, they are vulnerable to attack, there are many effective countermeasures available to the attacker, and generally their beams are degraded by scattering and absorption by the atmosphere and so they must function above it. Furthermore, each platform of any space-based system will be on station over the launch area of Soviet ICBMs only a small percentage of the time as it circles the globe in a low earth orbit. Many platforms will be needed if the defense is to provide continual protection against ICBM launches.[32]

The requirements for a system using chemical lasers can be illustrated by extrapolating this technology far beyond what has been demonstrated today to what is believed to be attainable in the future. Let us first calculate the kill potential of such a laser against an ICBM booster during second stage burn starting about one minute following launch when it has risen above most of the atmosphere. The kill range will increase with the output power of the laser and with the ability of the optical system to create a well-focused beam; and this range will decrease if the target is hardened against higher values of energy per unit area which are required in order to cause thermo-mechanical damage to the relatively thin skin of the booster.

In order to get an idea of the numbers involved, consider a hydrogen-fluoride (HF) chemical laser with a wavelength of 2.7 microns. High powers

---

32. In order to avoid this, it would be necessary to achieve effective operating ranges from geosynchronous orbits at an altitude of 36,000 km. This is not a practical prospect.

and efficiencies have been demonstrated for continuous wave operation of such lasers. For focusing, consider an optically perfect 4-meter-diameter mirror so that the diffraction limited value of the beam spread is $\theta_d = 0.54 \times 10^{-6}$ radians.

We envisage a power output of 100 megawatts (MW) as an ambitious and feasible goal for an earth-orbiting chemical laser in the infrared region of the spectrum. Our assumptions of 100 MW power output and of 4-meter-diameter main optics performing at the diffraction limit correspond to a system whose brightness is $4.4 \times 10^{20}$ watts per steradian. Such a high value of brightness is more than an order of magnitude beyond the level currently envisaged for technology demonstration during the 1980s[33] and will require major technical advances.

To continue this illustration we must specify the kill fluence, i.e., the energy that must be deposited per unit of area of the missile in order to destroy it. It is envisaged that boosters can be hardened to withstand the thermo-mechanical effects of a fluence of 20 kilojoules (kJ) delivered within a short time interval (several seconds) to one square centimeter of their surface.[34] At a range of 1,500 km, the beam spot size in the optically perfect laser system described above will be approximately 0.5 square meters, corresponding to a power density of 20 kW/cm² (kilowatt per square centimeter). It thus takes one second to deliver a kill fluence of 20 kJ/cm² at this range. During this second, the missile will travel about 7 km along its path and the laser beam will have to track it accurately in order to hold the beam spot on the same part of the booster as it advances. This will require an aiming and tracking accuracy better than used above for intercept.

With the above reference parameters we can now compute the total energy that must be orbited in space and the total number of laser battle stations that are required to form an effective defensive layer against a maximum ICBM attack from the Soviet Union. For each ICBM attacked at a range of 1,500 km, a total energy of about 100 MJ must be expended. In order to

---

33. A 15 MW laser with 10-meter diameter perfect optics has approximately the same value of brightness. It is intended to demonstrate a 2 MW chemical IR laser in 1987 and scale upward to 10 MW "almost immediately," and to demonstrate beam control of 4-meter segmented optics in 1988. See Clarence A. Robinson, Jr., "Panel Urges Defense Technology Advances," *Aviation Week and Space Technology*, October 17, 1983, p. 17.

34. Present-day boosters are typically rated as being able to withstand fluences of 1kJ/cm² delivered to their few-millimeter-thick skins. An increase in their skin thickness by a fraction of a centimeter layer of heat shield material such as carbon is one ready way to increase hardness to the value used in the text. The added weight of the heat shield (about 6 kg per square meter of surface area) could be offset by a relatively small reduction in missile payload (e.g., by removing one RV on an SS-18 ICBM).

defend against the current total Soviet force of 1,400 ICBMs in an all-out launch against U.S. targets, a minimum of $140 \times 10^9$ joules (140 GJ) is needed.

Much more energy than this must be lifted to space orbit, however. As they circle the earth, most of the lasers in a constellation of defensive satellites will be far beyond their maximum effective kill range from the ICBM launch areas. In addition, more than one laser must be within kill range in order to attack all 1,400 Soviet ICBMs, whenever and however they are launched, before they complete their boost phase. To illustrate, let us assume that our conceptual 100 MW laser system can, within 2 seconds, acquire, track, aim, attack, assess damage, attack again if necessary, and realign and stabilize its optics to its next target. This means that one laser at most can destroy something like 90 missiles, assuming that the boost phase lasts for three minutes at altitudes above most of the atmosphere, i.e., above 15–20 km. Thus, no fewer than 16 laser stations must always be "on station" above the Soviet ICBM launch area.

Since the earth is rotating under the orbit of the laser stations, many clusters will have to be orbited in order to ensure that there are always 16 within range whenever the Soviets might choose to launch their ICBMs. Characteristically, a laser with a 1,500 km kill range is on station 5 percent of the time[35] over the Soviet launch areas and therefore a total of $16 \times 20 = 320$ laser battle stations is the minimum number required for constant coverage against a massive ICBM attack.

With the assumptions of this example, it is necessary to lift a total energy into orbit for powering the lasers of $20 \times 140$ GJ. Thus, even with very optimistic assumptions about what can be achieved *in principle*, one finds that 6,000 tons of fuel must be lifted into orbit to power such a constellation of hundreds of lasers.[36] This requirement adds up to more than 250 shuttle loads of fuel alone.[37] If the engagement time is reduced from three to two

---

35. This is explained simply as follows: four planes, each with 13 satellites in low earth orbit separated from one another by 3,000 km, will keep each point of the earth's surface at the latitude of Soviet territory always within a 1,500 km operating range of at least one satellite. Since the Soviet ICBM fields are spread out, between 2 and 3 of these satellites are always on station, corresponding to an absentee ratio of roughly 20 or an availability ratio of 5 percent on station. These numbers are only approximate but provide a good idea of the necessary size of the laser constellation.

36. HF is a very efficient high-power laser. Its specific energy is 1.4 kJ/gm (kilojoules per gram) of fuel, or at 33 percent efficiency 500 J/gm. To provide 2800 GJ requires, therefore, 6000 tons of fuel.

37. The Fletcher Committee has recommended that the United States must expand its space transportation system to provide a capability to launch about 100 tons to medium altitude orbits tens of times a year. See Clarence A. Robinson, Jr., "Study Urges Exploiting of Technologies," *Aviation Week and Space Technology*, October 24, 1983, p. 50.

minutes, the required number of lasers increases from 320 to 480. The number would also increase with further hardening of Soviet boosters.

There are additional generic difficulties for any defensive systems whose components are predeployed in space. In addition to specific countermeasures that render them ineffective, their foremost difficulty is that they will most likely be vulnerable to direct enemy attack. Among the simplest direct threats are small and relatively cheap space mines with conventional explosives. These could be launched into orbit and detonated by radio command from ground to damage the large, delicate, and highly vulnerable optical parts. Of course, the Soviets could also put nuclear bombs in orbit. These would have an enormous range for damage of such systems and their necessarily "almost perfect optics." Ground-based laser beams pose another direct threat to the sensitive optical sensors of such a system. The space-based defenses would have to be prepared to "blink" if so attacked in order to avoid damage, particularly if the incident radiation comes in short, intense pulses.

Not all laser platforms would have to be attacked and put out of action, just a sizable fraction of that small percentage that is on station over the ICBM launch areas during boost phase. The offense can then be confident that a sizable fraction of its attack will penetrate the first layer of the defense. It is important to recognize that such space-based systems are much more delicate (and expensive) than the individual ICBMs against which they are deployed and therefore the task of protecting them is inherently more difficult than that of hardening the ICBMs against them.

Countermeasures other than direct attack include further hardening of the missiles against the incident beam energy to a level somewhat higher than the 20 kJ/cm$^2$ assumed in the above example. Two approaches are to coat the booster with a somewhat thicker (by a few millimeters) heat shield, or dispense an aerosol to absorb the incident fluence and disperse it harmlessly. Another useful technique is to spin the missile at the rate of a few turns a second during boost so that the beam energy is distributed at lower fluence around the booster surface.

There are additional problems that can be created for such a defensive system. Precursor nuclear bursts at high altitudes can precede an attack and disrupt the defensive system's operations, particularly its sensors and communication links. There is, of course, no spare time available for replacing or reconstituting these components since the entire system must operate within the first few minutes in order to destroy an ICBM during boost phase.

After boost is completed, the targets are or can be made harder and require a greater fluence to destroy; and once the MIRVs are deployed, they are not only much harder, but there are many more of them to destroy.

The defensive system can also be decoyed by false targets consisting of bright rockets and other hot sources simulating the missile exhaust. And, as the history of MIRVs shows, the Soviet Union could simply increase its arsenal of offensive missiles and warheads in order to maintain its deterrent in the face of a prospective defensive system.

In the final analysis, a very extensive and expensive constellation of chemical lasers predeployed in space seems to offer no credible prospect of forming an effective defensive layer against a large scale attack at the current high levels of the threat.

POP-UP SYSTEMS: GROUND-BASED X-RAY LASERS

To avoid most of these problems, the battle stations may be deployed on ground-based missiles that are ready to launch—i.e., "pop up"—on notification of an enemy attack. They must be substantially smaller and lighter than the infrared chemical lasers, each one of which would require about a shuttle load of fuel.

The most promising pop-up system consists of x-ray lasers, driven by nuclear explosives and mounted onto a missile that can be launched very rapidly. By itself, a nuclear explosion releases a very large amount of energy which is not focused, but which emerges in all directions. If a sufficiently large fraction of the energy from the nuclear explosion can be used to drive one or more lasers, and thus be focused into very highly collimated beams, it can cause severe impulsive damage to objects at very great distances.

Open U.S. and Soviet scientific literature has analyzed and discussed different materials and physical conditions for pumping x-ray lasers. The kinds of x-ray lasers the Soviets discuss include pumping transitions in zinc and iron in a plasma within appropriate physical conditions so that there will be a population inversion between several excited levels. A number of lasting transitions with energies from 10s of electron volts (eV) up to a kilovolt (keV) have been discussed.[38] The corresponding wavelengths of the coherent radiation vary between one-tenth and one-thousandth of a micron. With powerful and compact sources such as nuclear explosives, and with short

---

38. P.L. Hagelstein, "Physics of Short Wavelength Laser Design" (Ph.D. dissertation, Lawrence Livermore National Laboratory, 1981).

laser wavelengths in the x-ray region, it is feasible to mount this kind of system on a rocket to be boosted into space upon detection of a missile attack.

The kill mechanism in this case is impulsive damage due to ablative blow-off, caused by a very short intense pulse of incident energy, in contrast to the thermal heating by the chemical laser. The maximum kill range will depend on the gain that can be achieved in practice as well as the required kill fluence. This technology is still very immature,[39] but we shall assume here that there are no operational limits posed by limits in gain.

The most important operational issue is whether a pop-up system of this type can be deployed sufficiently rapidly even to attempt a boost-phase intercept. Modern ballistic missiles complete their powered flight, or boost phase, within three to five minutes after launch. At that point the remaining target becomes the post-boost vehicle consisting of the individual warheads and the "bus" on which they are mounted.[40] This means that a pop-up system has only a few minutes available to be boosted to a high enough altitude above the atmosphere in order to be able to initiate an attack. In practice an x-ray laser can only operate at altitudes above 100 km. X-rays of 1 keV energy and lower are absorbed by the atmosphere at lower altitudes.[41] Thus a defensive x-ray laser would have to be launched literally within seconds of the launch of an enemy attack.

Furthermore, such a pop-up system would have to be based far off shore from continental U.S. soil, and near the Soviet territory.[42] Otherwise, due to the curvature of the earth, it would be impossible for the x-ray laser beams to "see" the booster above the horizon before the end of burn.

A pop-up system would have to be entirely automated for quick response. This poses serious policy problems because the automated processes would necessarily include authorized release of nuclear weapons, as well as decisions as to whether and how to respond, depending on the intensity and

---

39. It is reasonable to assume that a comparable level of fluence is required to destroy a booster by impulsive kill as by thermo-mechanical damage. In order to deliver 20 kJ/cm$^2$ to a range of 1,500 km, the brightness of an x-ray laser will have to be increased by a factor of 10,000 beyond the goal reportedly recommended for demonstration in 1988 by the Fletcher Committee.

40. The bus for dispensing the MIRVs is considerably smaller and can be made harder than the booster; it can also be designed to release MIRVs very rapidly and therefore we focus here on boost-phase intercept.

41. The minimum altitude is even higher for lower energy x-rays since the absorption coefficient increases by a factor of 2$^3$ for each decrease by a factor of 2 in x-ray energy. The minimum altitude can be raised even higher by precursor nuclear explosions at high altitudes that are properly designed to heat the top of the atmosphere, thereby causing it to expand and rise.

42. This would contravene Article IX of the ABM Treaty.

tactics of the attack. For example, since an x-ray laser can fire only once, destroying itself in the act, should it be launched against a single attacking ICBM, or only against a suitably large barrage?

In addition to these formidable operational requirements, there are two countermeasures that can deny any possibility of a pop-up x-ray laser defense. First, the offense can deploy high-thrust "hot" missiles that complete their burn at altitudes below the top of the atmosphere. Second, the offense can alter the trajectory of the launch, depressing it so as to complete its burn below 100 km. Thus even if we assume that x-ray laser systems are developed with sufficiently high gain to look promising for destroying boosters at long range, the offense can use the atmosphere to defeat them.

HYBRID SYSTEM FOR BOOST-PHASE INTERCEPT

One widely discussed concept is that of a system of ground-based lasers whose beams are aimed up to a small number of large relay mirrors in synchronous orbits at 36,000 km altitude. These relay mirrors then direct the beams to various mission mirrors, orbiting earth at lower altitudes from which the beams are redirected onto their targets.

This hybrid system has the advantage of putting fewer parts in orbit than the space-based laser system, and it also avoids the very severe time constraints on a pop-up system for boost-phase intercept. However it too faces several severe and unavoidable technical and operational challenges. The directed light beams must travel very great distances, 36,000 km up and 36,000 km back from the relay mirrors in geosynchronous orbit. Large optics and short wavelengths are necessary to reduce the diffraction and keep the energy focused on such a long path from laser to target.

This concept necessarily relies on "active optics" to compensate for the effects of atmospheric turbulence which cause scattering and defocusing of the directed light beams. "Active optics" means the following: a weak laser beam from space shines to ground and its beam spread is analyzed as a measure of local atmospheric turbulence, which is then compensated for by use of deformable focusing mirrors for transmitting the beam from the ground-based laser. This technology while still immature has been progressing rapidly and, in principle, can achieve the goal of transmitting highly focused beams through the atmosphere.[43] In addition to atmospheric com-

---

43. Dr. George Keyworth II, the President's Science Advisor, claimed that "we've also seen very recent advances that permit us to compensate for atmospheric break-up of laser beams." See *Reassessing Strategic Defense, Remarks of Dr. G.A. Keyworth to the Council on Foreign Relations,* Washington, D.C., February 15, 1984, mimeo, p. 12.

pensation, there is the problem of weather, and in particular of cloud cover absorbing the laser energy. This requires replicating the ground-based lasers at widely distributed sites in order to have a high probability that an adequate number are free of cloud cover, or basing the lasers high near mountain tops above the clouds.

There remains a further, unavoidable operational problem: the few large focusing mirrors in space are themselves a small number of high value targets vulnerable to attack. They and the large ground-based laser stations are reminiscent of the large phased-array radars which proved to be the "Achilles heel" of the earlier generation of ABM systems. It has yet to be clarified how the large and delicate space mirrors or the ground-based installations could be protected with confidence.[44]

MID-COURSE INTERCEPT AND BATTLE MANAGEMENT

The mid-course phase lasts 20–25 minutes after completion of the booster burn. During the first few minutes the post-boost vehicles will be targets until they have dispensed the individual warheads, which then become the targets. Although the time constraints are less severe than in the boost phase, there are other factors that make midcourse interception difficult. The post-boost vehicles are generally more difficult to destroy than the boosters, since they are smaller and can be made harder, and designed to release MIRVs very rapidly. The warheads are smaller and harder still, since they must withstand extreme stresses caused by deceleration and atmospheric friction as they slam back into the top of the atmosphere. They are also far more numerous. In addition, each missile may dispense hundreds of light decoys which, in the absence of friction above the atmosphere, follow the same paths as the warheads.

Although the capacity to handle data has increased greatly in recent years, so has the size and sophistication of the offense, as well as its ability to confuse the defense. For example, the offense can use anti-simulation (the technique of making warheads look like decoys) to confuse the sensors and stress, if not saturate, the data-handling capacity of the defense. One means

---

44. Other concepts for boost-phase interception—particle beams and high voltage microwaves—are even less promising, for the technology is much less advanced than that of lasers. Schemes to use kinetic energy interceptors such as small missiles or pellet screens have been widely judged ineffective, on grounds of time constraints, countermeasures, and vulnerability. For further discussion, see the report from which this paper is taken.

to do this is to enclose the warheads in balloons with several thin metal-coated layers so that all balloons have the same appearance, whether or not there is a warhead inside. Additional difficulties can be caused by precursor detonations of nuclear weapons. The infrared radiation from the air heated by high altitude nuclear explosions creates a severe background—known as "red-out"—against which the sensors of a proposed mid-course ABM layer must operate and "see" the warhead.

In Defense Department documents[45] the different layers of a defensive system are described as operating semiautonomously. As part of the overall battle management—i.e., monitoring, allocating the available defensive systems, assessing the results of the attack, and refiring if necessary—data would be passed to successive layers in the defense. Input data from the sensors must be organized and filtered to see which objects can be discarded and which are candidates for further analysis. An effective boost-phase intercept that clears away close to 90 percent of the threat is thus crucial in making the battle management and data-handling problems more tractable.

No viable concept has yet been devised for a highly effective mid-course defense against a massive threat of many thousands of warheads plus many times more decoys.[46] The critical needs include not only a battle management software that far exceeds in complexity and difficulty anything accomplished so far,[47] but also the ability to protect all the critical space-based components against enemy attack.

Directed energy systems such as chemical lasers based in space, or pop-up ground-based x-ray lasers are also candidates for mid-course defense.

---

45. See, for example, *Defense Against Ballistic Missiles: An Assessment of Technologies and Policy Implications* (Washington, D.C.: Department of Defense, March 6, 1984).

46. Dr. Robert Cooper, Director of the Defense Advanced Research Projects Agency, said in Hearings on May 2, 1983, that "I think the single thing that we have not focussed attention on in the past, which may represent the most stressing technological problem, is the complexity of any comprehensive battle that we would have to wage against a large-scale strategic missile attack. . . . Currently we have no way of understanding or dealing with the problems of battle management in a ballistic missile attack ranging upward of many thousands of launches in a short period of time." See *Hearings before the Committee on Armed Services of the U.S. Senate: DoD Authorization for Appropriations for FY 1984 (March–May 1983)* (Washington, D.C.: U.S. Government Printing Office, 1983), p. 2892.

47. The first conclusion of the Battle Management, Communications, and Data Processing panel of the Fletcher Committee Report was: "Specifying, generating, testing, and maintaining the software for a battle management system will be a task that far exceeds in complexity and difficulty any that has yet been accomplished in the production of civil or military software systems."

Although they would not have the severe time constraint of a boost-phase intercept, they face the operational problems we have just described. The entire system—including intelligence, communications, and surveillance satellites and the optical and directed high energy components, whether on ground, in low-earth orbit, or at synchronous altitude—must survive and operate in a hostile environment for many minutes in order to engage the threat. Moreover, each individual warhead—as well as the many additional decoys—will have to be attacked, and the warheads are generally much harder targets requiring a substantially larger kill fluence or impulse. Hence the overall energy requirements are much greater than those already established in our earlier illustrations of boost-phase intercept.

TERMINAL DEFENSE

The terminal layer of the defense takes advantage of the atmosphere to slow down and strip away the lighter decoys that accompany the warheads during mid-flight. The requirements for a terminal defense of hardened military targets are much simpler than for a nationwide defense. Because the targets to be defended are small and hardened to withstand very high levels of overpressure, interception does not have to take place at a great distance from them. In addition, the goal of hard-site defense is not to destroy all incoming warheads, but only enough of them to cause the attacker to expend more of his force than he destroys.

Recent improvements in technology have enhanced the prospects for a cost-effective hard-site defense. Important advances include: interceptors that achieve much higher accelerations; improved accuracy, which raises the possibility of non-nuclear kill; and sensors that can discriminate warheads from decoys at higher altitudes. It should be emphasized, however, that hard-site defense is very different from the goal of a nationwide defense designed to render nuclear weapons "impotent and obsolete." As such, it should be discussed on its own merits.

The requirements for the terminal layer of a nationwide strategic defense are much more severe than for hard-site defense because the urban-industrial targets are much larger and more vulnerable, and have much higher value. Terminal defense alone offers no prospect of defending the nation against a massive attack. This conclusion was reached during the earlier ABM debates of 1969–70, and the new technologies do not alter it. If, however, a terminal defense were to operate behind effective boost-phase and mid-course defensive layers which remove all but a few percent of the attacking warheads, this conclusion might have to be reexamined.

ANTI-SATELLITE TECHNOLOGY (ASAT) AND DEFENSE VS. DEFENSE

ASAT presents a much simpler technical problem than exoatmospheric ABM, since the targets are softer, fewer, predictable both in their position and time, easier to discriminate, not easily replaced, and have vulnerable communication and control links from earth. There is no doubt that ASAT systems will be effective against satellites in low-earth orbit in the near future. Extending ASAT effectiveness to synchronous orbit altitudes of 36,000 km presents a technical challenge, but no fundamental problems.

The significance of ASAT for strategic defense lies in the threat it poses to the ABM's space platforms, in particular to the warning, acquisition, and battle management sensors. The significance of the SDI program for ASAT is that it will spur technical developments that will inevitably threaten the critical communication and early warning satellite links on which ABM must rely. Thus ASAT threatens ABM, but ABM developments contribute to ASAT.

More generally, one side's ABM will threaten the other's, especially the space-based components. ABM could be used as part of a first strike. A pop-up x-ray laser system, for example, could be employed to suppress both the ABM and the retaliatory strike capability of the other side when they emerge above the atmosphere.

## SDI: The Implications for Deterrence and Stability

The longing to escape from living indefinitely under the threat of nuclear destruction is both natural and strong. Escape through revolutionary technological developments cannot be ruled out *a priori*, but premature adoption of illusory concepts can make things worse, not better. We must ask whether the new technologies enhance deterrence and stability, or imperil them.

The technical and operational considerations laid out in the preceding section suggest that President Reagan's vision of escaping from deterrence cannot be realized with foreseeable technology. But behind the current SDI program lies a different concept, that of a mixed offense-defense strategic posture, with a layered ABM capable of destroying a large part of a massive missile attack. Faced with such a defense, it is argued, the potential attacker (who might in prudence assume that the defense was more effective than it actually was) would realize that the benefits of a first strike would be marginal, and so would not be tempted to launch one.[48]

---

48. See note 4.

It is not enough, however, to argue that the United States can build a defense that could destroy a large part of an attacking force. The inevitable Soviet response must also be taken into account. If the United States deploys a partially effective ABM, that will challenge, in the first instance, not just the Soviet incentive to strike first, but the efficacy and reliability of the Soviet retaliatory force. The Soviet leaders will look at the emerging posture to judge whether the United States remains vulnerable to, and deterred from initiating, nuclear war. In so doing, they will apply the test they have put vividly in several statements: does the United States still recognize that it would be suicidal for it to start a nuclear exchange?

The Soviet leaders will fear that the United States might be intending now—or might decide in a crisis—to launch a first strike, relying on its ABM to deal with a diminished Soviet response. Hence the Soviet Union, on a worst case analysis, is likely to give priority to countermeasures to ensure that its retaliatory forces can penetrate the defense and inflict an intolerable level of devastation on the United States. Soviet leaders have emphasized that they can and will maintain the capacity to retaliate in the face of any U.S. strategic programs.[49]

Given the number and power of nuclear explosives and the diversity and survivability of delivery means, it remains an illusion to think that either the United States or the Soviet Union can be protected by any means other than the prevention of nuclear war. Since this hard reality exists, and should be apparent to the leaders of both superpowers, a deliberate first strike could be grimly rational only if one side's leaders could persuade themselves that the other's retaliatory forces could be attacked and destroyed with sure effect and subsequent impunity. Is an extensive but imperfect U.S. ABM deployment the right way to reduce any Soviet temptation to act on such an expectation?

As the Scowcroft Commission noted, the plausibility of a Soviet disarming strike is small, in view of the diversity of U.S. strategic forces, and the high level of survivability of a substantial majority of the warheads they carry.[50]

---

49. There is no reason to suppose that they will not be able to do this. All analyses of the cost and time required for the retaliatory offensive forces to retain their damage capability *against soft targets* (cities, industrial areas, soft military targets) in the face of ABM deployment have shown that the advantage rests clearly with the offense. If such cost estimates were to shift in favor of defense, one factor in the equation would be changed, but there is no reason to anticipate that this is about to happen.

50. *President's Commission on Strategic Forces*, Washington, D.C., April 1983, pp. 12–14.

The real risk of nuclear war is not a cold-blooded decision to initiate one, but what might happen under the pressures and suspicions of a crisis—an accidental triggering nuclear incident, miscalculation, loss of control by responsible leaders. An effective but imperfect ABM on one side would exacerbate the risk because the side that did have an ABM might calculate that it would be better off if it struck first and used the ABM defense to deal with the weakened response. (This same calculation would affect both sides, if both possessed ABM.) Similarly, the side that did not have ABM might calculate that its situation would be better (however bad) if it struck first and avoided being caught trying to retaliate with a weakened force against the ABM defense.

In a crisis each side would wrestle with the fear that the opponent might strike first, and if war seemed inevitable a preemptive strike would be judged better than awaiting the adversary's initial blow. Soviet military writing suggests that Soviet strategic policy contemplates preemption in this kind of scenario.[51] If it does, the kind of "enhanced deterrence" under examination here would not weigh heavily, for the preemptive strike would be made not on any calculation of absolute advantage, but solely on one of *relative* advantage: since a nuclear attack is inevitable anyway, strike first in order to enhance one's own effectiveness and diminish the enemy's as much as possible.

No definitive conclusion can be reached in the abstract about "enhanced deterrence" from a mixed defense-offense strategic force. If one looks at the history of Soviet–American strategic competition, it is difficult to believe that either side would let the other deprive it of a retaliatory deterrent capability. If the concern is with deterring a first strike, it is equally difficult to be confident that a constructive difference would result. A mixed defense-offense posture designed to deprive the opponent of a first strike capability is likely to look—in motivation and in capability—uncomfortably like a first strike posture. This would be particularly true of a space-based ABM system, where the fragility and vulnerability of the components would make them unreliable except in support of a preemptive strike.

A further argument has been advanced in favor of an effective but imperfect defense: that it would lessen damage and save lives in the event of a

---

51. In the 1950s, some Soviet military theorists advocated a strategy of preemption; in the 1960s, this seems to have been supplemented with, but not displaced by, a launch-under-attack policy of some kind.

nuclear war. But not many nuclear warheads need penetrate the defense in order to inflict massive devastation and casualties on urban areas. If only a fraction of the attacking warheads gets through, the consequences for people and property will be catastrophic.

Proponents of ABM argue that, even if it is only partially effective, a potential attacker will be conservative and attribute substantial capability to the defense which faces him. This is the argument for enhancing deterrence. But if deterrence fails and an attack is made, the attacker will calculate his requirements conservatively, and program more warheads than are actually needed to achieve whatever level of destruction is judged necessary. It is thus a grim prospect that, for soft targets, at least as many casualties and as much damage will follow a partially effective ABM defense effort as without it.[52]

The only way to save lives is to prevent nuclear war. The prime test of ABM is how it contributes to that goal, and it does not pass that test well.

ABM DEFENSE OF HARD SITE TARGETS

Until recently United States ABM R&D has given priority to technology for defense of ICBMs and other hard targets. The technology for this mission has matured considerably. Such hard-site ABM deployment is not *in principle* inconsistent with the ABM Treaty, the main concern of which is that local ABM defenses not be so numerous or so configured as to constitute a base for a nationwide defense. If unambiguous hard-site defense seemed attractive, some modifications or agreed interpretations of the 1972 Treaty and 1974 Protocol might be needed. These might not be too difficult to work out if agreement could be reached on the longer-range implications of resuming limited deployments for hard-site defense. If this could not be done, the United States and the Soviet Union would face difficult decisions about the relative advantage of the ABM *status quo* as against unilateral changes which might result in military and political instabilities.

MOVING TO ABM AND DEFENSE DOMINANCE AFTER START

If the Soviet Union and the United States could agree to sturdy limitations on offensive systems, and were both confident of the soundness of the agreement, one basic criticism of the SDI would be voided. But so long as

---

52. Note that these reservations regarding the value of a mixed offense-defense posture for enhancing deterrence apply equally to a nearly leak-proof defense while it is being constructed.

deeply rooted rivalry persists between the Soviet Union and the United States, even successful arms control agreements will be subject to constant tension and strain. Each side will place ultimate reliance for prevention of nuclear attack on its residual deterrent offensive forces. Even if it should prove possible to shift from offense dominance to defense dominance, the two sides would find it hard to abandon their deterrent forces completely.

A strategic relationship in which defense was dominant would require a framework of effective restraints on offensive forces. It would also require prior agreement on the ABM systems to be deployed. Even then, agreement on ABM deployments would not be easy to reach, because of the inherent ambiguity of "defense." ABM deployments would impose other strains on the strategic balance too: for example, their inherent ASAT capability would threaten satellite warning and communications systems.

Even if a broad arms control regime were put in place for strategic offensive forces, it is far from clear that the shift to ABM would be judged reassuring and stabilizing, or worth the cost. In particular, it will be very difficult to manage the passage through a period of uncertain deterrence and high instability that connects today's condition of offense dominance with the goal of defense dominance. It will remain an intriguing possibility for that uncertain future, but a possibility rather than a clearly desirable goal. The sensible procedure would be to give priority to reducing offensive systems, and introduce ABM only by agreement at a later stage if it then appears warranted.

### The Prevention of Nuclear War

The Scowcroft Commission report, which was endorsed by President Reagan on April 29, 1983, recommended that the United States: (a) maintain and improve the survivability of deterrent retaliatory forces; (b) reduce the value and attractiveness of vulnerable targets such as MIRVed ICBMs or large Trident submarines by replacing them in due course with individual units of lower value such as Midgetman and smaller submarines; and (c) assure the penetration effectiveness of the retaliatory missiles.[53]

The last goal is partly to be achieved by unilateral measures such as diversified forces and penetration aids. It is also made substantially easier by

---

53. *President's Commission on Strategic Forces*, pp. 20–21.

the ABM Treaty, which for twelve years has guaranteed the penetration of missiles by drastically limiting ABM defenses. It is the prime example of how arms control can complement and reinforce national strategic policies and programs. If progress is made in ensuring the survivability of retaliatory forces and in reducing the value and attractiveness of vulnerable targets, it will become easier for each side to accept the penetration capability of the other's offensive missiles. If such progress is not made, ABM will continue to be considered to the extent that technical and cost factors allow.

Note that ABM would add primarily to the deterrent value of ICBMs by protecting them, whereas the ABM Treaty, by ensuring penetration, adds to the deterrent value not only of ICBMs but also of SLBMs, the most survivable component of the U.S. strategic triad and the one with the most retaliatory warheads. SLBMs at present face no significant defensive barrier to execution of their mission. Assessing deployment of an effective but partial ABM defense must weigh not only the hope of reducing the threat to survivability of ICBMs (and to some extent bombers ) but also what price might have to be paid for lessening the deterrent value of SLBMs and surviving ICBMs as the Soviet ABM system expanded in response.

ARMS CONTROL

For an ABM system to be effective, offensive systems will have to be limited. Under Secretary of Defense DeLauer has stated that "with unconstrained proliferation, no defensive system will work."[54] Administration spokesmen assert that one purpose of accelerated ABM activities is to bring additional pressure on the Soviet Union to treat arms reduction negotiations seriously.

The general rationale for this approach seems to be that the Soviet Union is driving the offensive strategic arms race by its unremitting buildup and its pursuit of strategic superiority. Only if the Soviet leaders see that this effort is doomed to frustration by present U.S. strategic programs, reinforced perhaps by major ABM deployments, will they abandon their dangerous ambitions and bargain seriously for reductions in their threatening and destabilizing ICBMs. Dr. George A. Keyworth, Science Advisor to the President, said in a recent speech:

"Now Soviet leaders are pragmatic—and smart. When confronted with mounting evidence that they're facing a future in which the ICBM will lose its previously unchallenged position as a devastating first-strike weapon,

---

54. Richard Halloran, "Higher Budget Foreseen for Advanced Missiles," *The New York Times,* May 18, 1983, p. 11.

they'll shift their strategic resources to other weapons systems. *They'll* change *their* perception of strategic war. Critics cite this as a failing, but I see it as a major plus. If we can reduce the effectiveness of the ICBM, we can make it far easier to negotiate its reduction and elimination. *Let* the Soviets move to alternate weapons systems, to submarines, cruise missiles, advanced technology aircraft. Even the critics of the President's defense initiative agree that *those* weapons systems are far more stable deterrents than are ICBMs. . . . if *we* can be the first to develop defensive deterrent capabilities, we would have a persuasive negotiating posture for arms reductions. We could then approach the Soviets with the mutual knowledge that their immense ICBM fleet no longer has the intimidating effect it once had. Under those circumstances, we could propose to join them in methodically eliminating the ICBM as the premier weapon of strategic war."[55]

To date, however, the pressure of U.S. strategic modernization seems to have been counterproductive for arms control. Since President Reagan's speech, the Soviet Union has broken off indefinitely both Euromissile and START negotiations, rather than moving towards accommodation on what the United States considers sound lines.

There are undoubtedly tactical and political considerations behind the Soviet actions. But there is good reason also to take seriously the Soviet Union's statement of December 8, 1983, at the adjournment of the START talks, that it feels compelled "to re-examine all the issues which are the subject of discussion at the talks. . . ."[56] It would be logical for the Soviet Union to be reexamining the START proposals of both sides in the light of the SDI. The Soviet proposal envisages an overall reduction of 25 percent in strategic delivery systems and continuing restraints on MIRVs. The U.S. position concentrates on MIRV and throwweight reductions, and possible subsequent movement away from large MIRVed missiles to smaller single-warhead missiles.

If the Soviet Union must be prepared to counter and penetrate a U.S. ABM defense, how will it regard the American proposal for deep cuts in delivery systems and warheads, and a move away from MIRVs to single warheads? This has been an issue for the United States too. One justification for the MX missile advanced by the Scowcroft Commission relates to the ABM Treaty:

it is important to be able to match any possible Soviet breakout from that treaty with strategic forces that have the throw-weight to carry sufficient

---

55. *Reassessing Strategic Defense,* p. 18.
56. *The New York Times,* December 9, 1983, p. 16.

numbers of decoys and other penetration aids; these may be necessary in order to penetrate the Soviet defenses which such a breakout could provide before other compensating steps could be taken.[57]

Will the Soviet Union feel any differently about the value of its heavy MIRVed missiles, which the United States wishes it to reduce or give up?

President Reagan's firm commitment to ABM is more likely to complicate START than to give it new and healthy impetus. The increase in U.S. ABM R&D will be matched by accelerated Soviet R&D in both offensive and defensive systems. In turn, the United States will feel compelled to counter Soviet ABM activities by stepping up work on penetration aids. Moreover, one has to wonder how long serious commitment to deep START reductions, or to shifting away from large MIRVed missiles to single warhead missiles, will survive in the United States as Soviet R&D takes on added impetus and ABM deployment is begun or is suspected.

The ABM Treaty was a path-breaking attempt to break the old action-reaction cycle in weapons development and deployment, and to reduce suspicion and fear as to the motives behind these activities. Escape from that pattern has been halting and partial at best, but complete reversion to the old pattern is not to be encouraged.

U.S. ABM R&D will go ahead at some substantial level, as a hedge against Soviet deployments and for improved understanding of major weapons technologies. That in itself should provide adequate indication to the Soviet leaders that they cannot expect to gain by refusing to hold to the ABM Treaty or to negotiate appropriate updating of its provisions. But unilateral movement by the United States towards deployment is as likely as not to reduce their readiness to bargain. Moreover, it will give impetus not only to Soviet ABM programs but also to offsetting offensive programs (penetration aids and delivery systems), which will take on their own momentum. Not everyone will view with the same complacency as Dr. Keyworth the prospect of a Soviet surge in cruise missile, advanced bomber, and submarine technology and deployments, either as making the U.S. more secure or as auguring well for deep cuts in strategic weapons.

MAINTAINING THE VIABILITY OF THE ABM TREATY

Too much is at stake in the present tense state of U.S.–Soviet relations to make it prudent to undermine remaining elements of stability and common

---

57. *President's Commission on Strategic Forces*, p. 17.

understanding before we have something in which we can have more confidence as a replacement. Deterrence has been basic to stability and the
prevention of nuclear war. An essential guarantee of deterrence, as recognized and defined in the ABM Treaty, has been reciprocal limitation of ABM
defenses. Gradual erosion of the Treaty could imperil deterrence and
heighten the risk of nuclear war more quickly than ABM deployment, should
it live up to even moderate expectations, could restore the balance. The
Treaty therefore needs to be understood and supported, by positive measures
rather than merely by passive adherence.

Arms control has not progressed as well as was hoped when the ABM
Treaty was signed in 1972. SALT II limitations and reductions were negotiated, but the Treaty was not ratified by the United States. The more ambitious
START negotiations are in recess with an uncertain future, and the INF talks
in Europe are at an impasse. The ABM Treaty, which claimed at most to
facilitate the limitation and reduction of strategic offensive arms, is not to
blame for this state of affairs. It was one foundation of the abortive SALT II
agreements, and is a precondition of the continuing adherence to the SALT
limits on launchers, warhead fractionation, and force modernization, without
which START would face even more formidable obstacles. Without the ABM
Treaty, the pace of modernization, diversification, and growth of offensive
arsenals on each side could well have been more rapid. Large-scale ABM
deployment over the past decade could have heightened suspicions and
tensions, and fostered mutual fears of a first strike, thus increasing the risk
of nuclear war by accident, miscalculation, or loss of responsible control. As
one of the few instruments by which the superpowers record their determination to avoid nuclear war and take significant action to that end, the
ABM Treaty has validated its main premise and purpose—even if precari-
ously so.

The future of the Treaty is now in doubt. The mutual suspicions of the
superpowers extend to questioning each other's desire to avoid nuclear war.
Each finds evidence that the other is pursuing strategic superiority and
preparing for a first strike. Each suspects the other of preparing to break out
of the Treaty. Both nations have continued active R&D, professedly within
Treaty provisions, but with ambiguities on either side.

The ABM Treaty is affected by advances in technology. The issue of upgrading air defenses which the Treaty sought to deal with under 1972 conditions is being complicated by the deployment of cruise missiles in air-,
ground-, and sea-launched modes. The effort on both sides to develop defenses against tactical or intermediate-range ballistic missiles enhances the

component and system capabilities of potential ABM systems and exacerbates this problem.

In these circumstances it is not enough to defend the letter of the existing treaty or to bicker about compliance and suspected infringements. It is essential to act to keep emerging technology within existing treaty provisions where possible, and to reach new understandings where necessary. Simply challenging suspected violations will not help maintain the viability of the treaty, and may lead to a situation in which technology has moved on and the treaty is progressively a monument to past hopes. It would be ironic if this were to happen just as we reached the conclusion that restraint on ABM remains essential to strategic stability, the prevention of nuclear war, and the reduction of nuclear arsenals. To let events run blindly to this end would indeed be an abdication of responsibility.

To prevent this from happening, the ABM Treaty should be carefully protected while the role of ABM is being reconsidered and strategic arms reductions are being pursued. Procedurally, this calls for making full use of the Standing Consultative Commission. This body has the charge (under Article XIII) to "promote the objectives and implementation of the provisions of this Treaty" in ways going far beyond the working out of dismantling procedures for obsolete facilities, or debating compliance issues. Especially, it is to:

(d) "consider possible changes in the strategic situation which have a bearing on the provisions of this Treaty; . . .
(f) "consider, as appropriate, possible proposals for further increasing the viability of this Treaty; including proposals for amendments . . .

The United States and the Soviet Union ought in the SCC to identify technological and strategic changes since 1972, and try to reach agreement on their implications for the treaty. The United States should take the initiative to explain President Reagan's vision, and to solicit and explore Soviet concerns. Changes in the treaty should by no means be excluded. They might take the form of an additional protocol, or amendment of the treaty text.[58]

---

58. The managers of the U.S. ABM R&D program state repeatedly that it is being conducted scrupulously within treaty bounds. It is no criticism of their good faith to suggest that this is not a judgment to be left to them. It would be valuable to set up a review panel to ensure that R&D activities are indeed fully consistent with the treaty. This would place the United States in a better position to discuss compliance issues with the Soviet Union. Such a panel ought to include, along with officials of the current administration, people of broad strategic experience such as those on the Scowcroft Commission.

Cooperative action by the United States and the Soviet Union could reinforce confidence that the two nations see the purposes and value of the ABM Treaty in consistent ways, and that each is determined to act separately and jointly towards the fundamental aim of preventing nuclear war.

## Conclusion

Our analysis raises grave doubts, on technical and strategic grounds, about the wisdom of accelerating or expanding ABM R&D. Deliberation and restraint are imperative not simply because of the enormous costs involved, but because the strict limitation of ABM deployments is one of the few points of real agreement reached in the U.S.–Soviet dialogue about nuclear war, and has important practical consequences for the U.S. deterrent and for such fragile stability as now exists.

We do not now know how to build a strategic defense that can render nuclear weapons impotent and obsolete. A long-term R&D program is required to determine whether we can move to realistic systems concepts. Each of the critical technologies must make advances by orders of magnitude beyond anything thus far demonstrated.

There are, moreover, great barriers to operational effectiveness. Space-based components would be vulnerable to destruction by direct attack, and could be rendered impotent by countermeasures. Pop-up systems for boost-phase intercept face severe time constraints, and could be denied engagement altogether if the offense moved to high-thrust boosters. The multi-layered defense with its many sensors and severe battle-management requirements could not be fully tested, but would have to work perfectly in a hostile environment the first time it was turned on.

The United States may view its defenses as enhancing deterrence by increasing Soviet uncertainty about the potential effectiveness of striking first. But the Soviet Union may (as it says it does) see American ABM R&D as evidence of an attempt to deprive it of an effective retaliatory capability.

The initiation of an intensified R&D program looking towards a declared deployment goal is not a harmless step. Even if no system is ever deployed, increased instability can result as both sides build up their forces over necessarily long lead times to preserve their retaliatory capability and try to match each other's anticipated ABM capability. To avoid this, the SDI must be closely coupled with effective arms control. If defensive systems are to contribute to a safer and more stable strategic relationship between the

United States and the Soviet Union, they will have to be embedded in a strict arms control regime that limits offensive systems. Technology alone will not solve the political problem of managing the strategic relationship with the Soviet Union.

The ABM Treaty has a central bearing on U.S.–Soviet relations and prospects for cooperative pursuit of peace and security. Reliance on deterrence, with its accompanying threat of mutual devastation, stirs profound feelings of anxiety. Nevertheless, so long as nuclear weapons exist (even at a fraction of present numbers), deterrence through clear common awareness of the consequences of initiating nuclear war will be prudent and indispensable.

It is the combination of deterrence with harsh and uncompromising confrontation that is ominous. While deterrence will not easily be dispensed with, it can be *subordinated* to the identification of mutual interests, so that areas of cooperation can be found, even if only warily. Arms control has a special prominence and urgency in the present situation. Because the catastrophic and uncontrollable character of nuclear war is recognized by responsible national leaders on both sides, preventing nuclear war has been accepted as one of the few clear areas of mutual interest even under current harsh circumstances.

In this context mutual deterrence takes the form not of a simple reciprocal threat but rather of mutual self-restraint on any disposition to initiate the use of nuclear weapons, and this is made possible by the perception that the determination to avoid nuclear war is present on both sides. However grim the continuing existence of nuclear arsenals makes this prospect, it is superior to the vision of continuing confrontation behind opposing defensive shields.

These conclusions lead to several recommendations. First, the SDI program should be limited to research for exploration of scientific possibilities, as a hedge against technological breakthroughs or Soviet ABM deployments. Engineering development should be deferred, in order to avoid violating the ABM Treaty and giving the program unstoppable technological momentum.

Second, consultations with the Soviet Union should be initiated in the SCC under Article XIII of the ABM Treaty. The United States should make clear the following points: the limited nature of its SDI program; its determination to comply fully with the ABM Treaty, and to insist on Soviet compliance; its readiness to discuss any aspect of the program that raises questions in Soviet minds; its readiness to explore modifications or clarifications of the Treaty in the light of developments since 1972; its appreciation of the link between ABM limitations and the START negotiations on offensive systems. This would help to maintain the viability of the ABM Treaty in a practical sense, and to allay suspicions raised by some of the recent rhetoric about strategic defense.

Third, in seeking to enhance deterrence, increase strategic stability, and improve the prospects for arms control, the United States should recognize that, for the foreseeable future, the active pursuit of strategic defense will not contribute to those goals as surely and clearly as the ABM Treaty does; and that other measures (improved survivability of strategic forces, reduced ratio of warheads to missiles) promise a more timely improvement in its deterrent posture than a dangerous and destabilizing movement towards ABM deployment.

# Do We Want the Missile Defenses We Can Build?

*Charles L. Glaser*

**O**n March 23, 1983, President Reagan delivered his famous "Star Wars" speech in which he called for development of defenses capable of making nuclear weapons "impotent and obsolete."[1] The President's speech raised the hope that U.S. cities could be made invulnerable, a task requiring near-perfect defense against massive Soviet ballistic missile attack, and set off a national debate on the feasibility of such highly effective defenses.[2] Studies of the "Star Wars" concepts now under research show there is virtually no hope that they will provide near-perfect defense in the foreseeable future.[3]

The author would like to thank Albert Carnesale, Ashton Carter, Lynn Eden, Philip Sabin, Steven Miller, and the members of the Avoiding Nuclear War working group for their helpful comments on earlier drafts of this article.

*Charles L. Glaser is a Postdoctoral Fellow on the Avoiding Nuclear War Project at the John F. Kennedy School of Government, Harvard University, and a Research Fellow at the School's Center for Science and International Affairs.*

1. *The New York Times*, March 24, 1983, p. 20.
2. The term "near-perfect" refers to systems capable of significantly reducing the costs the Soviet Union can inflict by attacking the United States' cities, population, and economic capabilities. How well a defense must perform to be considered near-perfect depends on three factors. First is the level to which a Soviet attack against U.S. cities must be reduced before its costs are judged significantly reduced. Analysts often use the levels of damage associated with "assured destruction"—25 percent of the population and 50 percent of industrial capability—as a benchmark. There is, however, substantial disagreement: some analysts argue that defenses are of little value until potential damage is reduced far below these levels; others believe any reduction in the potential damage to U.S. cities is worth pursuing. This issue is discussed in Ashton B. Carter, *Directed Energy Missile Defense in Space—A Background Paper* (Washington, D.C.: U.S. Congress, Office of Technology Assessment, April 1984), pp. 66–67. Carter, for the sake of discussion, uses 100 megatons as the level of penetration at "which a defense would be judged near-perfect." Second, for a given U.S. defense, the Soviet ability to inflict damage depends on the size and penetrability of its force. The third factor is the type of Soviet attack against which the defense is measured. For example, a defense that could provide some protection of U.S. cities following a U.S. counterforce attack might be unable to do so if the Soviet Union launched a first strike against U.S. cities.
3. Carter concludes in *Directed Energy Missile Defense in Space*, p. 81, that the prospect of developing near-perfect defense "is so remote that it should not serve as the basis of public expectation or national policy about ballistic missile defense (BMD). This judgment appears to be the consensus among informed members of the defense technical community." The studies done outside the government concur with this assessment: Sidney D. Drell, Philip J. Farley, and David Holloway, *The Reagan Strategic Defense Initiative: A Technical, Political and Arms Control Assessment* (Stanford, Calif.: Center for International Security and Arms Control, Stanford Uni-

*International Security*, Summer 1985 (Vol. 10, No. 1) 0162-2889/85/025-33 $02.50/1

This discrediting of the notion of near-perfect defense is shifting the ballistic missile defense (BMD) debate within the defense community to less-than-near-perfect defense, i.e., to BMD incapable of protecting U.S. cities, but having other strategic goals.[4] These include defense of military targets and "light" area defenses that might reduce the damage to cities in small attacks even though penetrable by a large dedicated attack. Defense capable of satisfying some of these goals might be available in the near future.

Debate over BMD is likely to focus on less-than-near-perfect defense for many years. Although President Reagan directed his speech at defenses that would radically change the nuclear situation by making U.S. cities invulnerable, the speech and the strategic defense initiative (SDI) research and technology program have restored legitimacy to strategic defenses in general. The combination of this renewed legitimacy with the dubious prospect for near-perfect defenses will lend political and budgetary support to BMD capable of achieving only more modest objectives.

---

versity, 1984); and Union of Concerned Scientists, *Space-Based Missile Defense* (Cambridge, Mass., 1983). See also Sidney D. Drell and Wolfgang Panofsky, "The Case Against Strategic Defense: Technical and Strategic Realities," *Issues in Science and Technology*, Fall 1984, pp. 45–65; James R. Schlesinger, "Reckless Rhetoric and Harsh Realities in the Star Wars Debate," *International Security*, Vol. 10, No. 1 (Summer 1985); and Harold Brown, "The Strategic Defense Initiative: Defensive Systems and the Strategic Debate," *Survival*, forthcoming.

The two studies ordered by President Reagan following his announcement of the Strategic Defense Initiative are less pessimistic. The study of BMD technologies and systems, directed by James C. Fletcher, is the more optimistic, concluding: "The technological challenges of a strategic defense initiative are great but not insurmountable. . . . The scientific community may indeed give the United States 'the means of rendering' the ballistic missile threat 'impotent and obsolete.'" *The Strategic Defense Initiative: Defensive Technologies Study* (Washington, D.C.: Department of Defense, March 1984), p. 23. The study of the policy implications of BMD [Fred S. Hoffman, Study Director, *Ballistic Missile Defenses and U.S. National Security: Summary Report*, prepared for the Future Security Strategy Study (Washington, D.C., October 1983), hereinafter cited as *The Hoffman Report*] is quite cautious, stating that "nearly leakproof defenses may take a very long time, or may prove to be unattainable in a practical sense against a Soviet effort to counter the defense" (p. 2); and "Such [highly effective] defenses may result from the R&D programs pursuant to the President's goal, but it is more likely that the results will be more modest" (p. 9).

4. During the 1960s the U.S. experienced a similar shift in the objectives of BMD as it became clear that the available systems could not protect U.S. cities from Soviet ballistic missile attack. There is a striking resemblance between the debate over anti-ballistic missile (ABM) systems of the late 1960s and early 1970s and the current BMD debate. (ABM and BMD are used here interchangeably to refer to defense against ballistic missiles.) Many of the arguments examined in this article also played a role in the earlier debate. Representative arguments against ABMs are found in Abram Chayes and Jerome B. Wiesner, eds., *ABM: An Evaluation of the Decision to Deploy an Antiballistic Missile System* (New York: Harper and Row, 1969). Representative arguments in favor are found in Johan J. Holst and William Schneider, Jr., eds., *Why ABM?: Policy Issues in the Missile Defense Controversy* (New York: Pergamon Press, 1969).

This article examines the following arguments for deploying BMD that is capable of performing certain limited missions but incapable of making U.S. cities invulnerable:

—To increase ICBM (intercontinental ballistic missile) survivability
—To protect command, control, and communications and other military targets
—To defeat small Soviet nuclear attacks, thereby raising the nuclear threshold
—To increase the uncertainty confronting the attacker
—To encourage and support arms control, especially reduction of offensive forces
—To protect against accidental, unauthorized, $n^{th}$ country, and terrorist attacks
—To counter Soviet violations of the ABM Treaty
—To gain the political and economic benefits of U.S. technical superiority
—To enhance the United States' offensive capabilities

Are these arguments cogent and analytically sound? How well do they apply to the current nuclear situation? What counterarguments might be made? On balance, do the benefits of BMD deployment suggested by these arguments exceed the costs?

Proponents of BMD recommend an evolutionary deployment strategy based on the arguments stated above: the U.S. should develop and deploy "intermediate" systems, i.e., BMD capable of performing these less demanding missions, as soon as possible. The U.S. would first deploy BMD capable of performing the least demanding missions, probably beginning with the defense of ICBMs. Then, in the hope that BMD technologies improve faster than penetration techniques, defenses would take on more demanding missions with the final objective remaining perfect defense. Proponents argue that an evolutionary strategy is not risky because at each stage of deployment the benefits provided by BMD exceed its costs.[5]

I argue, however, in sharp contrast to the arguments of SDI proponents, that deployment of limited or intermediate BMDs would, on balance, reduce U.S. security. The benefits of deploying these systems are found upon examination to be much less than the proponents' arguments at first suggest. And, unless one desires to provoke an intensified competition with the Soviet

---

5. See, for example, *The Hoffman Report*; and Keith B. Payne and Colin S. Gray, "Nuclear Policy and the Defensive Transition," *Foreign Affairs*, Vol. 62, No. 4 (Spring 1984), pp. 820–842.

Union, the costs of BMD deployment are much more impressive: the ABM Treaty would have to be either amended or, more likely, terminated, thereby eliminating restraints on Soviet BMD; each superpower, believing that the other's BMD threatens its security, would likely react by expanding and improving its offensive force in order to have confidence in its ability to overcome the adversary's defense; this competition between offense and defense would exacerbate U.S.–Soviet relations; these changes in the political and strategic situation would make arms control agreements much harder to reach than today; and, finally, the economic costs of U.S. strategic nuclear forces would increase immensely. Thus, deployment of less-than-near-perfect BMD and the associated evolutionary strategy should be rejected.

*Deploy BMD to Increase ICBM Survivability*

This section draws on the extensive debate over the vulnerability of U.S. ICBMs in order to evaluate the benefits of protecting ICBMs with BMD.[6] Three lines of argument are generally used to support the need for invulnerable ICBMs as part of the mix of U.S. strategic forces.[7]

---

6. The benefits of defending ICBMs are discussed in Payne and Gray, "Nuclear Policy and the Defensive Transition," pp. 823–825; George A. Keyworth II, "The Case for Strategic Defense: An Option for a Disarmed World," *Issues in Science and Technology*, Fall 1984, pp. 38–42; and "The President's Strategic Defense Initiative" (Washington, D.C.: U.S. Government Printing Office, January 1985), p. 3.

Renewed interest in protecting ICBMs with BMD preceded the SDI. Addressing this issue are "ABM Revisited: Promise or Peril," *The Washington Quarterly*, Vol. 4, No. 4 (Autumn 1981), pp. 53–85; G.E. Barash et al., *Ballistic Missile Defense: A Potential Arms-Control Initiative* (Los Alamos, N.M.: Los Alamos Scientific Laboratory, January 1981); Albert Carnesale, "Reviving The ABM Debate," *Arms Control Today*, Vol. 11, No. 4 (April 1981), pp. 1–2, 6–8; Colin S. Gray, "A New Debate on Ballistic Missile Defense," *Survival*, Vol. 23, No. 2 (March/April 1981), pp. 60–71; Jan M. Lodal, "Deterrence and Nuclear Strategy," *Daedalus*, Vol. 109, No. 4 (Fall 1980), pp. 166–174; Michael Nacht, "ABM ABCs," *Foreign Policy*, No. 46 (Spring 1982), pp. 155–174; and Office of Technology Assessment, *MX Missile Basing* (Washington, D.C.: U.S. Government Printing Office, September 1981).

7. The following discussion draws upon more extensive analysis in Albert Carnesale and Charles Glaser, "ICBM Vulnerability: The Cures Are Worse Than the Disease," *International Security*, Vol. 7, No. 1 (Summer 1982), pp. 70–85.

In addition to the three arguments discussed in the text, some proponents of BMD argue that the U.S. should deploy defenses now to defend its ICBMs in order to maintain crisis stability during the transition from the current nuclear situation, in which both superpowers' cities and industry are highly vulnerable to attack, to a situation in which defenses radically reduce this vulnerability. The crisis instability would result when the Soviet Union could partially defend itself against a retaliatory attack, but not against a U.S. first strike. See, for example, Payne and Gray, "Nuclear Policy and the Defensive Transition," p. 284. There are two problems with this argument. First, it does not support the near-term deployment of BMD to defend ICBMs. The

First, some argue that such a force is needed to discourage a Soviet first strike, especially during a crisis. However, others answer that even if the ICBM force were completely vulnerable, the other legs of the triad ensure the United States' ability to retaliate massively. Taking this into account, proponents of increasing ICBM survivability suggest instead that if U.S. ICBMs are highly vulnerable the Soviet Union could gain a coercive position by initiating a counterforce exchange that increases the relative size of its force to the U.S. force.[8] BMD, then, is one way of eliminating this danger.

Other analysts present persuasive counterarguments. They argue that the deterrent value of surviving forces is best reflected by absolute force size, not relative force size, since the former determines the amount of damage that could be inflicted. The U.S. now has more survivable weapons than are required to achieve reasonable retaliatory objectives, including especially destroying a large fraction of Soviet society. So, independent of its relative size, the U.S. force would be able to inflict essentially the same amount of damage following a Soviet attack against its ICBMs as before the attack. Thus, even if the Soviet Union could improve the relative size of its force, it could not gain a significant advantage, e.g., the ability to blackmail the U.S.

This first line of argument also ignores the enormous risks confronting a Soviet leader. Although a counterforce attack that left U.S. cities largely undamaged might encourage the U.S. not to retaliate against Soviet cities, no Soviet leader could be confident that the U.S. would respond with a purely counterforce attack. In addition, any incentive for Soviet leaders to attack U.S. ICBMs is reduced further by their awareness of the danger of undesired escalation: the difficulty of maintaining command and control of nuclear forces following a large attack and the "fog of war" make all-out war likely, even if both superpowers prefer to terminate the conflict at a lower

---

Soviet Union will not be able in the foreseeable future to defend its cities against U.S. retaliation: even if U.S. ICBMs were completely destroyed in a first strike, the Soviet Union would face a large retaliatory force composed of SLBMs, strategic bombers, and cruise missiles. The Soviet defense that could be deployed in the near future would not be able to protect Soviet cities from this retaliatory force and, therefore, could not significantly decrease crisis stability. Second, if the superpowers ever deploy defenses that significantly reduce the vulnerability of their home-lands, then these defenses would also make their forces highly survivable, thereby greatly reducing any preemptive incentives.

8. See, for example, Paul H. Nitze, "Deterring Our Deterrent," *Foreign Policy*, No. 25 (Winter 1976–77), pp. 195–210; Paul H. Nitze, "Assuring Strategic Stability in an Era of Détente," *Foreign Affairs*, Vol. 54, No. 2 (January 1976), pp. 207–232; and Secretary of Defense Harold Brown, U.S. Department of Defense, *Annual Report*, Fiscal Year 1982 (Washington, D.C.: U.S. Government Printing Office, 1981), pp. 41, 52–59.

level.[9] Thus, while "pure" counterforce attacks and exchanges are possible, they are probably quite unlikely, and the conclusions that follow from their consideration should be weighted accordingly.

The second line of argument holds that the U.S. needs survivable ICBMs because they possess capabilities unavailable in the other legs of the triad. Secure communications, high accuracy, and the speed of ballistic missiles enable ICBMs to carry out reliably prompt attacks against Soviet ICBMs and other time-urgent targets. However, again, the counterarguments identify serious weaknesses. A highly survivable ICBM force is not required for more limited conflicts since a large fraction of the U.S. ICBM force would remain available following such a limited Soviet attack. Yet examination of less limited conflicts in which the Soviet Union launches an attack against all of the United States' ICBMs shows that, in these cases, there would be few, if any, time-urgent targets following the Soviet attack.[10] Those scenarios that do require prompt second strikes are incredible and unlikely. Moreover, even if there were important time-urgent second-strike targets, deployment of BMD by both superpowers might not increase the United States' prompt second-strike capability: although U.S. BMD would make its ICBMs harder to destroy in a first strike, Soviet BMD would make it harder for surviving U.S. ICBMs to destroy Soviet targets in a second strike.

The third argument in favor of survivable ICBMs is that an invulnerable triad is more resistant to improvements in Soviet forces than an invulnerable diad. Of the three arguments this is the strongest: a survivable strategic force forms the core of the U.S. deterrent and should not be compromised. The survivability of U.S. ballistic missile submarines (SSBNs), however, reduces the importance of possessing an invulnerable triad. The threat to SSBNs remains highly theoretical.[11] Should advances in anti-submarine warfare capabilities occur that could not be handled with countermeasures, there would likely be time for the U.S. to restructure its land- and sea-based forces.

---

9. For a complete statement of these counterarguments see Robert Jervis, "Why Nuclear Superiority Doesn't Matter," *Political Science Quarterly*, Vol. 94, No. 4 (Winter 1979–80), pp. 617–633; and *The Illogic of American Nuclear Strategy* (Ithaca: Cornell University Press, 1984).
10. See Carnesale and Glaser, "ICBM Vulnerability," pp. 78–82. The importance of prompt second-strike counterforce is also questioned by Richard K. Betts, "Elusive Equivalence: The Political and Military Meaning of the Nuclear Balance," in Samuel P. Huntington, ed., *The Strategic Imperative* (Cambridge, Mass.: Ballinger, 1982), p. 130.
11. See Richard L. Garwin, "Will Strategic Submarines Be Vulnerable?," *International Security*, Vol. 8, No. 2 (Fall 1983), pp. 52–67.

Moreover, if the U.S. decides to increase the survivability of its ICBM force, BMD will have to be compared to the many other ways this can be achieved.[12]

In short, then, although survivable ICBMs are preferable to vulnerable ones, the benefits of increasing ICBM survivability, with BMD or other means, are relatively small.

### Deploy BMD to Protect $C^3$ and Other Military Targets

ICBMs are not the only U.S. military capability vulnerable to Soviet nuclear attack. The other legs of the triad, i.e., long-range bombers and ballistic missile submarines, are partially vulnerable; the command, control, and communications ($C^3$) systems for the strategic forces can be severely degraded by even relatively small attacks; and U.S. conventional military forces are especially vulnerable to nuclear attack. The forces could be subjected to attacks of varying intensity: at one end of the spectrum are nuclear attacks against a few conventional force targets, e.g., critical ports needed to support forces fighting in Europe, while at the other end of the spectrum are full-scale attacks against the United States' strategic nuclear forces, including $C^3$.

Proponents of BMD tend to favor protecting military targets against the entire range of attacks because they believe that increasing the survivability of military targets reduces the adversary's ability to achieve his objectives, thereby strengthening deterrence.[13] In contrast, opponents of BMD do not believe there is much value in defending these targets against any type of attack: the U.S. already has large retaliatory capabilities, and defense can add little to its ability to deter such attacks by threatening retaliation. This section examines the arguments for defending $C^3$. The two sections following examine arguments that apply to all types of military targets.

Some proponents of BMD argue that the U.S. should defend its $C^3$ against a large Soviet attack, maintaining that less-than-near-perfect defenses can deny the Soviet Union high confidence in its ability to destroy these targets.[14] The counterargument has two components. First, the U.S. need not be able to defeat a Soviet attack on its $C^3$ to deter it. A Soviet leader would believe

---

12. BMD is only one of many ways to increase ICBM survivability; others include deception, mobility, hardening, launch under attack, and arms control. For analysis of the various ways of basing the MX, see Office of Technology Assessment, *MX Missile Basing*.
13. See, for example, *The Hoffman Report*, p. 1.
14. Ibid., pp. 9–10.

that any full-scale attack on U.S. $C^3$ targets would result in full-scale retal-iation: the attack would result in a large number of U.S. casualties; it would likely destroy the United States' ability to control, and therefore limit, its retaliation;[15] and it would not be launched as part of a bargaining strategy since the attack would destroy the capabilities required for damage assess-ment, communication, and war termination. Consequently, if the Soviet Union attacked the United States' command and control, it would almost certainly also attack U.S. strategic nuclear forces. The damage then would be much higher than if only $C^3$ were attacked, thereby further reducing any U.S. incentives for restraint. In short, the only rational reason for launching a full-scale attack against $C^3$ is to reduce the damage the U.S. could inflict in retaliation.

This, however, is infeasible for the Soviets. Although $C^3$ is considered the weak link in U.S. strategic capabilities, analysts believe that the U.S. would almost certainly be able to retaliate massively after a Soviet attack against its command and control.[16] So, the Soviet Union cannot reasonably hope to reduce the United States' ability to inflict damage by attacking its $C^3$ and, therefore, has virtually no incentive to launch such an attack. In this case, therefore, denial of the ability to attack $C^3$ targets adds little, if anything, to deterrence. Thus, the benefits of protecting $C^3$ with BMD would be small.

Still, if nuclear war appeared inevitable, a Soviet leader might, because of the slight chance of significantly reducing U.S. retaliation, rationally decide to launch a preemptive attack against U.S. $C^3$. Thus, the importance of the second component of the counterargument: defending $C^3$ would not signif-icantly increase its ability to survive a dedicated Soviet attack. There are fewer than 100 critical fixed $C^3$ targets.[17] Assume that U.S. BMD were quite effective and could raise the "attack price" to 20 (i.e., the Soviet Union would have to allocate 20 warheads to the target to have confidence that it would be destroyed).[18] In this case the Soviet Union would have to allocate 2000

15. Desmond Ball, *Can Nuclear War Be Controlled?*, Adelphi Paper No. 169 (London: International Institute for Strategic Studies, Autumn 1981); and John D. Steinbruner, "Nuclear Decapitation," *Foreign Policy*, No. 45 (Winter 1981–82), p. 18.
16. Ball, *Can Nuclear War Be Controlled?*, p. 37; and Steinbruner, "Nuclear Decapitation," pp. 18–19.
17. Ball, *Can Nuclear War Be Controlled?*, p. 35; Steinbruner, "Nuclear Decapitation," p. 18. Also describing the small number of $C^3$ targets is John J. Hamre, Richard H. Davison, and Peter T. Tarpgaard, *Strategic Command, Control and Communications: Alternative Approaches for Modernization* (Washington, D.C.: Congressional Budget Office, 1981), p. 13.
18. An attack price of 20 is quite optimistic for systems that might be available in the next two decades, so this argument is weighted toward the defense. Most disagreement on the attack

warheads to $C^3$ targets instead of the 100 to 200 required without BMD. However, such an attack would leave the Soviet Union with approximately 4000 ICBM warheads for attacks against U.S. ICBMs and other military targets. So, BMD could raise the price of an attack on $C^3$, but even very effective BMD could not deny the Soviet Union the ability to destroy these targets with a first strike. Moreover, raising the total attack price would significantly discourage the Soviet Union from starting an all-out war only if the price exceeded the forces available: in an all-out war, holding forces in reserve would be of little value. If this point were ever reached, the Soviet Union could expand its offensive force to offset the United States' BMD.

*Deploy BMD to Defeat Small Soviet Attacks, Thereby Raising the Strategic Nuclear Threshold*

Some proponents of BMD assert that a moderately effective BMD could reduce the probability of nuclear war by protecting the U.S. from a small Soviet nuclear attack. Without BMD, nuclear war could start with the use of only one or a few nuclear weapons. Therefore, goes the argument, because a thin defense might completely intercept a small attack, it would require the attacker to launch a larger attack. But, the argument continues, the Soviet Union might be unwilling to launch a large attack because it would constitute too large an escalation. So, even if the U.S. could not prevent the Soviet Union from destroying certain targets, there is value in requiring the Soviet Union to attack with a large number of weapons. Donald Brennan argues:

[I]t is very likely that a government that would otherwise plausibly consider escalating an intense crisis to the strategic nuclear level would have second thoughts about the matter if it was obliged to fire a large-scale salvo rather than one or a very few weapons.[19]

---

price that might be achieved with traditional endoatmospheric systems (those that target reentry vehicles in the atmosphere) is between two and eight, with many analysts skeptical of estimates above five. It is difficult to estimate the attack price that might be achieved by adding exoatmospheric systems (those that target incoming warheads before they reenter the atmosphere) because the performance of these systems depends heavily on the specific system architecture and its ability to defeat a wide variety of potentially devastating penetration tactics. See Ashton B. Carter, "BMD Applications: Performance and Limitations," in Ashton B. Carter and David N. Schwartz, eds., *Ballistic Missile Defense* (Washington, D.C.: Brookings, 1984), pp. 110–130.
19. Donald G. Brennan, "BMD Policy Issues for the 1980s," in William Schneider, Jr. et al., *U.S. Strategic-Nuclear Policy and Ballistic Missile Defense: The 1980s and Beyond* (Cambridge, Mass.: Institute for Foreign Policy Analysis, 1980), p. 27.

It is far more likely, however, that both attacker and defender would care more about the number of targets destroyed than about the number of warheads that were launched. A thin defense could increase the attack price, but could not prevent the Soviet Union from destroying a relatively small number of targets. BMD would not force the Soviet Union to attack more targets, only to use more weapons. Since this larger Soviet attack would not do significantly more damage than would an attack not increased by the need to overcome BMD, the probability of U.S. counter-escalation should not increase significantly. So, BMD should not significantly influence Soviet decisions unless the attack price were high enough to severely deplete the Soviet force, thereby placing the Soviet Union at meaningful military disadvantage. However, a moderately effective BMD protecting a relatively small number of targets could not have this effect.

Thus, less-than-near-perfect BMD would not deny the Soviet Union the ability to use limited nuclear attacks for military or bargaining purposes. The Soviet Union, as part of a bargaining strategy, for instance, could still escalate a war with a small nuclear attack to demonstrate its resolve and to increase the United States' assessment of the probability that the war would escalate to yet higher levels.

More important in the context of the current debate, the Soviet Union would still be able to mount limited nuclear strikes for military purposes. For example, the Hoffman Report argues that during a large conventional war in Europe the Soviet Union might launch a limited strategic nuclear attack against U.S. force projection targets to deny the U.S. the ability to provide military support to Western Europe, and that BMD could play a key role in deterring this type of attack. BMD of moderate capability, according to this argument, by forcing the Soviet Union to increase the size of its attack, would reduce Soviet confidence in its ability to destroy these targets without running too high a risk of further escalation.[20]

---

20. The Hoffman Report maintains that: "In the event of imminent or actual large-scale conflict in Europe, another high-priority Soviet task would be to prevent quick reinforcement and resupply from the United States. . . . In the absence of defenses, the Soviets . . . could also accomplish this task with higher confidence by means of quite limited nuclear attacks on such [reception] facilities in Europe and on a restricted set of force projection targets in CONUS. . . . [A]n intermediate ballistic missile defense deployment of moderate capabilities could force the Soviets to increase their attack size radically. This would reduce or eliminate the Soviets' confidence that they could achieve their attack objectives while controlling the risks of a large-scale nuclear exchange." *The Hoffman Report*, pp. 10–11.

However, the Soviet Union can maintain high confidence of destroying the small number of U.S. force projection targets by increasing the number of weapons directed at each target. Increasing the total number of weapons, while holding the number of targets fixed, would not significantly increase the damage to the U.S. Therefore, the larger attack seems to be hardly riskier than the smaller attack. Moreover, a Soviet leader willing to attack the U.S. with nuclear weapons presumably is prepared to run extremely large risks. BMD would not increase significantly the risk of attacking a few U.S. targets and, therefore, would be unlikely to change such a leader's decision. Thus, the Hoffman Report's argument appears quite weak.

### Deploy BMD to Increase the Uncertainty Confronting the Attacker

Proponents maintain that BMD would strengthen deterrence by increasing Soviet uncertainty about the success of an attack, thereby making it less attractive. The Hoffman Report argues, "Uncertainty about the offense-defense engagement itself contributes to deterrence of attack by denying confidence in the attack outcome."[21] An Administration description of the SDI states:

Effective defenses against ballistic missiles have potential for enhancing deterrence. . . . they could significantly increase an aggressor's uncertainties regarding whether his weapons would penetrate the defenses and destroy our missiles and other military targets. It would be very difficult for a potential aggressor to predict his own vulnerability in the face of such uncertainties.[22]

This argument, however, overstates both the potential of defenses to increase uncertainty and the benefits if uncertainty were increased. First, the argument that increasing uncertainty strengthens deterrence is incomplete: it overlooks the effect of the Soviets' defenses on their deterrence calculation. If both superpowers deployed defenses, an attacker would face greater uncertainty about both the effectiveness of his attack *and* the effectiveness of the adversary's retaliation. The net effect of defenses is, therefore, indeterminate.

---

21. Ibid., p. 10.
22. "The President's Strategic Defense Initiative," p. 3.

The attacker's willingness to take risks determines how uncertainty would affect deterrence. A conservative Soviet decision-maker would likely overestimate the capability of the U.S. defense and underestimate the capability of his own defense. As a result, deterrence would be strengthened. However, an overconfident Soviet decision-maker would underestimate the capability of America's defense and overestimate the capability of his own defense; hence, deterrence would be weakened.[23] It may well be that, more often than not, decision-makers who face potentially catastrophic outcomes will act cautiously. However, uncertainty is not unambiguously good, and we should not overlock the possibility of a decision-maker who in a dire situation optimistically evaluates uncertainties about force capabilities.[24]

Second, the argument that BMD will increase uncertainty significantly applies only to large attacks against military targets. As discussed above, if the number of targets is not large, then the attacker can overcome the defense by increasing the size of the attack. Uncertainties about the effectiveness of the defense could require the attacker to increase further the size of the attack to ensure high confidence of success. However, for BMD available in the next two decades, compensating for uncertainties could make the overall attack price prohibitively high only for a large target set. Moreover, the Soviet Union could reduce the significance of the increased attack price by expanding its offensive force—which would likely cost less than the U.S. BMD.

Finally, for the reasons discussed above, these large Soviet attacks against U.S. ICBMs, $C^3$, and other military targets are now adequately deterred. Supporting this conclusion, and especially germane to this argument, is the fact that large technical and operational uncertainties already exist about the effectiveness of a Soviet attack, especially one directed at hardened targets.[25] A decision-maker who could be deterred by uncertainties about attack out-

---

23. We should not overlook the fact that defenses shift the most probable outcome of an attack as well as the distribution that surrounds it. Thus, there are cases in which defenses would tend to strengthen deterrence, independent of the nature of the decision-maker. For example, if both countries deploy BMD that unambiguously protects only forces, then a decision-maker considering a counterforce attack and anticipating countervalue retaliation would find that BMD makes attacking less attractive. In this case, as in others, the effect of uncertainty about the adversary's BMD depends on the specific decision-maker.
24. Crises in which ill-founded expectations of military success may have led to riskier crisis policies and, as a result, wars are discussed in Richard Ned Lebow, *Between Peace and War: The Nature of International Crisis* (Baltimore: Johns Hopkins University Press, 1981), pp. 242–247.
25. See John D. Steinbruner and Thomas M. Garwin, "Strategic Vulnerability: The Balance Between Prudence and Paranoia," *International Security*, Vol. 1, No. 1 (Summer 1976), pp. 138–181; and Matthew Bunn and Kosta Tsipis, "The Uncertainties of a Preemptive Nuclear Attack," *Scientific American*, November 1983, pp. 38–47.

comes would likely be deterred by the uncertainties that already exist. In contrast, if these uncertainties in combination with the other reasons for not launching a counterforce attack are not adequate to deter a Soviet decision-maker, then the additional uncertainty created by BMD is likely to have little effect. In sum, *if* the uncertainty created by BMD is beneficial, it can at most strengthen deterrence only slightly.

*Deploy BMD to Encourage and Support Arms Control, Specifically Offensive Reductions*

Proponents of BMD maintain that defenses reduce the military utility of nuclear weapons, thereby making it easier to trade away existing offensive weapons and less attractive to build additional ones. Moreover, continues the argument, if the cost of building offenses to defeat defenses is greater than the cost of building defenses, i.e., the "cost-exchange ratio" favors the defense, then U.S. deployment of defenses might essentially force the Soviet Union to give up its offensive capability. Presumably President Reagan's statement that research and development of BMD "could pave the way for arms control measures to eliminate the weapons themselves" was based upon these reasons.[26]

Opponents, however, argue convincingly that deploying BMD will not facilitate arms control, warning that if BMD reduces the Soviet Union's ability to perform strategic missions, then it will simply increase the size and penetrability of its force to restore these capabilities. Moreover, because offensive reductions would further reduce Soviet capabilities, it will then be harder for the superpowers to limit offensive forces once the U.S. deploys BMD.[27] Five points support this counterargument.

First, U.S. BMD would not reduce the value the Soviet Union places on being able to perform certain missions with nuclear weapons; rather, BMD might increase the difficulty of performing these missions. Therefore, the utility of nuclear weapons would remain high, and the Soviet Union is likely

---

26. *The New York Times*, March 24, 1983, p. 20. See also "The President's Strategic Defense Initiative," p. 6; Gray, "A New Debate on Ballistic Missile Defense," pp. 68–69; Payne and Gray, "Nuclear Policy and the Defensive Transition," p. 839; Keyworth, "The Case for Strategic Defense," pp. 35–44; James C. Fletcher, "The Technologies for Ballistic Missile Defense," *Issues in Science and Technology*, Fall 1984, pp. 28–29; and *The Hoffman Report*, p. 11.

27. See, for example, McGeorge Bundy, George F. Kennan, Robert S. McNamara, and Gerard Smith, "The President's Choice: Star Wars or Arms Control," *Foreign Affairs*, Vol. 63, No. 2 (Winter 1984–85), pp. 264–278.

to react to U.S. BMD by increasing the size and penetrability of its offensive force to defeat the United States' defense.

Some analysts respond that if both superpowers deploy BMD, then the value of being able to carry out counterforce and countervalue attacks would be reduced. Both superpowers, therefore, would be willing to forgo their ability to perform these missions. For example, if U.S. BMD eliminates the Soviet ability to destroy U.S. forces, then the U.S. need not be able to destroy Soviet forces; so, continues the argument, Soviet defense of its forces would not threaten a necessary U.S. mission, and the U.S. would be willing to reduce the size of its offensive force.[28] This logic, however, fails to explain superpower behavior heretofore. Since presumably each superpower could reduce the other's offensive force requirements by reducing the size of its own offensive force, this logic predicts that negotiations would have produced large mutual reductions of the superpowers' offensive forces. Yet not even minor reductions in strategic nuclear offensive forces have occurred. Why should we expect deployment of BMD to totally reverse the way the superpowers define their security requirements and plan their forces?

Second, even if the superpowers followed this unrealistic logic, BMD would not spur the radical reduction of Soviet forces required to protect U.S. society. The BMD proponents' argument holds that the superpowers will give up the capabilities threatened by the adversary's BMD. However, in the coming decades each superpower could deploy at most moderately effective BMD, leaving its adversary with a redundant assured destruction capability and the ability to destroy a large fraction of its fixed military targets. There would be no more incentive than today to trade away these capabilities. In

---

28. D.G. Brennan, applying this logic, argues that BMD could lead to symmetric reductions in the homeland vulnerability of the superpowers; see "Post-Deployment Policy Issues in BMD," in D.G. Brennan and Johan J. Holst, *Ballistic Missile Defence: Two Views*, Adelphi Paper No. 43 (London: International Institute for Strategic Studies, November 1967), p. 9. Arguing in a similar vein, Payne and Gray maintain that the Soviet Union's reluctance to reduce its ICBM force stems, in part, from its doctrinal requirement for a damage-limitation capability. Currently, ICBMs form the core of this capability, providing an offensive counterforce capability. Therefore, continues the argument, because Soviet strategic defense could substitute for Soviet ICBMs in this damage-limitation role, Soviet deployment of defenses would make the Soviet Union more willing to reduce the size of its offensive force. (Payne and Gray, "Nuclear Policy and the Defensive Transition," p. 839.) This argument, however, appears logically inconsistent: at least in theory, offensive counterforce, strategic defense, and/or negotiated reductions of the adversary's offensive force could be used to reduce the adversary's ability to inflict damage. Thus, at this level of generality, it is unclear why the Soviet doctrinal priority given to reducing the U.S. ability to inflict damage does not now favor negotiated reductions of offensive forces.

addition, each superpower, looking to the future through a conservative planning lens, is likely to fear improvements in the other's defense that leave it at a strategic disadvantage, possibly even to the point of unease about its assured destruction capability. From this perspective, the adversary's defense actually *increases* the value of excess offensive forces since redundant weapons serve as a hedge against the adversary's development of a superior strategic defense. Fear of this possibility will encourage both superpowers to increase the size of their offensive forces. Offensive reductions would make the possibility of improvements in the adversary's defense more threatening.

Third, the cost-exchange argument of proponents of BMD, while not without merit, suffers serious weaknesses. According to this argument, the superpowers should be willing to reduce their offensive forces, or at least freeze them, if the cost of maintaining an assured destruction capability exceeds the cost of the defensive system required to deny this capability. However, the cost-exchange ratio of such highly effective defenses is likely to heavily favor the offense for the foreseeable future. For one thing, the only concepts proposed for highly effective area defense require expensive boost phase technologies, which still appear susceptible to defeat by relatively inexpensive countermeasures.[29] In addition, defenses become more costly as they are asked to perform more demanding missions since they must be able to defeat the full range of offensive countermeasures, which in turn makes the cost-exchange ratio more favorable to the offense.[30] Furthermore, because each superpower would believe that its fundamental security interests were threatened by the adversary's defense, it would likely be willing to pay a disproportionate sum to defeat it—even if a favorable cost-exchange ratio existed, it would not be sufficient to subdue competition between each superpower's offense and the other's defense.[31]

---

29. On the existence of countermeasures see Carter, *Directed Energy Missile Defense in Space*, pp. 45–52, 69–70.
30. Ibid., pp. 45–46. In addition to technical countermeasures, such as decoys, there are tactical countermeasures. A city defense must engage almost all of the attacking warheads. Thus, unlike a defense of silos, it does not benefit from the leverage provided by preferential defense. In fact, just the opposite applies, since the offense can defeat the defense by concentrating its attack on cities of especially high value. Carter, "BMD Applications," in Carter and Schwartz, *Ballistic Missile Defense*, p. 170.
31. For example, Brennan argues that the cost-exchange ratio would likely have to exceed 5 to 1 to dominate U.S. planning. See "Post-Deployment Policy Issues," in Brennan and Holst, *Ballistic Missile Defence: Two Views*, p. 7.

Fourth, the inclination to offset the adversary's defense would probably be reinforced by military institutions: no organization is likely to concede that its missions that are threatened by the adversary's defense are not vital to the country's security. For example, it is hard to imagine the U.S. Air Force accepting the argument that Soviet BMD was so effective that the U.S. should no longer maintain the capability to target Soviet forces or that U.S. ICBMs had become virtually worthless and should be retired. Indeed, past Soviet defense deployments have instead motivated American offensive programs: the American MIRV was spurred in part by Soviet BMD, and the current B-1, Stealth, and air-launched cruise missile programs are responding to Soviet air defense deployments.

Finally, and possibly most important, deploying BMD will not facilitate the limitation and reduction of offensive forces if it has a destructive effect on overall superpower relations. If the adversary's deployment of strategic defenses is understood to reflect aggressive intentions, as it almost certainly would be, then the superpowers are likely to be unable to pursue offensive limits or any other form of arms control.

Proponents also argue that BMD would make possible large reductions in offensive forces by reducing the difficulty of adequately verifying such a treaty. At force levels much lower than today's, the superpowers would require extremely effective verification capabilities because even small numbers of illegal nuclear weapons could be strategically significant. Defenses reduce the danger of cheating and breakout at these low levels by reducing the importance of an advantage in the number of weapons. Thus, BMD could increase U.S. confidence in its ability to detect significant violations by raising the level at which cheating becomes militarily significant. Therefore, continues this argument, BMD would allow the U.S. to relax its verification requirements for agreements that radically reduce offensive forces.[32]

Although this argument might apply in certain situations, it is not an argument for deploying defenses now. The superpowers have not made progress toward the successful negotiation of large offensive reductions. Verification and breakout are far from the most serious barriers to these agreements. Massive offensive reductions will not be negotiated in the near future, so this argument should not influence current U.S. policy. Moreover, for the reasons discussed above, deployment of BMD will reduce the superpowers' ability to negotiate offensive reductions.

---

32. See, for example, Payne and Gray, "Nuclear Policy and the Defensive Transition," p. 838.

*Deploy BMD to Protect Against Accidental, Unauthorized, N$^{th}$ Country, and Terrorist Attacks*

Some BMD proponents focus on the danger posed by certain small nuclear attacks and maintain that BMD could reduce the damage if one of these attacks occurred. The source of a small attack could be an accidental or unauthorized launch of Soviet missiles or some country other than the Soviet Union (an n$^{th}$ country) or a terrorist group.[33]

Clearly, if one of these small ballistic missile attacks occurs, then the U.S. would be better off with a ballistic missile defense that can reduce the damage than without one. The importance of this argument, however, also depends upon the likelihood of such attacks. As with all nuclear attack scenarios, estimating probabilities of occurrence, even relative probabilities, is highly speculative. The probability of an accidental Soviet ballistic missile attack, at least during normal peacetime conditions, is generally believed to be quite small.[34] In addition, improvements in Soviet command and control and maintenance of forces that need not be placed on high alert during crises or launched under attack could further reduce this probability. (These improvements would likely to be in the interest of both superpowers.)

The probability of n$^{th}$ country attacks depends upon the number of countries that threaten the U.S. with nuclear weapons carried on ballistic missiles. Referring to this as the n$^{th}$ country problem suggests there will be many countries that might launch small ballistic missile attacks against the U.S. In fact, of the countries that now have this capability—the Soviet Union, China, France, and the United Kingdom—only the Soviet Union is not a U.S. ally. During the late 1960s the possibility of a Chinese ballistic missile capability was presented as the principle reason for deploying a light area defense.[35] That capability has developed more slowly than then anticipated, and the

---

33. See, for example, "The President's Strategic Defense Initiative," p. 4; Fletcher, "The Technologies for Ballistic Missile Defense," pp. 26–27; Keyworth, "The Case for Strategic Defense," pp. 41–42; and Payne and Gray, "Nuclear Policy and the Defensive Transition," p. 825.
34. See Paul Bracken, "Accidental Nuclear War," in Graham T. Allison, Albert Carnesale, and Joseph S. Nye, Jr., eds., *Hawks, Doves and Owls: An Agenda for Avoiding Nuclear War* (New York: W.W. Norton, forthcoming).
35. A persuasive argument against deploying BMD in response to the Chinese threat was made by Allen S. Whiting, "The Chinese Nuclear Threat," in Chayes and Wiesner, *ABM*, pp. 160–170. The opposing argument was made by Frank E. Armbruster, "The Problem of China," in Holst and Schneider, *Why ABM?*, pp. 221–234.

Chinese have now deployed only a few ICBMs.[36] More importantly, U.S.–Chinese relations have improved significantly. Furthermore, not many countries are likely in the near future to acquire the capability to attack the U.S. with nuclear weapons delivered by ballistic missiles. In addition to the problem of acquiring nuclear weapons, countries face the more difficult task of building long-range ballistic missiles.[37] The U.S. could reduce the future size of $n^{th}$ country and terrorist threats by giving higher priority to slowing the proliferation of nuclear weapons and ballistic missiles.

Possibly most important in terms of this argument, BMD could not prevent a determined adversary from attacking the U.S. with nuclear weapons by some other means of delivery. The country or terrorist group could deliver nuclear weapons by aircraft, ships, and a variety of clandestine means much more easily than by building intercontinental missiles. In addition, although the proliferation of nuclear weapons and advanced means of delivery does increase somewhat the threat of nuclear attack against the U.S., the U.S. ability to retaliate should be sufficient to deter most countries—especially since most presumably would lack a second-strike capability, and thus would stand at America's mercy after they attacked.

This discussion does not, of course, prove that accidental or $n^{th}$ country ballistic missile attack is impossible, nor that a light BMD would be of no value. It does, however, suggest that these attacks are highly unlikely and that BMD is unlikely to make a significant difference in preventing $n^{th}$ countries from delivering nuclear weapons to U.S. cities if they want to.

*Deploy BMD to Counter Soviet Violations of the ABM Treaty*

Some proponents of BMD argue that the U.S. should withdraw from the ABM Treaty to counter Soviet violations of arms control treaties. Colin Gray holds that:

Soviet violations now work uniquely to their benefit, however, so the United States would have little to lose by abandoning treaties that only restrain its behavior.[38]

---

36. International Institute for Strategic Studies, *The Military Balance 1984–1985* (London: IISS, 1984), pp. 90–91.
37. See Aaron Karp, "Ballistic Missiles in the Third World," *International Security*, Vol. 9, No. 3 (Winter 1984–85), pp. 166–195.
38. Colin S. Gray, "Moscow Is Cheating," *Foreign Policy*, No. 56 (Fall 1984), p. 148.

He further argues that Soviet violations of the ABM Treaty have given them a superior near-term breakout potential. In addition, if the U.S. fails to respond to these violations, the Soviet Union may conclude that the U.S. lacks the determination necessary to achieve its foreign policy objectives.[39]

There seems to be little question that the large phased-array radar the Soviets have built near Krasnoyarsk violates the treaty's prohibition of early warning radars that are not located on the periphery of the Soviet Union and facing outward.[40] However, this violation does not threaten the United States' strategic capabilities, and, therefore, does *not* provide the Soviet Union with a significant military benefit. Moreover, the U.S. can deny the Soviet Union any breakout advantage that the combination of this violation with ongoing Soviet BMD programs allowed by the treaty might provide. First, by increasing the ability of its ballistic missile force to penetrate future Soviet BMD systems, the U.S. can reduce the benefits of Soviet breakout. Second, by bringing its BMD programs to the pre-deployment stage allowed by the ABM Treaty, the U.S. can ensure that the Soviet Union cannot gain a significant advantage in a BMD deployment race. Whether the U.S. actually needs to increase its ability to rapidly match potential Soviet BMD deployment is questionable, however, since the danger now posed by Soviet breakout is quite small. In short, Soviet violations do not create a critical military need to abandon the ABM Treaty.

Soviet violation of the treaty is therefore, at least for now, a political problem. Any treaty violation, even if not militarily significant, is politically significant, calling into question the wisdom of engaging in arms control agreements with the Soviet Union, and creating fears that, over time, Soviet violations really will make a military difference. Thus, assuming effective limitation on both superpowers' BMD is in its interest, the United States'

39. Ibid., pp. 141–152.
40. The U.S. has charged the Soviet Union with violations or probable violations of many arms control treaties, including SALT II and the Threshold Test Ban Treaty as well as the ABM Treaty. In addition to the radar, potential violations of the ABM Treaty are said to include development of mobile ABM components and the concurrent operation of SAM and ABM components. Problems of Soviet compliance with the ABM Treaty are reported in "The President's Unclassified Report to the Congress on Soviet Non-Compliance with Arms Control Agreements" (Washington, D.C.: The White House, February 1, 1985); "The President's Report to the Congress on Soviet Noncompliance with Arms Control Agreements" (Washington, D.C.: Office of the Press Secretary, The White House, January 23, 1984); and General Advisory Committee on Arms Control and Disarmament, "A Quarter Century of Soviet Compliance Practice Under Arms Control Commitments: 1958–1983, Summary" (Washington, D.C., October 1984). For an analysis of the Administration's findings, see *F.A.S. Public Interest Report*, Vol. 37, No. 3 (March 1984), and *Arms Control Today*, Vol. 14, No. 3 (March–April 1984).

response to the Soviet radar should be designed to preserve the ABM Treaty by establishing that the Soviet Union cannot gain an advantage by cheating on arms control agreements.

This argues against rushing to withdraw from the ABM Treaty to deploy counterbalancing defensive capabilities: since the treaty still holds the potential to significantly restrict Soviet BMD, withdrawing because of the Soviet radar would be like "throwing the baby out with the bath water." The U.S. should continue to demand that the Soviet Union provide satisfactory explanations of their apparent non-compliance with arms control treaties. Discussions of compliance should be pursued in the Standing Consultative Commission (SCC) to minimize unnecessary political posturing.[41] If Soviet explanations prove unsatisfactory, the U.S. will have to search hard for options that both preserve the ABM Treaty and make it clear that the Soviet Union cannot gain advantages by violating arms control agreements. Analysts, with these objectives in mind, have suggested a variety of ways this compliance problem might be resolved. One is for the Soviet Union to modify the radar to reduce its early warning capability—its orientation might be changed to make it look more like a radar designed to track satellites, which is what the Soviets now claim it is. Another suggests the U.S. try to reach a common understanding with the Soviet Union in the SCC to allow deployment of defensive systems that match or offset Soviet violations. For example, Michael Krepon suggests "a common understanding . . . limiting the number of large phased-array radars, regardless of their stated purpose."[42] While such an adjustment of the ABM Treaty to accommodate violations of its provisions would strike a symbolic blow to the treaty, and more generally to arms control, which requires U.S. confidence in Soviet treaty compliance, it would preserve the treaty's fundamental objectives and would demonstrate that the Soviet Union cannot gain an advantage by cheating on agreements. Alternatively, the U.S. might accept apparent Soviet non-compliance as the result of divergent interpretation of ambiguities in the ABM Treaty but insist

---

41. The SCC was established by Article XII of the ABM Treaty. It provides a forum to consider "questions concerning compliance"; "possible proposals for further increasing the viability of this Treaty, including proposals for amendments"; "proposals for further measures aimed at limiting strategic arms"; as well as other issues related to compliance with and implementation of the treaty.
42. Michael Krepon, "Both Sides Are Hedging," *Foreign Policy*, No. 56 (Fall 1984), pp. 170–171. See also Thomas K. Longstreth, John E. Pike, and John B. Rhinelander, *The Impact of U.S. and Soviet Ballistic Missile Defense Programs on the ABM Treaty (Third Edition)*, A Report for the National Campaign to Save the ABM Treaty, March 1985, pp. 67–74.

upon supplementing the treaty with clarifications that eliminate the possibility of similar problems in the future.

It is likely, however, that satisfactory resolution of this non-compliance issue depends more on the determination of the U.S. and the Soviet Union to preserve the ABM Treaty and on U.S.–Soviet relations more broadly than on the merits of these specific approaches. A strong U.S. commitment to preserving the treaty may be a prerequisite. If this commitment exists and is matched by comparable Soviet interest in preserving the treaty, then the U.S. will likely be able to design a response that it can negotiate with the Soviet Union and sell to its public and allies.

*Deploy BMD to Gain the Political and Economic Benefits of U.S. Technical Superiority*

Colin Gray argues:

If, as seems unavoidable, the United States must sustain military competition with the Soviet Union for many decades to come—since the political fuel for the competition cannot be cut off—it is cost effective to compete most vigorously in those areas wherein the structural basis for an enduring lead is present, and with regard to which the Soviet Union, for excellent reasons, harbors the deepest of anxieties.[43]

In other words, even if BMD cannot meet its military objectives, it can contribute more broadly to U.S. strategy.

In light of the preceding analysis, which suggests that the benefits of less-than-near-perfect BMD are quite small, this argument may reflect a key underlying source of disagreement on BMD policy. From this perspective, the BMD debate becomes a proxy for answers to questions like: Is intense military competition a necessary and/or desirable extension of ongoing political competition? Would U.S. security be increased by a competition that threatens the Soviet Union with the specter of technological, if not strategic, inferiority? Should the U.S. use military competition to drain the inferior Soviet economy, thereby weakening the Soviet Union overall? BMD is part of a general national security debate that, with some danger of oversimplification, can be characterized as occurring between analysts who stress the existence of common U.S. and Soviet interests and analysts who emphasize

---

43. Colin S. Gray in Carter and Schwartz, *Ballistic Missile Defense*, p. 407.

the existence of conflict and competition. Although complete analysis of this debate is beyond the scope of this paper, it is useful to make the potential role of BMD explicit.

First, while military competition between the U.S. and the Soviet Union will continue for the foreseeable future, the superpowers have some control over its intensity. Deploying defenses will almost certainly increase the competition: Soviet deployment of BMD will increase U.S. leaders' doubts about the adequacy of their offense, and vice versa. As a result, the current competition in offensive forces would be exacerbated; and, of course, a new full-fledged competition in defenses would likely be set in motion. Advocates of defenses often disagree with this prediction, pointing to the buildup of offensive forces since the signing of the ABM Treaty to support their case. Their argument, however, which compares the arms buildup we have experienced to no buildup, is misleading. The correct comparison is between the buildup we have experienced and the one we would have experienced had there been no ABM Treaty (and had extensive BMD been deployed).

Second, a competition in BMD combined with an increased competition in the offensive forces that the BMD challenged would greatly increase the economic cost of strategic nuclear forces to both the U.S. and the Soviet Union. Since this competition would not significantly increase U.S. security, it could be "cost effective" only in the sense that it places more of a strain on the Soviet economy than on the U.S. economy.[44]

Third, the use of BMD to achieve military and economic advantages is incompatible with the achievement of arms control agreements—the Soviet Union will not accept highly inequitable agreements. Analysts cannot argue consistently that BMD is a means of achieving both objectives.

In short, there is virtually no doubt that BMD will increase the intensity of nuclear weapons competition and exacerbate superpower relations. Thus, it is clear that disagreements about how to improve U.S. security could result in divergent conclusions about the deployment of BMD. Analysts who believe that intensified superpower competition and initiatives to weaken or

---

44. This argument assumes the Soviets will follow the U.S. in deploying BMD, which is likely to be the case. It is interesting to note, however, that the Soviet Union could respond by increasing and improving only its offensive force. It would likely cost the Soviet Union far less to offset the United States' BMD than it cost the U.S. to build the BMD, i.e., the cost-exchange ratio is likely to favor the offense heavily. In this case (which, granted, is unlikely to occur), the U.S. deployment of BMD would move the arms competition into an area of U.S. disadvantage—its security would not be increased, and its economy would be drained more than the Soviet Union's.

intimidate the Soviet Union are in the U.S. interest will tend to favor deployment of BMD, and vice versa.

*Deploy BMD to Enhance U.S. Offensive Capabilities*

Lastly, some proponents favor marrying BMD with air defense, civil defenses, and offensive counterforce in order to reduce the Soviet Union's ability to retaliate following a U.S. first strike. In this scheme, BMD would blunt Soviet retaliation already weakened by an American attack against Soviet strategic nuclear forces and command and control. This BMD mission is less demanding than protecting against a Soviet first strike: American offenses aid in limiting damage by providing a kind of "pre-boost-phase" or "silo phase" intercept that leaves the defense with fewer retaliatory warheads to destroy. Proponents present two reasons for pursuing this essentially offensive strategy. First, much like the familiar "Star Wars" arguments, this combination of offensive counterforce and active defense would enable the U.S. to reduce the cost to itself of an all-out war. Second, by denying the Soviet Union a similar capability, the U.S. would create a situation in which an all-out war would be much more costly for the Soviet Union than for the U.S. This clear superiority, continues the argument, would increase the United States' ability to protect third areas (such as Western Europe) from Soviet conventional attack, i.e., it would strengthen extended deterrence. This is especially true, continues the argument, because Soviet leaders view military conflict in these traditional military terms—they are more afraid of being defeated than of suffering retaliatory damage.

This argument falls uncomfortably between the argument for "Star Wars," which imagines both superpowers with impenetrable shields, and the arguments for less-than-near-perfect defenses: this defense is not good enough to protect the U.S. against a first strike against its cities, but is good enough to contribute significantly to defense against a second strike.[45] Moreover, because proponents of this damage limitation strategy want to gain a strategic advantage, this use of defenses, unlike the President's version of "Star Wars," cannot be cast as mutually beneficial to both superpowers.

---

45. This comparison actually includes two separable factors: first, does the defense provide the U.S. with an advantage or does it benefit both superpowers roughly equally; and second, can an offensive counterforce attack significantly increase the ability of U.S. defenses to reduce the costs of Soviet retaliation? At least in theory, therefore, there are four possible kinds of situations in which defenses reduce the vulnerability of the U.S. homeland.

This argument has not played a prominent role in the current public debate over the SDI, but it was an important argument for BMD for many years before President Reagan's 1983 initiative, and it likely remains important in the minds of many BMD advocates today. For instance, Colin Gray, who has forcefully framed arguments for less-than-near-perfect defenses, also argues that to satisfy its extended deterrence requirements the U.S. must acquire a significant damage limitation capability, while maintaining the ability to destroy the Soviet state. Specifically, he asserts that combining an offensive strategy with homeland defense should reduce U.S. casualties to about 20 million.[46]

The feasibility of offensive damage limitation has received less attention than the more benign image of a defense that does not require the U.S. to strike first. Analysis of "Star Wars" has focused on the competition between U.S. BMD and Soviet ballistic missiles, countermeasures, and tactics. A complete analysis of the feasibility of combining offense and defense to reduce Soviet retaliatory capabilities would also examine the competition between U.S. offensive counterforce and the survivability of Soviet retaliatory forces. Most analysts believe that the U.S. deployment of offensive counterforce weapons cannot provide the ability to reduce significantly the potential cost of Soviet retaliation: first, the cost-exchange ratio in a competition over the Soviets' ability to retaliate massively against U.S. cities and industry favors the offense; second, the Soviet Union values this capability highly and, therefore, will pay large sums to maintain it. These are the same reasons that U.S. BMD cannot win such a competition against Soviet ballistic missiles. Thus, the strongest argument against this damage limitation strategy is likely to be that it is technically and economically infeasible. This is especially true since defenses protecting Soviet forces would presumably reduce the effectiveness of the United States' first strike. Studies of the feasibility of this offensive damage limitation strategy that addressed potential synergisms between offensive counterforce and active defense, however, would fill a gap and help to clarify the debate.

This brief discussion suggests a new dimension of the debate over less-than-near-perfect defense. An evolutionary deployment of BMD by the U.S.

---

46. Colin S. Gray and Keith Payne, "Victory Is Possible," *Foreign Policy*, No. 39 (Summer 1980), pp. 14–27; and Colin S. Gray, "Nuclear Strategy: The Case for a Theory of Victory," *International Security*, Vol. 4, No. 1 (Summer 1979), pp. 54–87.

might, in theory at least, lead to two distinct outcomes: in one the U.S. gains superiority, while in the other both superpowers are presumed to benefit from the mutual reduction of the vulnerability of their homelands. The reader, then, is reminded that a larger debate about these final objectives looms in the background of the current debate over less-than-near-perfect defenses. Moreover, if the envisioned U.S. superiority requires offensive counterforce, then its desirability raises additional difficult questions: the safety of counterforce has been challenged on grounds that it invites preemptive war, which in turn makes the control and termination of conventional wars more difficult.

## The Costs of Deploying BMD

The ABM Treaty now severely restricts the BMD deployments of both superpowers. Extensive U.S. deployment of BMD can be obtained only by amending or terminating the treaty, thereby allowing Soviet deployment. Switching from a nuclear situation in which neither superpower has extensive BMD to one in which both do would have significant strategic, political, and economic costs for the United States.

The costs of such a shift depend upon whether the ABM Treaty is amended or terminated. The vulnerability of ICBMs has created interest in amending the treaty to allow defense of only ICBMs. Such a treaty would, in theory, continue to restrict the most important potential threat to U.S. capabilities from Soviet BMD, i.e., reduction of the U.S. ability to retaliate against Soviet value targets, including cities and economic and industrial capabilities. Therefore, the costs of amending the treaty are smaller than terminating it because the strategic capability of Soviet BMD would continue to be constrained. In practice, however, such an amendment is likely to represent a large step toward termination of the treaty. First, U.S. attempts to amend the treaty might be unsuccessful, with one possible outcome being the total loss of constraints on BMD. For example, the U.S., having committed itself to an amendment to allow deployment of ICBM defenses, might, following unsuccessful renegotiation, choose to withdraw from the Treaty altogether instead of allowing the Soviet Union to block its deployment of BMD. Negotiating an amendment that allows defense of only ICBMs promises to be especially difficult because the Soviet Union is likely to be more interested in protecting leadership, command and control, other military targets, and economic targets than in protecting ICBMs.[47]

Second, Soviet systems ostensibly deployed to protect ICBM silos might provide some protection of other targets as well. Even if the actual area defense capabilities of these systems were small, United States analysts and targeting staff would tend to give them the benefit of the doubt, creating fears about whether the amended treaty serves U.S. interests and spurring moves toward complete termination of the treaty. Finally, extensive Soviet deployment of BMD would increase concern that the Soviet Union could gain a strategic advantage through breakout. Concern about Soviet breakout potential is an important factor in the reassessment of U.S. strategic defense policy now taking place. These concerns would only be increased by ICBM defenses that provide a base for a nationwide defense. Moreover, amendments to the treaty that allow still more extensive deployment, e.g., defense of $C^3$ as well as ICBMs or a thin area defense, would suffer even more severely from these problems. In short, amending the ABM Treaty is likely to hasten its termination.

Soviet BMD, which would almost certainly be deployed following the amendment or termination of the ABM Treaty, would reduce the ability of U.S. reentry vehicles to penetrate to their targets. However, strategic defenses available in the foreseeable future could not deny the United States the ability to perform its key deterrent missions. First, the Soviet Union will remain unable to protect its cities and industry. The U.S. now has a very large survivable strategic nuclear force, only a small fraction of which must penetrate Soviet defenses to destroy a large fraction of Soviet value targets. The Soviet Union, like the U.S., will be unable to deploy anything approaching the necessarily near-perfect strategic defense in the foreseeable future. Moreover, to eliminate any possible reduction in its ability to retaliate, the U.S. could, and almost certainly would, increase the number and penetrability of its ballistic missiles; increase the survivability of its ICBMs with BMD or by other means; and continue to maintain bombers and cruise missiles to penetrate Soviet air defenses. Furthermore, by maintaining an unquestionable ability to retaliate massively against cities and other targets of value the U.S. could ensure a high degree of crisis stability: the Soviet Union would be unable to reduce significantly the costs of a U.S. attack, so striking first

---

47. On Soviet BMD programs, see Sayre Stevens, "The Soviet BMD Program," in Carter and Schwartz, *Ballistic Missile Defense*, pp. 182–220.

and incurring U.S. retaliation would be essentially as bad as being struck first.[48]

Second, whether Soviet BMD could protect military targets would depend upon the value the U.S. places on being able to destroy these targets. Soviet BMD could make the total price of attacking large target sets, for example ICBMs and other military targets, quite high. However, an expanded and improved U.S. ballistic missile force could offset the Soviet BMD. The question, then, is would the U.S. be willing to pay potentially vast sums to maintain its full menu of retaliatory options against Soviet strategic forces and other military targets? There is substantial disagreement about the need for such a capability. Ironically, it is the advocates of deploying BMD to strengthen deterrence who believe that counter-military retaliation is important for deterrence, especially extended deterrence; they are likely to see Soviet defense of military targets as more threatening than do opponents of BMD.

A large expansion of Soviet BMD deployments might reduce significantly the threat posed by the French, British, and Chinese nuclear forces. Most analysts believe these independent nuclear forces increase U.S. security and,

---

48. My argument here differs from that of opponents of BMD who argue that mutual deployment of BMD will undermine the United States' deterrent capabilities and reduce crisis stability. In fact, at least to first order it appears inconsistent to argue both that the offense will prevail in an offense-defense competition and that the superpowers' deployment of BMD will significantly reduce their strategic capabilities. It is possible, however, to make such an argument by including what might be considered second order considerations. For example, one could argue that uncertainties and/or misperceptions created by BMD could lead a risk-taking decision-maker to launch a nuclear attack that would otherwise be deterred. However, while this argument has merit, it would not apply if the superpowers reacted to each other's defenses with offenses that offset not only the most probable performance of the defense, but also offset worst-case estimates that far exceed the most likely estimate. In this case, which is the one BMD opponents use to predict an endless expansion of offensive forces, the uncertainties created by BMD would be quite small, especially when compared to the uncertainties that already exist.

The argument that less-than-near-perfect BMD will decrease crisis stability suffers from essentially the same weaknesses. The strongest incentive to preempt would exist when a country could significantly reduce the potential costs of an attack against its cities by launching a counterforce first strike. The BMDs that might be available in the next two decades could not provide this damage-limitation capability against a competitive offensive threat. This conclusion is reinforced by the fact that defending forces is less demanding than defending cities. Thus, mutual deployment of BMD could increase both countries' ability to retaliate against cities and industrial capabilities, tending to increase crisis stability. If defenses in combination with offensive counterforce ever begin to provide a significant damage-limitation capability (an outcome that available analyses suggest is unlikely), then crisis stability would likely be reduced. Working against crisis stability is the fact that defenses would be more effective against a ragged retaliatory attack, one not optimized to defeat the defense, and against a smaller retaliatory attack than a larger first strike. Even then, however, the decrease in crisis stability might not be large since such highly effective defenses would greatly increase the survivability of retaliatory forces.

therefore, would consider this effect of Soviet BMD to be negative. In addition, because their nuclear capabilities would be threatened by amendment or termination of the ABM Treaty, U.S. efforts to change the treaty would likely strain its relations with these allies.

While the U.S. would be able to maintain its deterrent capabilities, the political costs of this offense-defense competition would be large. Soviet defenses will appear threatening to the U.S., and vice versa. The Soviet Union's BMD, like the ability it now has to attack U.S. strategic nuclear forces, will be understood by many analysts to reflect aggressive intentions. Analysts will likely see deployment and improvement of Soviet BMD as attempts to gain a strategic advantage, to undermine U.S. deterrent capabilities and, in the long run, to disarm the U.S. Soviet defenses will raise U.S. force requirements, making agreements to limit offensive forces harder to achieve than today and making U.S. security seem harder to maintain. Uncertainties about the effectiveness of Soviet defenses will continue to drive up U.S. force requirements. The natural tendency of U.S. force planners to assume the best about Soviet defenses and the worst about their own would likely generate unending fears that U.S. offense and defense were both inadequate. In this environment of increased mutual fear and hostility, the superpowers are more likely to overlook their common interests and to see only hostility in each other's actions. The result could well be a reduction in their ability to avoid and to manage crises.

The economic costs of deploying BMD will depend on the objectives of the U.S. defense and the size and penetrability of the Soviet offense. Any militarily significant BMD will cost tens of billions of dollars.[49] No one knows how much a highly effective multi-layer BMD including boost-phase intercept

---

49. Very few estimates of the cost of less-than-near-perfect defenses are now available. Although analysts have raised the possibility of deploying BMD to perform these less demanding missions, they have not specified in any detail the goal of the defense deployment, the BMD technologies that would be deployed, or the Soviet threat the defense would confront. Without this information, realistic cost estimates cannot be developed. A useful point of reference may be the Army's 1980 cost estimate for its Low-Altitude Defense System (LoADS) to defend the MX missile based in multiple protective shelters (MPS). A LoADS system to defend 200 MX missiles based in 4600 shelters was projected to cost $8.6 billion in 1980 dollars. This was a low estimate, which did not include the cost of additional warning and threat assessment systems and C³ systems required to support LoADS. See Office of Technology Assessment, *MX Missile Basing*, p. 125. A traditional BMD of 1000 fixed ICBMs would likely be many times as expensive: first, the number of missiles to be defended is much larger; and, second, the defense, lacking the leverage provided by MPS basing, would have to be able to charge a much higher attack price than LoADS. Defense of ICBMs is discussed in detail in Carter, "BMD Applications," in Carter and Schwartz, *Ballistic Missile Defense*, pp. 122–146.

(if ever developed) would cost; rough estimates fall in the 500 billion to one trillion dollar range.[50] Whatever the goal of the defensive system, the cost of deploying BMD would be far greater than the cost of the initial BMD system. The U.S. would improve its BMD to offset changes in the Soviet ballistic missile force. In addition, there would be the cost of improving and/or expanding the U.S. offensive force to offset Soviet BMD. Moreover, certain strategic defense goals, e.g., defeating small Soviet attacks, require the U.S. to deploy expensive air defenses against Soviet bombers and cruise missiles.

*Conclusion*

This analysis finds that the deployment of less-than-near-perfect defenses would provide, at most, small benefits on concerns of second or third order importance; deterrence would not be strengthened significantly. Yet the costs of deploying BMD are found to be extensive: U.S. offensive force requirements would increase, U.S.–Soviet competition would intensify, serious arms control efforts would likely disappear, and the economic costs of strategic nuclear forces would increase immensely.

This conclusion undermines the argument for an evolutionary deployment of BMD that hinges on the belief that less-than-near-perfect defenses would provide net benefits at each stage of deployment. My analysis reaches the opposite conclusion: U.S. deployment of the ballistic missile defense that might be available in the foreseeable future, assuming the Soviet Union also deployed BMD, will reduce U.S. security. Thus, an evolutionary strategy can be defended only with the hope that perfect or near-perfect defenses will someday be developed.[51] However, given the extremely low probability of such an effective defensive system ever being developed and deployed, the U.S. should reject the evolutionary deployment strategy.

Because deployment of less-than-near-perfect BMD is not in the United States' interest, priority should be given to preserving the ABM Treaty. The SDI conflicts with this objective. Although the Administration stresses that the SDI is only a research program and is therefore permitted by the ABM

---

50. See, for example, R. Jeffrey Smith, "Schlesinger Attacks Stars Wars Plan," *Science*, November 9, 1984, p. 673.
51. Although feasibility is a necessary condition for pursuing an evolutionary strategy, it is not a sufficient condition. For an analysis of the desirability of mutually deployed near-perfect defenses, see Charles L. Glaser, "Why Even Good Defenses May Be Bad," *International Security*, Vol. 9, No. 2 (Fall 1984), pp. 92–123.

Treaty, its enthusiasm for and commitment to the deployment of BMD conveys a willingness to terminate the treaty. Moreover, development and testing now planned in the SDI strain the limits of the treaty, and the BMD deployments envisioned by the SDI will require its termination. The Soviet Union must wonder whether they should expect the U.S. to withdraw from the treaty by the early 1990s.

Most important for preserving the ABM Treaty is for the Reagan Administration to change its course. Ideally, President Reagan would reverse himself completely, announcing that new studies show the SDI is a mistake. More realistic steps include reductions in funding for the SDI and a relaxation of commitment to the long-term goal of near-perfect defense. Unfortunately, the Administration appears unwilling to rein in its SDI even slightly.

Somewhat ironically, the U.S. should continue research and development of BMD that is *permitted* by the ABM Treaty in order to reduce the potential benefits to the Soviet Union of withdrawing from the treaty. Such a policy, designed to deter Soviet deployment of BMD, requires a careful balancing of conflicting pressures. Pursuit of BMD activities allowed by the treaty could have two negative effects. First, U.S. R&D could provide support in the Soviet Union for increased BMD R&D. This would be counterproductive if the net effect was to increase pressure in the Soviet Union to amend or withdraw from the treaty. Second, the larger the U.S. R&D effort, the stronger the domestic pressure will be to loosen the treaty constraints: larger budgets will increase the influence of those responsible for BMD, and successful R&D will create pressures for deployment. The momentum of R&D could threaten the objective it was initiated to achieve.

To avoid undermining the ABM Treaty, the U.S. should be absolutely clear about its reasons for pursuing R&D: 1) to deter Soviet withdrawal from the treaty; and 2) to hedge against Soviet breakout from the treaty. Providing these reasons for R&D cannot, however, be convincing to the Soviet Union or to domestic constituencies if U.S. leaders are simultaneously talking about the substantial future benefits of BMD. Thus, R&D designed to preserve the ABM Treaty should be supported by policy statements declaring the United States' long-term interest in preserving a nuclear situation in which neither superpower has deployed BMD extensively.

R&D that approaches the constraints imposed by the ABM Treaty should be understood to have served its purpose: the Soviet Union would no longer be able to achieve an advantage in this area of BMD technology. The treaty should not be amended to allow R&D to continue since this would not be necessary to achieve either of the objectives stated above. Moreover, we should recognize that there are significant advantages to pursuing R&D at less than the fastest possible rate since this will reduce the negative effects of R&D. The "best" rate requires a balance between keeping reasonable pace with the Soviet Union and controlling the pressures stimulated by R&D for amendment of the ABM Treaty.

Beyond these changes in its current policy, the U.S. should explore policies to strengthen the ABM Treaty. First, because certain treaty constraints are open to interpretation, the U.S. and the Soviet Union should, in the SCC, try to develop a shared understanding of the treaty's boundaries. Where permitted research ends and prohibited development begins is not absolutely clear; whether certain technologies are components of an ABM system, and therefore banned, or are something less than components, and therefore allowed, can be even less clear. The U.S. and the Soviet Union are likely to undermine the treaty if they press the limits of possible interpretations.[52] Working through these issues will require that each superpower believes the other is interested in preserving the treaty. Changes in current U.S. policy are, therefore, a prerequisite for strengthening the treaty.

Second, the ABM Treaty is threatened by potentially dual-capable weapons, including anti-tactical ballistic missiles, surface-to-air missiles, and anti-satellite weapons, which are not banned and could reduce its effectiveness. The ABM Treaty covers only defense against strategic ballistic missiles and therefore does not ban anti-tactical ballistic missiles (ATBM). ATBMs could, however, contribute to the unraveling of the ABM Treaty. Drawing a precise line between ABMs and ATBMs is necessarily difficult: certain strategic weapons, e.g., short-range SLBMs, have trajectories similar to long-range tactical ballistic missiles. An extensive ATBM deployment would have some capability against strategic offensive systems and would be certain to create fears that its ABM capability could be upgraded.[53] In addition, there is an important asymmetry: Soviet ATBMs can be deployed on Soviet territory. Because,

---

52. These and other legal issues are discussed in Abram Chayes, Antonia Chayes, and Eliot Spitzer, "Space Weapons: The Legal Context," in Jeffrey Boutwell, Donald Hafner, and Franklin Long, eds., *Weapons in Space* (New York: W.W. Norton, forthcoming); and Alan B. Sherr, *Legal Issues of the "Star Wars" Defense Program* (Boston: Lawyers Alliance for Nuclear Arms Control, 1984).

53. Stephen Weiner, "Systems and Technology," in Carter and Schwartz, *Ballistic Missile Defense*, p. 73. The relationship between ABM and ATBM is cast in a positive light by *The Hoffman Report*, which states, "The advanced components, though developed initially in an ATM mode, might later play a role in continental United States (CONUS) defense" (p. 3).

by definition, the U.S. homeland is beyond the reach of tactical missiles, U.S. deployment of ATBMs on its territory would violate the ABM Treaty. The combination of this asymmetry with the ATBM's ability to intercept certain strategic missiles could place the U.S. in a situation where the Soviet Union appears to have an extensive ABM capability and the U.S. has none. A ban on ATBMs would close a significant loophole in the ABM Treaty.[54]

There is also substantial overlap between BMD and ASAT technologies.[55] The point at which an ASAT system qualifies as an ABM system, i.e., is capable of intercepting strategic ballistic missiles in flight trajectory, is unclear. Thus, the potential dual capability of ASAT weapons may provide an opportunity to circumvent the ABM Treaty. Agreement to limit or ban ASAT testing and deployment would reinforce the treaty. Although a complete assessment of ASAT arms control is beyond the scope of this paper, it is important to note that the BMD issue is of much greater moment than the ASAT issue.[56] This assertion is not intended to minimize the significance of ASATs, but rather to highlight the central importance of strategic defenses. Area defenses could, at least in theory, threaten the adversary's capability to retaliate, which is essential for deterrence. By contrast, satellites provide extensive military support, but their role is not entirely irreplaceable.[57] In this sense the BMD and ASAT issues are incommensurate: the importance of the area BMD decision overwhelms the ASAT decision. This weighs heavily in favor of limiting ASAT capabilities.

The Reagan Administration now faces something of the opposite choice: successful ASAT arms control negotiations would place severe restrictions on BMD development and deployment. This would be hard to reconcile with the Administration's commitment to strategic defense. Independent of the Administration's beliefs about the value of mutual ASAT restraints, the close

---

54. Analysis of whether the net benefit of mutual deployment of ATBMs exceeds the risk that such deployment poses to the ABM Treaty is beyond the scope of this paper. Potential benefits of NATO deployment of ATBMs are discussed in *The Hoffman Report*, pp. 3, 10–11.

55. See Ashton B. Carter, "The Relationship Between ASAT and BMD," in Boutwell, Hafner, and Long, *Weapons in Space*.

56. Arguing that ASAT arms control is not in the overall interest of the United States is President Reagan, *Report to the Congress on U.S. Policy on ASAT Arms Control*, Washington, D.C., 31 March 1984. Opposing opinions are found in William J. Durch, *Anti-Satellite Weapons, Arms Control Options, and the Military Use of Space*, Report prepared for the U.S. Arms Control and Disarmament Agency, July 1984; Richard L. Garwin, Kurt Gottfried, and Donald L. Hafner, "Antisatellite Weapons," *Scientific American*, June 1984, pp. 45–55; and John Pike, "Anti-Satellite Weapons and Arms Control," *Arms Control Today*, Vol. 13, No. 11 (December 1983), pp. 1, 4–7.

57. Office of Technology Assessment, *Arms Control in Space: Workshop Proceedings* (Washington, D.C.: U.S. Government Printing Office, 1984), p. 11

relationship between BMD and ASAT makes success in ASAT arms control unlikely.

While this article argues against the deployment of BMD and in favor of strengthening the ABM Treaty, the foregoing analysis also suggests a general observation about BMD policy: because none of the options for BMD deployment are without costs, we must address difficult trade-offs. Arguments in the BMD debate that suggest that deploying BMD brings only benefits are incomplete; they advocate, but do not analyze. Analyses of the benefits of defending U.S. targets tell only part of the story and will always favor deployment of BMD. Making the U.S. invulnerable to Soviet attack has undeniable appeal, but is an unrealistic objective and only distorts U.S. defense policy. U.S. security requires complete analysis of the more realistic limited BMD options. Such analysis, however, will direct the U.S. away from BMD and toward policies to maintain and strengthen the ABM Treaty.

# Ballistic Missile Defense and the Atlantic Alliance

*David S. Yost*

$\mathbf{T}$he Atlantic Alliance may be at the threshold of a new debate on the implications of ballistic missile defense (BMD) for European security. Secretary of Defense Caspar Weinberger and several U.S. Senators and Congressmen support a thorough review of U.S. BMD options, including possible revision of the 1972 Anti-Ballistic Missile (ABM) Treaty and its 1974 Protocol. Although active defense of intercontinental ballistic missiles (ICBMs) seems the most likely application for BMD, other strategic defense options are reportedly under consideration. European-based BMD against theater ballistic missiles such as the SS-20, SS-21, SS-22, and SS-23 is being examined as well. Such defenses are known as anti-tactical ballistic missiles (ATBM) or anti-tactical missiles (ATM). The term "ATM" is preferred in that it implies capability against cruise as well as ballistic missiles.

The political and strategic issues that BMD programs could raise within the Alliance should be explored as deliberately as possible before economic resources are committed. Material for preliminary analysis resides in previous Alliance deliberations on BMD and in the informal discussions recently provoked in Europe by obvious U.S. interest in BMD options, including ATM. The issues go to the heart of NATO's established theory of deterrence and offer an opportunity for fundamental reassessment.

This essay is based on extensive interviews in Europe in 1980 and 1981. Despite obvious risks of over-simplification, owing to the diversity of views in each country on most issues, I have chosen to conform to standard practice by referring to the "Europeans" as a shorthand for what appear to have been and to remain dominant trends in West European opinion. Special thanks are owed to Colin Gray, who first encouraged me to investigate this topic, and to various observers in government and industry (including Benson Adams, Guy Barasch, Charles Kupperman, and Richard Nuttall) who commented on earlier drafts. The views expressed are nonetheless mine exclusively, and should not be construed to represent those of the Department of the Navy or any U.S. government agency.

*David S. Yost is an Assistant Professor at the U.S. Naval Postgraduate School, Monterey, California. He is the author of* European Security and the SALT Process, *Washington Paper, Number 85 (Beverly Hills and London: Sage Publications, 1981) and editor of* NATO's Strategic Options: Arms Control and Defense *(New York: Pergamon Press, 1981).*

*International Security,* Fall 1982 (Vol. 7, No. 2) 0162-2889/82/010143-32 $02.50/0

*Alliance BMD Deliberations, 1967–1968*

The previous Alliance deliberations on BMD helped to form European attitudes that have become firmly entrenched over the past fifteen years. The
principal deliberations took place in Nuclear Planning Group (NPG) meetings
from April 1967 through April 1968. The two key issues were U.S. plans for
the Sentinel ABM system and the possibility of BMD in Europe.

McNamara's September 1967 speech announcing the decision to deploy
the Sentinel ABM system for defense of the United States against projected
Chinese strategic capabilities "created considerable resentment among the
allies" for several reasons, including convictions that "the announcement
had been made without sufficient consultation and that the United States
had failed to honour its obligations to the NPG."[1] The anti-Chinese orientation of Sentinel was seen in Europe as based on "hysterical and dangerous"
American fears of China, so that "the dangers that are thought to arise from
BMD deployment seem to be incurred for no good reason."[2]

These presumed dangers were partly those thoroughly articulated by the
American opponents of BMD at approximately the same time—above all,
that strategic stability and prospects for arms control and détente would be
needlessly endangered by highly expensive technology that probably would
not be reliably effective. Few indeed were the Europeans of that era who
supported a U.S. BMD program as in the West European interest because it
might promote coupling by assuring the continued invulnerability of American retaliatory forces.[3] The most frequently offered West European argument
against U.S. ABM deployment was that it would promote a neo-isolationist
"Fortress America" concept, allowing Western Europe to stand alone and
vulnerable. The West European reaction was admirably summarized by Johan
Holst, recently the Norwegian state secretary for foreign affairs:

It is, on the whole, surprising to note the extent to which European opinion
has been so unanimously unfavorable to any deployment of ballistic missile
defenses. The generally critical attitude does not differentiate between various alternative U.S. BMD deployment configurations. . . . The expectation

---

1. Paul Edward Buteux, *The Politics of Nuclear Consultation in NATO, 1965–1974: The Experience
of the Nuclear Planning Group* (Ph.D. dissertation, London School of Economics and Political
Science, 1978), p. 114.
2. Laurence W. Martin, *Ballistic Missile Defence and the Alliance*, Atlantic Paper Number 1 (Bou-
logne-sur-Seine, France: Atlantic Institute, 1969), p. 31.
3. An example was Elizabeth Young in *Survival*, Vol. 12, No. 4 (April 1970), p. 149.

[is] that any BMD deployment is likely to generate an arms race which, in turn, will increase tensions between the two superpowers. . . . If, however, we assume a similar Russian BMD deployment, the threat the United States could mobilize on part of her allies might look less impressive the more the Soviet BMD promised to reduce the damage of any American retaliation. Hence, a bilateral BMD deployment might on balance also be perceived as reducing the validity of the guarantee.[4]

Still further "overwhelmingly hostile" arguments were expressed in Europe regarding BMD. Even ICBM defenses might lead to limited area defenses and thus to virtual "decoupling" of the U.S. guarantee, with an enhanced "possibility of nuclear war at Europe's expense." Superpower BMD deployments could even serve the American purpose of "elimination of independent centres of nuclear power in the West," and at the least would reinforce West European feelings of political-military inferiority and subordination. Finally, the European perception that "BMD seems to be entirely concerned with fighting wars rather than with deterrence" guaranteed the concept a "chilly reception."[5]

European distaste for the concept of BMD based in Europe was even more emphatic. In April 1968 the NPG decided that Europe-based BMD "would be too costly, not totally effective, and might compromise arms limitation discussions between the United States and the Soviet Union."[6] Denis Healey, then British secretary of state for defense, reportedly stressed "its prohibitive cost and lack of effectiveness against a Soviet attack."[7]

In contrast, Holst felt that the "technical problem of providing some reasonably effective defense at a meaningful level in Europe is probably surmountable," owing to the slower re-entry speed of medium-range ballistic missiles (MRBMs) (compared to ICBMs). Holst, however, did note three political obstacles to ABM deployment in Europe: 1) Since the kill mechanism would probably have to be nuclear, he foresaw "a political problem in terms of convincing a suspicious audience about the reliability of the design against

---

4. Johan J. Holst, "Missile Defense: Implications for Europe," in Johan J. Holst and William J. Schneider, Jr., eds., *Why ABM? Policy Issues in the Missile Defense Controversy* (New York: Pergamon Press, 1969), pp. 190, 194. See also Theodore Sorenson, "The ABM and Western Europe," in Abram Chayes and Jerome B. Wiesner, eds., *ABM: An Evaluation of the Decision to Deploy an Antiballistic Missile System* (New York: Harper and Row, 1969), pp. 179–83.
5. Martin, *Ballistic Missile Defense*, pp. 29–36.
6. Buteux, *Politics of Nuclear Consultation*, p. 123.
7. Benson D. Adams, *Ballistic Missile Defense* (New York: American Elsevier Publishing Co., 1971), pp. 137–138, 179.

accidents and abuse." 2) Disagreements about which localities would be defended "could have disruptive rather than integrating effects" in the Alliance. 3) Most important was the risk to détente, the "danger that a BMD in Western Europe might tend to perpetuate a posture and atmosphere of confrontation."[8] Such political arguments, especially the latter, were probably as important as the technical and financial ones offered by Healey.

*General European Views on BMD*

In 1969, the NPG took relatively little notice of the U.S. decision to revise the anti-Chinese orientation of Sentinel to a Safeguard system dedicated to protection of U.S. retaliatory forces and to providing "thin" area defenses against accidental or small attacks, Soviet or Chinese. This contrast in NPG reactions may be explained in part by a recognition by Europeans that theater nuclear weapons analyses would be "the area of greatest allied input into alliance nuclear policy," with strategic force decisions mainly a U.S. responsibility.[9] An additional factor may have been the imminent commencement of the Strategic Arms Limitation Talks (SALT) and the prospect of negotiated constraints on BMD.

The ABM Treaty of 1972 and its 1974 Protocol were welcomed in Western Europe for all the reasons why BMD was recently opposed. The main benefit was seen as stabilization of the arms race and East–West relations generally, with a firm foundation for continuing détente. Ian Smart suggests three more specific reasons for West European approval: 1) The continued credibility of the British and French deterrents was enhanced. 2) The United States insisted that Article IX of the ABM Treaty (which prohibits the transfer of ABM technology to third countries) would not prevent the transfer of offensive weapons technology. 3) The United States did not make itself less vulnerable to ballistic missile attack than its Allies.[10]

In the intervening years, Europeans have generally become even more sensitive to détente considerations, and the ABM Treaty has assumed special importance as a surviving "keystone" of détente. U.S. interest in renegotiating the ABM Treaty therefore appears dangerous and potentially destabilizing to many in Western Europe, and abrogation still more so. An important

---

8. Holst, "Missile Defense," pp. 200–201.

9. Buteux, *Politics of Nuclear Consultation*, pp. 162–163.

10. Ian Smart, "Perspectives from Europe," in Mason Willrich and John B. Rhinelander, eds., *SALT: The Moscow Agreements and Beyond* (New York: The Free Press, 1974), pp. 187, 191, 194.

example of this view is the statement of the Palme Commission, which includes such influential West European politicians as Egon Bahr, Jean-Marie Daillet, Gro Harlem-Bruntland, David Owen, and Joop den Uyl:

If the ABM Treaty were abrogated and an unbridled offense/defense arms race ensued, the consequences would be severe. . . . as continued development of ABM systems buttressed the illusion that nuclear wars could be fought and survived in some meaningful sense, the risk of the use of nuclear weapons would multiply. Each side, fearing that the other might perceive advantage in a nuclear first-strike, might be tempted to act first. The instabilities and dangers in such a situation are obvious.[11]

These well-established general attitudes regarding BMD are reinforced by views in specific countries that illustrate the political obstacles within the Alliance that U.S. homeland BMD options and ATM might face.

FRANCE

Approximately fifteen years ago, when the prospects for BMD deployment for area defense in both the United States and the Soviet Union appeared serious, analysts predicted that the emerging French strategic nuclear force program would be deprived of all credibility by Soviet BMD; France could scarcely hope to build enough submarine-launched ballistic missiles (SLBMs) and intermediate-range ballistic missiles (IRBMs) to saturate Soviet defenses. The 1972 ABM Treaty therefore provided a very welcome opportunity for the French to continue the expansion of their strategic nuclear force program. The French government could have reasonable confidence that its deterrent's political utility would not be rendered ineffective without at least some advance warning through public abrogation of the ABM Treaty by either superpower, or through intelligence regarding clandestine Soviet research and development in ABM that might offer the Soviets an option of rapid ABM deployment.

What would the French do if more extensive and effective deployments take place? The official position for many years has been that such a threat is genuine, but that France is fully prepared for the eventuality.[12] One of the

---

11. *The SALT Process: The Global Stakes* (Vienna, Austria: Independent Commission on Disarmament and Security Issues, February 1981), pp. 2–3.
12. Alain Bru and Lucien Poirier, "Dissuasion et défense anti-missiles," *Revue de défense nationale*, December 1968, p. 1828. Hugues de l'Estoile, Lucien Poirier, and Didier le Cerf, "Les implications stratégiques de l'innovation technologique," *Revue de défense nationale*, January 1968, pp. 23–33, and February 1968, pp. 238–239. General Poirier, then a member of the Defense Ministry planning staff, was the key author of the 1968 documents that became the foundations of the still-valid 1972 defense white paper.

highest officials in the French Defense Ministry's planning department, writing under a pseudonym, explains that the ABM Treaty could be abrogated or circumvented (through air defense missile upgrades) at any time:

Nevertheless our principal guarantee resides in the reciprocal surveillance the two superpowers maintain over each other; each is in fact most interested in assuring that the other will respect the ABM Treaty. . . . So long as ABM defenses remain at the current level, the multiple-warhead system [MRV] that will be in service with the M4 [SLBM] should be able to exhaust these defenses without too much difficulty and to assure the penetration of a significant portion of our strategic missiles. Moreover, what is called the "hardening" of warheads and missiles can make our missiles more invulnerable to the effects of ABM warhead explosions. But still other solutions exist . . . [e.g.,] increasing the number of our missiles . . . [and] cruise missiles, which pose difficult problems for enemy defenses.[13]

When he was deputy director of the planning department, Colonel Guy Lewin added that the number of warheads on existing missiles could also be increased.[14] The director of military applications at the Commissariat à l'Energie Atomique has declared that re-entry vehicle (RV) separation will be such that, in conjunction with hardening, no enemy interceptor will be able to destroy more than one RV.[15] Decoys and other penetration aids may be under consideration as well.

At the same time, expressions of official concern have also been made. Foreign Minister Jean-François Poncet in February 1981 reportedly "cautioned the Reagan administration against building large-scale anti-ballistic missile systems . . . on grounds that this would create instability in Europe."[16] The director of the Foreign Ministry's planning department has stated that revising the ABM Treaty, even for ICBM defenses, "would weaken the technical credibility of our striking force with respect to the USSR."[17]

To date no relevant statements by the Mitterrand government have been made. It nonetheless seems plausible to assume that its officials would also oppose any revision of the ABM Treaty. Jean-Pierre Cot, now minister of

---

13. Ivan Margine, "L'avenir de la dissuasion," *Défense Nationale*, April 1978, p. 10.
14. Guy Lewin, "L'avenir des forces nucléaires françaises," *Défense Nationale*, May 1980, p. 18.
15. Jacques Chevallier, "Les armes et les ripostes mises en oeuvre par la défense française," in *La France face aux dangers de guerre: Actes du Colloque*, Vol. 1 (Paris: Fondation pour les Etudes de Défense Nationale pour l'Association des Anciens Elèves de l'Ecole Nationale d'Administration, 1980), pp. 175–176.
16. *International Herald Tribune*, February 26, 1981.
17. Jean-Louis Gergorin, "Menaces et politiques dans la décennie 1980," in *La France face aux dangers de guerre*, p. 65.

cooperation and development, once pointed out that Soviet BMD could undermine the French deterrent as part of his advocacy of French participation in SALT negotiations.[18] Morever, the chief of staff of the armed forces, in discussing the decision to construct a seventh SSBN by 1994, added that it would not be reasonable, in view of France's "sufficiency" needs for deterrence, to have more than seven SSBNs by the end of the century; more than seven or eight SSBNs would lead France away from the "sufficiency" principle.[19] Evidently this concept of sufficiency, influenced in part by France's economic capacity, assumes that Soviet BMD will not be upgraded beyond manageable limits.

It is most improbable that France would reopen the decision announced in the 1972 White Paper against any French BMD program on the grounds of cost and probable ineffectiveness,[20] given the short time-of-flight from the Soviet Union to France (around 10 to 12 minutes, of which 4 at most would be useful for interception).[21] Even if France had the resources to pursue BMD, a French decision to do so could legitimize Soviet interest in BMD and thereby severely undermine the credibility of the French forces. In the meantime, the French insist, as they have since 1967 when the question of IRBM vulnerability was first raised, that their fixed IRBMs are protected by their SLBMs. An attack against French IRBMs would be the plainest proof of aggression, and would justify strategic retaliation.[22] In Giscard d'Estaing's words in June 1980: "Any nuclear attack on French soil would automatically provoke strategic nuclear retaliation."[23] The French would therefore probably oppose Europe-based U.S. BMD against Soviet theater missiles—the antitactical missile (ATM) concept—as likely to legitimize Soviet BMD. Any form of U.S. homeland BMD beyond the ABM Treaty regime's limits would be seen as even more certain to provoke the expansion of Soviet BMD programs.

## BRITAIN

British sensitivity regarding the ABM Treaty, and possible improved Soviet BMD within its confines, can be seen in the Chevaline program for hardened,

---

18. Jean-Pierre Cot, "Plaidoyer pour l'intérêt national," in *La France face aux dangers de guerre*, p. 202.
19. Général Jeannou Lacaze, "La politique militaire," *Défense Nationale*, November 1981, pp. 13–14. Cf. Mitterrand in *Le Monde*, July 26–27, 1981, p. 6.
20. *Livre blanc sur la défense nationale*, Volume 1 (June 1972), p. 18.
21. General François Maurin in *La France face aux dangers de guerre*, pp. 389, 421.
22. Defense Minister Pierre Messmer in *L'Express*, December 11, 1967.
23. Giscard d'Estaing in *Le Monde*, June 28, 1980.

maneuvering and early separating re-entry vehicles with advanced decoys as penetration aids. The British government has acknowledged that, because of the Galosh ABM system protecting the Moscow region, the targeting list of "key aspects of Soviet state power" for British SLBMs may include targets outside Moscow—cities and "many high value targets, such as major dams and waterways, major oil refineries, major naval shipyards, major iron and steelworks, and major nuclear reactor establishments."[24] Since these targets are presumably unprotected today, why did the British go to the expense of the long-secret Chevaline program (nearly £1 billion over the 1973–1980 period) to harden the SLBM RVs and equip them with penetration aids? Apparently both Chevaline and the July 1980 Trident SLBM decision assume possible improvements in Soviet BMD. Soviet construction of significant BMD would require enhanced British penetration capabilities. This concern is expressed delicately in public British documents:

Though the Chevaline programme will keep our Polaris missiles able to penetrate anti-ballistic missile defences into the 1990s, continuing Soviet effort in research and development, allowed by the 1972 ABM Treaty, might in time reduce our assurance of this. . . . [Trident I's] MIRV capability and long range give excellent margins of long-term insurance against further advances in Soviet ABM and ASW capability. . . .[25]

In other words, Britain's position on BMD is similar to that of France. Precautionary steps have been taken in case Soviet BMD improves, while preservation of the ABM Treaty's constraints is emphatically preferred. Some observers have speculated that Britain would find superpower ICBM defenses more acceptable than any area defenses. This might be tolerable in theory for the maintenance of the credibility of Britain's deterrent, but even such ICBM defenses could promote the deployment of Soviet BMD that could be oriented to defense of population centers and other targets of potential British interest. No official British preferences have been expressed other than continuance in force of the existing ABM Treaty and Protocol provisions.

British interest in BMD is even less likely than French, given that the British have no hardened retaliatory forces to protect and greater financial constraints. British officials would probably oppose the ATM concept as

---

24. Official evidence in the *Twelfth Report from the Expenditure Committee*, cited in Lawrence Freedman, *Britain and Nuclear Weapons* (London: Macmillan, 1980), p. 47. See also Freedman's useful discussion of Chevaline, pp. 48–51.
25. *The Future United Kingdom Strategic Nuclear Deterrent Force* (London: Ministry of Defense, July 1980), pp. 7, 20. This statement would also apply to the even longer-range and more effective Trident II (D-5), which is to be purchased in lieu of Trident I.

likely to endanger the ABM Treaty. Because of Britain's close relationship with the United States in strategic nuclear matters, some observers have speculated that the British might be more likely than the French to accept readily a U.S. decision to seek extensive revisions in the ABM Treaty regime if BMD came to seem overwhelmingly necessary to preserve the credibility of the U.S. guarantee; that credibility might be seen as a higher priority than maintenance of the ABM Treaty for the sake of the penetrability of British RVs. Although such arguments might be adopted as rationalizations if Britain had no other choice, this speculative distinction between British and French views would almost certainly prove unfounded in practice. Britain would oppose revision of the ABM Treaty regime as firmly as France.

FEDERAL REPUBLIC OF GERMANY

The government of the Federal Republic of Germany (FRG) is perhaps more keenly aware than other European governments that the continuing credibility of the British and French deterrents is in the interest of Western Europe in general. This reason for favoring continuation of the ABM Treaty regime is, however, probably secondary to the Treaty's perceived importance for the future of détente and stability in East–West political relations. Favoring continuation of the ABM Treaty regime is implicit in the FRG's support for the continuing SALT/START arms control process, which is deemed "of permanent importance in all political efforts aimed at safeguarding peace and achieving stabilization in the East–West balance of power."[26]

At a time when U.S. strategic arms control policy remains under review, the ABM Treaty appears to be one of the few surviving pillars of détente— a link to the optimistic early years of the *Ostpolitik* initiated in 1969 by the current SPD–FDP government. Unilateral U.S. action to alter the ABM Treaty regime would be more upsetting than Soviet–American agreement on a revision, but no revision at all seems preferred. Perhaps even more than elsewhere in NATO Europe, the negative impression prevails that BMD is a technology more oriented to nuclear war-fighting than deterrence.

Given these general attitudes, the FRG would probably not welcome U.S. interest in the ATM concept. Yet West Germany would be the key European ally regarding ATM. A major share of the new intermediate-range nuclear forces (including all the Pershing IIs) are to be based in the FRG. Moreover, since Britain and France, who in any case have a special nuclear status, are

---

26. *White Paper 1979: The Security of the Federal Republic of Germany and the Development of the Federal Armed Forces* (Bonn: Federal Ministry of Defence, 1979), p. 69.

unlikely to take any initiative on ATM, the rest of the Allies (and the U.S. Congress) would await a West German decision with the keenest interest.

A positive ATM decision from the current West German government would be likely only under some combination of the following conditions: 1) further obvious deterioration of détente; 2) a strong and consistent U.S. commitment to ATM; 3) non-nuclear kill (NNK) mechanisms in the ATM system; 4) favorable financing and production arrangements; 5) deployment of ATM on the soil of at least two other non-nuclear Continental members of NATO—probably Belgium and Italy; 6) a NATO Council endorsement of the ATM concept; and 7) however paradoxical, the establishment of an appropriate arms control negotiating context that would demonstrate West Germany's interest in the continuation of détente and preference for an arms control solution. These conditions may be predicted from the recent history of West German participation in NATO theater nuclear modernization decisions; most, if not all, would probably apply with a CDU/CSU-led government as well.[27]

Some doubt that the ATM concept would be accepted even by a CDU/CSU-led government, even with these conditions—if they could all be fulfilled. Hans Rühle of the CDU-sponsored Konrad Adenauer Foundation has, for example, published a view of "profound skepticism" regarding the ATM option: ". . . this option appears neither technologically practicable within acceptable financial limits at present, nor is there any prospect of an optimized combination of strategic defense systems in the foreseeable future."[28]

The Alliance difficulties U.S. BMD programs might face become even clearer when both limited U.S. homeland options (ICBM defenses and "thin" area defenses) and potentially extensive U.S. homeland BMD are examined from a European perspective.

## U.S. BMD Alternatives

### LIMITED U.S. HOMELAND BMD OPTIONS

ICBM defenses constitute the least controversial option for three reasons.

---

27. See David S. Yost and Thomas C. Glad, "West German Party Politics and Theater Nuclear Force Modernization Since 1977," *Armed Forces and Society*, Vol. 8, No. 4 (Summer 1982), in press.

28. Hans Rühle, "A European Perspective on the U.S.–Soviet Strategic-Military Relationship," in William Schneider, Jr., et al., *U.S. Strategic-Nuclear Policy and Ballistic Missile Defense: The 1980s and Beyond* (Cambridge, Mass.: Institute for Foreign Policy Analysis, 1980), p. 51.

First, the technical feasibility of defense of hardened ICBM launchers seems increasingly credible, especially if leverage could be provided through mobile and/or deceptive basing. Second, ICBM defenses would not violate the long-standing offense-dominant verities of the "assured destruction" outlook. Pre-launch survivability of retaliatory forces would be enhanced with no degradation of their ability to penetrate to countervalue targets; the assumed stability of mutual counter-society threats would be unimpaired. Third, precisely because Soviet cities and other "countervalue" targets would remain undefended under this hypothetical revised ABM Treaty regime, British and French nuclear forces would retain their ability to penetrate to their presumed targets. Nonetheless, any attempt to revise the ABM Treaty regime to accommodate ICBM defenses could become controversial in Europe. Fears of an arms race destabilizing East–West relations would probably surpass more technical concerns about possible expansion and reorientation of ICBM defenses to area defenses, and so forth.

Limited area defenses would probably be even more controversial in Europe. Jan Lodal has made probably the strongest case for limited U.S. area defenses, partly on the grounds that they would strengthen the credibility of U.S. guarantees to Allies. Lodal reasons that

an active defense would eliminate any Soviet incentive to carry out "limited" nuclear attacks against U.S. territory, even if the United States had used tactical nuclear weapons to stop a Soviet invasion. The defense would be capable of intercepting a small-scale attack; a Soviet leader would have to launch a large attack (several thousand warheads) to penetrate these defenses. A rational Soviet leader ought to be deterred from launching such an attack, realizing that an assured destruction response is a much more credible reaction to an attack of several thousand warheads than to an attack of a few warheads.[29]

In Lodal's proposal for limited area defenses, mutual Soviet–American "assured destruction" capabilities would remain the guarantors of strategic stability. But what Lodal calls "a second 'firebreak' in the ladder of escalation" (in addition to the NATO-assumed conventional/nuclear firebreak) would be created: the United States could use battlefield nuclear weapons with less risk of catastrophic Soviet retaliation against the U.S. homeland.

Lodal rightly points out probable European objections to such area de-

---

29. Jan M. Lodal, "Deterrence and Nuclear Strategy," *Daedalus*, Vol. 109, No. 4 (Fall 1980), p. 167.

fenses. Even more than with ICBM defenses, the main European concerns would include fear of an arms race undermining prospects for détente and the legitimization of Soviet area defenses possibly reducing the effectiveness of British and French nuclear forces. Above all, limited U.S. area defenses "would make it relatively more likely that a war could be fought in Europe alone, without involving U.S. territory—a result that would be decried as decoupling." Although the ongoing debate about the prospective deployment of intermediate-range nuclear forces (INF) has shown great resonance among European publics of the false argument that the U.S. purpose in INF modernization is to "confine a nuclear war to Europe," Lodal's proposal is intended to maximize that possibility. Lodal judges that implementation of his proposal would nonetheless be "healthy for the alliance," because of the increased credibility of U.S. guarantees and thus the reduced likelihood of any conflict, even at the conventional level, in Europe: "No possible strategy can fully satisfy the European countries. . . . our European allies continue to look for an easy solution where none exists."[30]

EXTENSIVE U.S. HOMELAND BMD

In rejecting such proposals for "thin" area defenses, Colin Gray rightly points out that "small-scale nuclear strikes are not much in keeping with what is known about Soviet military style." Accordingly, Gray suggests that

a "thick," or truly serious, multi-level [BMD] deployment would usefully reduce American self-deterrence and so enhance the credibility of the extended deterrent. . . . in the absence of substantial homeland protection, U.S. strategic nuclear forces lack both credibility as an extended deterrent threat and ability in the event of need. The Soviet Union cannot be certain that this is so (even incredible threats deter to some extent) but the required quality of deterrence, its robustness in periods of very acute political stress, could well be lacking if the U.S. homeland continues to be totally at nuclear risk.[31]

More credible extended deterrence guarantees could thus be a by-product of extensive homeland defenses.

A true "damage-limiting" posture could, however, also include a theory of "escalation dominance" oriented toward controlling the Soviet Union's

---

30. Ibid., p. 171. Herman Kahn has expressed a similar attitude toward probable West European objections to U.S. BMD programs: "They won't like it, of course, but they are sensible people when they're forced to be sensible." (*U.S. News and World Report*, September 21, 1981, p. 54).
31. Colin S. Gray, "A New Debate on Ballistic Missile Defence," *Survival*, Vol. 23, No. 2 (March/April 1981), p. 68.

power projection advantages in Eurasia as well as denying the Soviets opportunities to play upon the probability of self-deterrence in an undefended America. In other words, while the "assured destruction" verity of near-total population vulnerability would be discarded, U.S. operational capabilities for strategic nuclear war, both offensive and defensive, would be improved in order to make implausible any Soviet theory of victory. Stability would be derived from a U.S. ability to dominate any escalation process by limiting damage to its population centers as well as its military assets. The United States could thus extend and honor guarantees—including, if necessary, first use of nuclear weapons—with less risk of self-deterrence because there would be less risk of homeland damage. Because deterrence would be improved, war would be less probable, and less catastrophic, if it occurred.

All the premises of this compelling and logically consistent strategic prescription are rejected by partisans of the "assured destruction" model of deterrence and stability, who are even more predominant in Western Europe than in the United States. It is assumed that reliable population defenses are infeasible in an offense-dominant world, and that any Soviet–American competition in defensive measures would dangerously destabilize the strategic nuclear balance, in addition to being self-defeating and extremely costly.[32] Even more than with limited defenses, Europeans would almost certainly deplore U.S. interest in extensive area defenses as undermining, if not destroying, the peace-preserving structure of deterrence; as possibly sliding from proper control of self-deterrence to a "first-strike" posture; as attempting to confine any future nuclear war to Europe; and as encouraging the Soviet Union to construct similar defenses.

Even if Europeans could generally accept Gray's judgment that, in "the context of U.S. BMD deployment, Soviet BMD would not be a destabilizing development,"[33] extensive Soviet defenses could make the need for costly improvements in conventional and battlefield nuclear forces in Europe more obvious. NATO's strategic and intermediate-range nuclear forces could not be as readily applied to deterring Soviet attacks with conventional or battle-

---

32. Calling this model of deterrence and stability "assured destruction" is admittedly somewhat unfair in that, despite popular perceptions to the contrary, U.S. targeting and operational doctrine has for many years included numerous counterforce and counter-military options. "Assured destruction" nonetheless became the shorthand characterization of U.S. declaratory policy in the late 1960s, and retains a certain descriptive merit owing to grave U.S. and West European deficiencies in active and passive defenses. Those who believe such deficiencies technologically unavoidable and strategically stabilizing carry forward the "assured destruction" logic of the late 1960s.
33. Gray, "New Debate on Ballistic Missile Defense," p. 65.

field nuclear forces because Soviet defenses would directly counter them. The recent political trials of sustaining an approximate three percent annual real increase in defense spending in NATO Europe suggest how welcome new programs for extensive conventional force improvement would be.

Moreover, extensive Soviet area defenses, more than ICBM or limited area defenses, would tend to reduce the deterrent value of the British, French, and (incidentally) Chinese nuclear forces. Putting aside the question of how great that deterrent value is in U.S. (or, more importantly, Soviet) eyes, the continued technical credibility of these nuclear forces is meaningful to more Europeans than those in the British and French governments. The United States was especially hostile to the French nuclear effort during the 1960s, when Robert McNamara was secretary of defense. For over a decade, however, U.S. policy has accepted, even vaguely approved, maintenance of the French forces (in, for example, the 1974 NATO Ottawa communiqué), though still not with the degree of active cooperation accorded to the British. While damage-limiters could argue that the greater good of the Alliance would be served by the improved deterrence derived from extensive U.S. area defenses, and that smaller independent deterrents would be less necessary, many Europeans would probably be skeptical, to say the least.

Still another West European argument against U.S. homeland BMD is that the resultant Soviet homeland BMD would undermine the U.S. ability to execute limited strategic options. While U.S. penetration technology could perhaps overcome Soviet defenses, the Soviets could nonetheless defend against limited strategic nuclear strikes more readily than against greater ones and could thus oblige the United States to consider more extensive options—increasingly less distinguishable from general nuclear response—in order to honor the guarantee. This is the obverse of the Lodal argument for "thin" area defenses, and it applies with even greater force if more extensive BMD programs in the United States and the Soviet Union are envisaged. The United States may well be self-deterred from executing any limited strategic options. Moreover, even if the United States were not self-deterred from employing such options, what value could they have when the United States is defenseless against the virtually inevitable Soviet strategic nuclear responses?

Finally, West Europeans remain skeptical about the arms control arguments for U.S. homeland BMD set forth by some Americans. Such arguments presume that Soviet and U.S. BMD programs would permit both sides to limit or even sharply reduce offensive strategic nuclear forces because ICBMs and other targets would be defended. Abrogation or revision of the ABM Treaty would be necessary, but extensive superpower homeland BMD could theoretically also promote strategic stability by minimizing the effects of cheating on negotiated offensive force levels and, more importantly, by enhancing uncertainty as to the cost and feasibility of offensive strike plans. Crisis stability could be improved by reductions in retaliatory force vulnerability, and by "allowing for nonnuclear interceptor launch under real or apparent attack."[34] European doubts derive from their judgment that superpower BMD would in practice not result in limitations or reductions in offensive forces, and would probably promote instability through intensified competition instead.

## U.S. Interest in ATM

Defense Secretary Harold Brown apparently did not even consider the feasibility of ATM defenses for the planned new intermediate-range nuclear forces (INF)—the ground-launched cruise missiles (GLCMs) and Pershing IIs (P-IIs) scheduled to begin deployment in 1983—when he made the following assessment:

If TNF are to provide a credible deterrent, they must be highly survivable in the aggregate, at least against conventional or limited nuclear attack. To a large extent, force survivability against these threats depends on mobility and concealment from Warsaw Pact target acquisition systems. Given the relatively limited deployment area for NATO land-based systems and short time of flight for Soviet ballistic missiles, absolute survivability against large-scale, bolt-out-of-the-blue nuclear attacks is *probably infeasible and certainly excessively costly*.[35] [Emphasis added]

This view is similar to the general European view expressed by RAF Marshall Sir Neil Cameron: "We can, of course, do nothing against a ballistic missile attack but dig deep. . . ."[36]

---

34. G.E. Barasch et al., *Ballistic Missile Defense: A Potential Arms-Control Initiative*, LA–8632 (Los Alamos, New Mexico: Los Alamos National Laboratory, January 1981), p. 23. While this study presents perhaps the most complete version of such arguments, similar ideas about BMD's potentially beneficial effects for strategic stability have been expressed by Herman Kahn, James R. Schlesinger, and others.

35. Harold Brown, *Department of Defense Annual Report Fiscal Year 1981* (Washington, D.C.: U.S. Government Printing Office, 1980), p. 146.

36. Neil Cameron, "Defense and the Changing Scene," *RUSI Journal*, Vol. 25, No. 1 (March 1980), p. 26. For a similar American view on the indefensibility of Europe, see Lodal, "Deterrence and Nuclear Strategy," p. 171.

In contrast, Principal Deputy Undersecretary of Defense Research and Engineering James Wade has made the following evaluation:

The question of active defense for theater nuclear forces is being looked at quite carefully. . . . it is reasonably clear that such a course could have merit. . . . both the GLCM and P-II are designed to achieve survivability against a number of threats through covert field deployment, frequent relocation in the field, and the reduction of signatures associated with field deployment. This mode of operation assumes enough warning to disperse to covert field sites prior to an attack.[37] An ATM could reduce the importance of warning time. . . .[38]

Since Wade's statements, it has been reported in more specific terms that arming the Patriot missile with a nuclear warhead for defense against Soviet theater missiles is under consideration. "A separate study contract is expected from the Army for development of a non-nuclear warhead for theater BMD to avoid the problem of obtaining release authority if Patriot is equipped with a nuclear warhead."[39] This report is consistent with other unofficial American discussions of theater ATM capable of neutralizing the threat posed by the SS-20 and other Soviet theater missiles, which could have conventional, nuclear, or chemical warheads.

It is not yet clear whether reported research activities will result in actual ATM programs. No plans currently exist to replace or supplement the nuclear-armed Nike-Hercules air defense system with new nuclear-warhead active defenses,[40] or with conventional-warhead systems capable of theater ATM defenses. If ATM programs were pursued, they could face serious opposition in Western Europe, given established views on BMD in general and special factors in Britain, France, and the FRG. Six problematic issues could inhibit or even frustrate a U.S. initiative in favor of ATM: technological credibility, Soviet countermeasures, Alliance cohesion, military rationales, implementation of INF modernization, and arms control issues (including INF negotiations as well as SALT—re-named START by the United States in November 1981—and the ABM Treaty).

---

37. Wade in U.S., Senate, Armed Services Committee, *Department of Defense Authorization for Appropriations for Fiscal Year 1981*, Hearings, Part 5, Research and Development, March 13, 1980 (Washington, D.C.: U.S. Government Printing Office, 1980), pp. 3013–3014.
38. Wade cited in "Protection for Europe-Based Nuclear Missiles," *Flight International*, October 18, 1980, p. 1496.
39. *Aviation Week and Space Technology*, June 22, 1981, p. 89.
40. *The FY 1983 Department of Defense Program for Research, Development, and Acquisition*, Statement by the Honorable Richard D. DeLauer, Under Secretary of Defense Research and Engineering to the 97th Congress, March 2, 1982, pp. VII–14.

*Technological Credibility of ATM*

European experts, to say nothing of politicians and the general public, will not be easily convinced of the technical feasibility of ATM defenses. The tendency is to assume that the infeasibility of reliable BMD was long ago established at the intercontinental level, and that shorter distances, shorter warning times, and lower trajectories make theater BMD even more difficult. BMD technology has in fact advanced, particularly in such areas as discrimination, computerization, data processing, radar, and other—i.e., optical—detection systems. Moreover, because the distance and duration of their boost flight phases are relatively short, theater ballistic missiles (especially the SS-21 and SS-23) have significantly slower re-entry speeds than SLBMs, ICBMs, and longer-range theater ballistic missiles like the SS-22 and the SS-20. Shorter flight-times tend to aggravate (or, as BMD specialists say, "stress") the intercept problem, while slower re-entry speeds tend to simplify it.

Skepticism has focused initially on the reported idea of using Patriot in an ATM role. When the Patriot study program (then called SAM-D) was started in the mid-1960s, an ATM role was envisaged, in addition to a capacity against high-performance aircraft at high and low altitudes. U.S. policy (as reflected in the NPG deliberations) then favored consideration of ATM. However, the ATM requirement was later dropped because of costs and the challenges of defending against heavy nuclear attacks, in addition to the Alliance recommendation against theater BMD. Patriot is now intended to serve as a replacement for the Hawk and Nike-Hercules air defense systems. Patriot radars are presumably designed only for anti-aircraft operations. Even if equipped with radars and data processing for ATM-capable discrimination and responsiveness, Patriot might not be prompt and accurate enough for an ATM role unless a nuclear warhead were used as the kill mechanism. Even then some observers would have grave doubts, especially concerning the higher re-entry speeds of longer-range Soviet missiles (e.g., the SS-20 and the SS-22). Developing a low-performance range ATM (on the basis of the Patriot, or, in the Soviet case, the SA-10) capable of intercepting cruise missiles or slower, shorter-range ballistic missiles (e.g., the SS-21 and SS-23) would be less challenging than developing a new high-performance missile. Probably only an entirely new high-performance ATM missile, if equipped with effective sensing and homing devices, could avoid the requirement for nuclear warheads.

The potential necessity for nuclear warheads, which would pose the political problem of introducing new nuclear weapons systems in Western Europe, underlines the many advantages of non-nuclear kill (NNK) mechanisms. These advantages would include reduced manpower requirements; simplified logistics, security, and command, control, and communications; minimized risk to allies; no self-inflicted nuclear effects—i.e., electromagnetic pulse (EMP) or blackout—hindering radars and communications systems; no requirement for nuclear weapon materials; simplified release authority; and confidence in system reliability, because total system testing is feasible, including destruction of incoming target warheads. Even though NNK warheads, designed for either direct impact or high explosive detonation near the RV, are cheaper than nuclear warheads, skepticism about NNK cost-effectiveness and reliability will persist until research (including operations research) leads to more definitive conclusions and the publication of authoritative assessments.

Although some observers, perhaps correctly, deem ATM "the only effective option" for "a reasonable degree of survivability" for NATO theater nuclear forces, including INF,[41] one recent official discussion of ATM technology noted the need to consider its costs and effectiveness in the context of alternatives to ATM:

The technology required to defend against an IRBM attack includes: Acquisition and tracking radar capable of picking up and tracking an incoming warhead; rapid, sophisticated signal processing equipment to allow firing an interceptor within a few seconds; a high-speed, high acceleration missile which can reach the incoming warhead in time to kill it at a sufficient range to preclude damage to the defended target; adequate terminal homing; and an interceptor warhead capable of destroying the incoming reentry body. In addition, if a system is to be used to defend a mobile target such as a Pershing launcher, all of the interceptors and supporting equipment must have mobility consistent with that of the target to be defended. This technology is attainable; much of it exists from our ABM development work, although there are differences between defending fixed targets against ICBMs, and defending mobile ones against IRBMs. What will be at issue is the degree to which ATM hardware contributes to survivability in the aggregate, what other active and passive measures can enhance survivability, arms control consideration, cost of ATM alternatives, and the best technical approach should we elect to field such a system. . . .

---

41. Wayne R. Winton, "Applications of BMD Other Than ICBM Defense," in *U.S. Arms Control Objectives and the Implications for Ballistic Missile Defense*, Proceedings of a Symposium held at the Center for Science and International Affairs, Harvard University, November 1–2, 1979 (Cambridge, Mass.: Puritan Press, 1980), p. 96. Cf. Carnes Lord, "The ABM Question," *Commentary*, Vol. 69, No. 5 (May 1980), p. 38.

We are not far enough along in our examination of this complex issue to have the answers to all these questions. . . . While we are not yet prepared to estimate what such a system might cost, I am certain that the cost would be substantial. . . .[42]

Alternatives to ATM would presumably include improved dispersal planning and mobility, deception, redundancy, and signature reductions.

The ABM development work that may be of greatest relevance to ATM is that on the Low Altitude Defense System (LoADs), which consists of small radars and interceptors designed for possible mobile deployment with the MX ICBM. In conjunction with emerging technology for endoatmospheric NNK, LoADs development might be directly applicable to ATM.[43] Nonetheless disagreement within the technical community persists as to the feasibility of reasonably effective BMD, owing to technical and operational problems, including the challenge of reliable NNK.[44]

Uncertainties would persist, even with reliable NNK. Endoatmospheric interception with NNK of the chemical warheads the Soviets have reportedly deployed on theater ballistic missiles could be less than satisfactory, since such an interception over Allied territory could spread the chemicals, depending on the altitude of the interception and other factors. Prompter interceptions might therefore provide better solutions against Soviet chemical warheads over the long term.

High-level NNK interceptions over Allied territory could also represent a problem if the Soviets designed their warheads with the mechanism of "salvage-fusing," whereby incoming RVs might be detonated by the impact of the interceptor's kill mechanism. While the obvious advantage of intercepting the Soviet warhead several miles away from its intended target would remain, a "salvage-fusing" nuclear explosion could interfere with subsequent Allied defenses because of its effects on radar and communications systems. However improbable the "salvage-fusing" possibility seems—owing to its great cost and difficulty, and risk of catastrophic failure—it could serve as a

---

42. Wade, in *DoD Authorization . . . FY 1981*, p. 3014.
43. Winton, "Applications of BMD"; Jonathan E. Medalia, *Antiballistic Missiles*, Issue Brief, Number IB81003 (Washington, D.C.: Congressional Research Service, September 1, 1981), p. 15.
44. For somewhat contrasting assessments of BMD technology, see the Los Alamos study cited in note 34; Guy Barasch, Nikki Cooper, and Ray Pollock, *Ballistic Missile Defense: A Quick-Look Assessment*, LA–UR–80–1578 (Los Alamos Scientific Laboratory, June 1980); Chapter 3 (on BMD) of U.S., Congress, Office of Technology Assessment, *MX Missile Basing* (Washington, D.C.: U.S. Government Printing Office, 1981); and the articles by William A. Davis, Jr., Deputy Ballistic Missile Defense Project Manager, in *National Defense*, September/October 1979 and December 1981.

basis for European technical skepticism. Such uncertainties as the technical effectiveness of NNK ATM would not necessarily deprive ATM of all deterrent value. Resultant Soviet uncertainties as to the effectiveness of their offensive strikes could still be helpful to deterrence.

*Soviet ATM Countermeasures*

Soviet responses to a U.S. ATM program for NATO could be offensive, defensive, and political.

Offensive responses would be systems designed to destroy, overwhelm, or circumvent Western ATM systems. Given the deficiencies of NATO's current air defenses, the Soviets could attack the ATM radars and other components with air-breathing systems. Improved air defenses and ATM defenses would therefore both be required in a serious damage-limiting effort by NATO.

ATM systems could also be saturated by the Soviets at specific points of interest, at costs partly dependent upon the sophistication and number of Soviet penetration aids as opposed to the ATM's discrimination capability and cost-effectiveness. Even without penetration aids, the numbers of RVs on Soviet theater ballistic missiles are very high:

The number of NATO military installations which the Soviets might target with nuclear weapons is, at most, approximately three hundred. The Soviets presently have ten delivery systems for each target, and when the SS-20 has been fully fielded they will have ten weapons for each target in this system alone. . . . The Soviet motivation for this tremendous capability, enabling them to destroy every military installation in NATO ten times over, continues to be a mystery in the West.[45]

The Soviet capability seems excessive, even allowing for redundancy to compensate for reliability uncertainties and the hypothetical contingencies of Western pre-emption or intra-war attrition, and to cover an even more ambitious target set in Western Europe. Some Western analysts have offered deceptively reassuring explanations in speculating that extra-rational factors (e.g., bureaucratic politics or cultural tradition) may account for the high numbers of Soviet deployments. The Soviets may also be deploying militarily

---

45. Francis X. Kane, "Safeguards from SALT: U.S. Technological Strategy in an Era of Arms Control," in Paul H. Nitze et al., *The Fateful Ends and Shades of SALT* (New York: Crane, Russak and Co., 1979), p. 116.

redundant INF for negotiating purposes, i.e., no loss of required target coverage, even if negotiations result in reductions. Whatever the explanations, the seemingly redundant warheads in effect constitute double insurance against potential future ATM capabilities—an impressive capacity to overwhelm ATM, which in turn works to discourage NATO from pursuing such systems. If an ATM were to cast doubt on the effectiveness of systems such as the SS-22 or SS-20, the Soviets would probably use it as a new rationale to deploy even more numerous and effective INF; this would make technical and cost-effectiveness arguments for ATM even more difficult for Western proponents.

Although Soviet redundancy in theater ballistic missile numbers makes it improbable, another Soviet offensive countermeasure could be attacking targets defended by European-based ATM with SLBMs and ICBMs. Knowledgeable observers assume that approximately 120 of the SS-11 ICBMs, as well as some SS-19 ICBMs, may have targets in Europe.[46] This means that a fully credible ATM would require an ability to intercept the very rapid ICBM and SLBM RVs, while the latter might come from any direction. (ICBM RVs could also come from any direction if the Soviets used the Fractional Orbital Bombardment System [FOBS], or orbited missiles; but these delivery techniques seem improbable as concerns targets in Europe.) An imperfect ability to defend against certain types of Soviet intermediate-range missiles would nonetheless constitute an improvement over the current situation, in which no defense exists against Soviet ICBMs, SLBMs, or intermediate-range missiles.

Defensive Soviet countermeasures would consist of expanded BMD systems. Since the ABM Treaty was signed, Soviet BMD research and development has been more intensive than that of the United States. Moreover, in contrast to the United States, the Soviets retain active BMD capabilities around Moscow, permitted by the ABM Treaty and its Protocol, and continue to perfect radars and air defense interceptor missiles for possible future upgrading to BMD roles. Whether the United States retains an advantage in the key areas of BMD technology is no longer clear; at the least, comparative U.S. advantages in certain areas have probably been reduced since 1972, given the contrasting levels of investment effort under the ABM Treaty regime. Although the United States perhaps retains an edge in some areas,

---

46. Lawrence Freedman, "The Dilemma of Theatre Nuclear Arms Control," *Survival*, Vol. 23, No. 1 (January/February 1981), p. 5.

for example, battle management and discrimination technologies and exoatmospheric nonnuclear kill mechanisms, it is reasonable to suspect that the dimensions of the U.S. lead in ABM technology have been significantly reduced. The Soviets may be better placed than the U.S. to deploy effective BMD in a timely fashion.[47]

Even if new BMD programs could be restricted through Soviet–American negotiations to capabilities against theater ballistic missiles, Soviet ATM, like U.S. ATM, could be virtually indistinguishable in practice from systems capable of intercepting ICBM and SLBM RVs. For full technical credibility, ATM systems would almost have to be capable of such interceptions. Geographical asymmetry would favor the USSR, in that a Soviet ATM, whether capable of ICBM and SLBM RV interceptions or not, could defend the homeland as well as Allies. U.S. ATM in Western Europe would defend Allies alone, not the U.S. homeland. Some analysts speculate that the recent consolidation of the Soviet Troops of National Air Defense (PVO Strany) with the troops of Air Defense of the Ground Forces (PVO SV) may be related to new problems posed by overlapping strategic and theater BMD challenges, though the centralization of air defenses could also be explained by other managerial aims.[48] Moreover, there remains the possibility that Soviet testing of air defense systems in an ABM mode, possibly in violation of the ABM Treaty, includes development of an ATM.[49]

On the other hand, one wonders how to interpret the repeated Soviet complaints that Pershing II would allow them only 6 to 8 minutes of warning time. (The Soviets naturally never indicate how much warning time Europeans could expect prior to the impact of their INF RVs, or indeed how much warning time U.S. coastal, or West European, targets could have prior to the

---

47. Indeed, the Soviets may surprise many Western observers by choosing themselves to propose revisions in the ABM Treaty regime or to end it. This essay is concerned primarily with current Alliance issues posed by prospective U.S. BMD decisions, and therefore reflects the widespread Western assumption that the Soviets will be reacting to U.S. and NATO decisions, not vice versa. Soviet decisions for BMD could substantially change the current climate of opinion regarding BMD in the United States and Western Europe. Soviet incentives (e.g., protecting key assets from a U.S. second strike) and disincentives (e.g., prospective alleviation of U.S. ICBM vulnerability) for BMD deployments constitute a large subject distinct from the purposes of this essay. See the DoD assessment of Soviet BMD in U.S., Senate, Committee on Appropriations, *Department of Defense Appropriations for Fiscal Year 1982*, Hearings, Part 5, June 1981 (Washington, D.C.: U.S. Government Printing Office, 1981), p. 466.
48. William F. and Harriet Fast Scott cited by Henry Bradsher in *Washington Star*, July 16, 1981, p. 10.
49. Winton, "Applications of BMD," pp. 96–97; cf. Senator Jake Garn, "Soviet Violations of SALT I," *Policy Review*, Number 9 (Summer 1979), pp. 24–28.

impact of Soviet SLBM RVs.) If sincere, the complaints could imply lack of confidence in their ATM upgrade capability. The Soviet ATM research and development program against Pershing II has been described as "aggressive";[50] but the United States officially attributes Pershing II "a high assurance of penetrating future Soviet defenses,"[51] partly because the GLCM-Pershing II combination stresses Soviet defenses and both have potential for penetrability measures upgrade.[52] The Soviet complaints about Pershing II could, moreover, also be part of the Soviet political strategy of portraying NATO INF modernization as "aggressive."

The Soviet political response to ATM would be an extension of the bargaining posture the Soviets have already assumed. U.S. interest in an actual ATM program would be seized and exploited for a variety of media themes: the unmasking of the truly aggressive "war-fighting" intentions of the West, the threat to strategic stability and world peace in violating the ABM Treaty with ATM, the initiation of a new "arms race" by the capitalist military-industrial complex, and so forth. More importantly, a U.S. ATM initiative could be perceived by the Soviets as an opportunity to promote antagonism between the United States and Western Europe.

*Alliance Cohesion and ATM*

The Soviet opportunity to promote antagonism would reside in the potentially divergent U.S. and West European appreciations of the utility of any BMD programs, including ATM. While Britain and France would have their own national reasons for opposing anything that might alter the ABM Treaty regime, these reasons would be endorsed by others in Western Europe and reinforced by the general tendency to see BMD as destabilizing and likely to promote an expensive and futile "arms race" that could end in war. If the United States determined that ATM could be cost-effective and militarily useful, and should be pursued as an active program, ATM could provide another example of the broad dichotomy in U.S.–NATO European views that was noted by Robert W. Komer, when President Carter's undersecretary of defense for policy: "Indeed, we Americans are increasingly asking whether

---

50. Senator John Warner in *DoD Authorization . . . FY 1981*, p. 3013.
51. Caspar W. Weinberger, *Department of Defense Annual Report Fiscal Year 1983* (Washington, D.C.: U.S. Government Printing Office, 1982), pp. III–72.
52. William J. Perry, then Under Secretary of Defense for Research and Engineering, in *DoD Authorization . . . FY 1981*, p. 3018.

Europe is as interested in its own defense as is the United States, or perceives the same threat."[53] In contrast, Europeans might view the American concerns as immoderate.

Secondary Alliance cohesion problems could arise if ATM programs were accepted by the Alliance. Intra-Alliance disputes could concern what systems and localities would be entitled to ATM protection, while ATM protection could be opposed for specific localities or for specific purposes.

*Military Rationales for ATM*

The precise military purposes ATM might serve have yet to be fully clarified. The statement by James Wade cited above (one of the few official comments on ATM's potential utility) stresses the prospective gain in survivability for NATO's new INF if one did not have to count on the Soviets cooperatively providing warning time to NATO before engaging in strikes against the INF. If INF survivability were more thoroughly assured, the INF deterrent threat to the Soviet Union would be more formidable—for pre-war deterrence and, depending on the thickness of the defenses and their endurance potential, for intra-war deterrence as well. While INF survivability may be adequately assured through warning time, mobility, and dispersal, fixed targets would remain vulnerable. If fixed targets such as airfields, nuclear weapons storage sites, and certain command, control, and communications ($C^3$) centers were also equipped with ATM defenses, the damage-limiting capabilities might significantly increase Soviet uncertainty as to the prospects of successful attack against NATO. ATM might, in particular, obstruct probable Soviet plans for pre-emptive nuclear strikes against theater nuclear weapons targets in Western Europe.

This concept, which one might call "theater damage-limiting for defense and deterrence," would be most practical and convincing if used in defense of hardened sites, especially INF and fixed $C^3$ centers. Defense of $C^3$ centers could in particular have a "force multiplier" effect, while allowing current $C^3$ vulnerabilities to persist simply offers the Soviets a lucrative opportunity to degrade the effectiveness of all types of forces.[54] The Soviets could overwhelm almost any defenses, if determined to do so, yet obliging the Soviets to increase the scale of their attack could be seen as raising the probability of

---

53. Robert Komer cited in *Aviation Week and Space Technology*, March 3, 1980, p. 57.
54. William R. Graham, "Reducing the Vulnerability of Retaliatory Forces and Command, Control and Communications: A Question of Balance," in David S. Yost, ed., *NATO's Strategic Options: Arms Control and Defense* (New York: Pergamon Press, 1981), pp. 170–178.

bringing about retaliation by U.S. strategic nuclear forces against the Soviet Union.

ATM could thus be considered a non-provocative and defensive means of denying the Soviets any opportunity they might wrongly perceive of relatively low-cost victory through selective theater nuclear strikes. Deterrence and stability would be enhanced, because obliging the Soviets to use far more nuclear warheads would raise the risks to the Soviet Union. If truly "thick" and cost-effective ATM defenses could be constructed, the Soviet potential for nuclear blackmail against Western Europe might be so severely eroded that doubts about the credibility of the U.S. strategic nuclear guarantee would become an almost secondary concern.

Another possible military rationale for ATM might be escalation control. At present, it is assumed that NATO would receive warning time sufficient for dispersal of the new INF, and that the warning time would have to be used to assure the survival of the INF. In a crisis situation, it seems likely that some Western politicians would argue that actual dispersal of the INF would be provocative, i.e., likely to aggravate the crisis and make war more unavoidable; and yet failure to disperse could equal the destruction of the INF. If ATM defenses were available, warning time and prompt dispersal would be less necessary. Political control over escalation processes might be enhanced if there were less military operational incentive to engage in seemingly provocative behavior.

None of the above military rationales is likely to have much appeal in Western Europe. Such rationales—especially the "protracted war" concept of "enduring" survivability—would appear more oriented toward actual war-fighting than toward deterrence. West Europeans generally are unwilling to accept the Soviet view (increasingly respected in the United States) that deterrent capabilities are a product of operationally effective war-fighting capabilities. Instead, West Europeans (even more than Americans) tend to favor a "deterrence-only" perspective based on threatening strategic nuclear retaliation against Soviet society. The U.S. threat to retaliate against the Soviet homeland is enough, they generally feel, to deter any Soviet invasion. In Ian Smart's words, "West European political leaders and their electorates have rarely, if ever, been willing to devote serious attention to what would happen if the deterrence of initial attack by threat of intolerable penalty should fail."[55]

The favorable reception in West European circles of McGeorge Bundy's

---

55. Smart, "Perspectives From Europe," p. 186.

keynote address to the 1979 Conference of the International Institute for Strategic Studies is a bit of anecdotal evidence for the same point. Above all, West Europeans drew comfort from Bundy's insistence that deterrence of any Soviet attack is well assured:

. . . no one *knows* that a major engagement in Europe would escalate to the strategic nuclear level. But the essential point is the opposite; no one can possibly know that it would not.[56]

Bundy's affirmation as to the genuine possibility of escalation was reassuring to Europeans because they prefer a concept of deterrence without intra-war escalation boundaries. West European faith in strategic deterrence is often associated with the assumption that more credible theater war-fighting capabilities would undermine strategic nuclear deterrence. The threat to punish Soviet society is the bedrock of deterrence in their view, not an ability to defeat a Soviet offensive against Western Europe.

The possible military rationales for ATM outlined above would, however, not only sound intolerably bellicose to many West Europeans; such rationales could also seem subtly designed to decouple the U.S. guarantee and to confine war to Europe. Protecting the new INF could be seen as creating a distinct "Eurostrategic" level of potential conflict, something the December 1979 NATO decision on INF was intended to avoid. Given the abiding concerns of West Europeans, many would suspect the United States of improving conditions for successful war-fighting in Europe out of a desire to confine a war to that region. What one might call the "incalculability of escalation" could be undermined in European perceptions if ATM promised to increase prospects for holding conflict to the theater level.

Using ATM to avoid premature dispersal of INF in order to enhance political control over the escalation process would not necessarily be an appealing argument in Western Europe because it would underline the potential vulnerability of the INF to Soviet attack. Because ATM defenses could be seen as guaranteeing intensive Soviet strikes intended to overwhelm them and destroy the INF, West European officials are likely to prefer to stress the probability that deterrence will not fail and the adequacy of dispersal through mobility and warning time as means of survivability for the INF. The risk of appearing "provocative" by dispersing the INF would have to be set against 1) the risk of appearing too frightened and vulnerable to do so; 2) the contrasting message of firmness and readiness to act—constructive for "crisis

---

56. McGeorge Bundy, "The Future of Strategic Deterrence," *Survival*, Vol. 21, No. 6 (November/ December, 1979), p. 271.

management"—that dispersal might usefully transmit; and 3) the risk of the INF being destroyed in their peacetime basing areas.

Strategic rationale arguments against ATM could also be derived from the "flexible response" doctrine of NATO. First, it could be argued, at present NATO assumes that whatever nascent ATM capability the Soviets have is inadequate for defense against the Pershing II. However, if the Soviets built up their own ATM capabilities, the option within "flexible response" of what might be called "deliberate limited escalation" would be undermined by Soviet defenses. NATO's strikes would have to be more extensive to achieve similar effects, a fact which might be seen as harmful to escalation control.

A second strategic rationale argument against ATM would apply if the kill mechanism were nuclear. An ATM system would be used as necessary to destroy incoming RVs threatening defended targets. If the kill mechanism were nuclear, this would amount to NATO's using nuclear weapons reactively against an "accidental" target. This situation would contradict the West European preference, for deliberately controlled political use of nuclear weapons if nuclear weapons ever have to be used. It is still assumed (despite mounting evidence of the nuclear orientation of Soviet theater forces) that, unless the Soviets initiate their aggression with a pre-emptive nuclear strike, NATO would precede the Soviets in making decisions on initial use of nuclear weapons, and that the initial use should be planned primarily for political effect with the resultant military effects of secondary importance.

A third strategic rationale (as well as an Alliance cohesion) argument directly follows. A nuclear warhead ATM could not be effective unless it could respond to incoming RVs automatically. This would require an agreement in advance among the Allies to use the weapons, with a pre-delegation of release authority, unless the United States were to insist on a strict interpretation of the 1962 Athens guidelines on consultation only "time and circumstances permitting." It would be politically very awkward for the United States to so insist. It would be no less difficult for the United States to obtain advance approval of nuclear release from the Allies.[57] Even if it

---

57. Some observers consider the nuclear-armed Nike-Hercules air defense system to be virtually unusable for these reasons. There is a linkage between Nike-Hercules and the Patriot in that the NNK Patriot for air defense is scheduled to replace the Nike-Hercules, and in that an ATM with a nuclear kill mechanism might be more readily accepted in Western Europe if presented as a successor to the Nike-Hercules system. It has been reported for years that West Europeans would like to retain the nuclear high-altitude anti-air capability that Nike-Hercules represents, and for this reason at least a nuclear successor system with ATM potential might be acceptable. (Cf. Walter Pincus in *The Washington Post*, November 1, 1981; and *Aviation Week and Space Technology*, August 29, 1977, pp. 47–48.)

were a case of defending against incoming SS-20 RVs, many West European officials might argue that warning system malfunctions could lead to a rapid and unnecessary escalation of a crisis if any nuclear explosion took place. The advantages reliable NNK would offer are again apparent.

*INF Modernization and ATM*

The impact an actual U.S. ATM program might have on the ongoing INF modernization program is indeterminate, but some European officials are concerned that the impact could be harmful in a number of ways. If the ATM required a nuclear warhead, it would constitute a new nuclear system for possible introduction into Western Europe, and hence a new focus of controversy. Moreover, whether nuclear or NNK, ATM would represent a sufficiently dramatic development to "overload" the West European decision-making process. European governments are reluctant to see any new dramatic issues raised (such as enhanced radiation battlefield weapons, chemical weapons, new mid-range nuclear missiles, etc.) that might make implementation of the December 1979 decisions on INF arms control and modernization even more difficult.

Some West Europeans might see ATM as a justification for not proceeding with the INF modernization decision, even if ATM were presented as a long-term necessity for defense of the new INF. Various arguments might be made to this effect:

—Given the ATM possibility, should the INF decision and its rationale not be re-examined in order to find a more optimal mix of systems and basing for ATM defenses?
—Why introduce new offensive systems at all if the problem posed by the SS-20 and other Soviet theater ballistic missiles can be solved effectively and directly through defensive systems that would not pose a threat the Soviets might perceive as "aggressive"?
—Given the likelihood that the Soviets would try to overwhelm any ATM, should any land-based systems be deployed at all?

A number of influential West Europeans are already concerned that land-based INF would constitute attractive targets in Soviet eyes, and an ATM program might tend to underline that probability. In the words of Carl Friedrich von Weiszäcker, "In case of a crisis, these weapons would naturally be the targets of a Russian first strike. The necessity of avoiding such crises would make Europe more vulnerable to blackmail."[58]

*ATM and Arms Control*

Would ATM violate the ABM Treaty? The ATM concept is distinct from the purpose of the ABM Treaty, which defines an ABM system as "a system to counter strategic ballistic missiles or their elements in flight trajectory" (Article II). Some observers therefore argue that ATM could be developed and deployed while complying with the ABM Treaty. After all, the purpose of an ATM would be to protect theater military assets.[59]

On the other hand, Article IX of the ABM Treaty states that "each Party undertakes not to transfer to other States, and not to deploy outside its national territory, ABM systems or their components limited by this Treaty." This provision of the Treaty could be used by the Soviets and by Western European (and American) opponents of ATM to argue that ATM represents an attempt to circumvent the ABM Treaty. It could rapidly become apparent that, even if the Treaty language does not explicitly exclude ATM, politically and in terms of public perceptions, ATM is covered by the Treaty's limitations.

Even if the United States were to point out that the ATM would not be capable of protecting continental U.S.-based assets, the Soviets would respond that this proves that U.S. "forward-based systems" (FBS) in Europe—mostly aircraft—must be limited by arms control measures. FBS are more threatening to the Soviets, if more survivable; the Soviets would therefore insist even more emphatically that FBS be included in the INF negotiations. The ABM Treaty issue could raise "arms race" and "destabilization" specters that might be almost impossible to exorcise in the social democratic circles of Western Europe, even with an NNK ATM. The more effective an ATM is, the more it will look like an ABM, even if incapable of defense against SLBM and ICBM RVs. Ironically, therefore, the less effective the ATM, the easier it might be to deploy in terms of public relations. While the ideal ATM would also be capable of intercepting SLBM and ICBM RVs, even an ATM capable

---

58. Carl-Friedrich von Weiszäcker, "Can a Third World War Be Prevented?" *International Security*, Vol. 5, No. 1 (Summer 1980), p. 204.
59. See, for example, Winton, "Applications of BMD," p. 97. Cf. Medalia, *Antiballistic Missiles*, p. 17.

only of intercepting Soviet INF could be part of a set of measures enhancing deterrence and security in Europe.

General arms control and détente issues follow directly from this situation. The Soviets might very well propose inclusion of ATM in the INF negotiations or in some other arms control forum, and the suggestion would probably be heartily approved by many sectors of West European opinion. It would be hard for the United States to extricate itself from such an arms control negotiation offer without being portrayed as a "warmonger." ATM's inclusion in the INF negotiations would tend to protract an already complex and difficult set of negotiations, and could make useful results even less likely. These negotiations already promise to disappoint many in Western Europe with unrealistic expectations about arms control, and to strain further the fabric of the Alliance.[60]

*Conclusion*

The sensitivity of West European governments and publics regarding all types of BMD is a factor the United States will have to consider as it examines options that might require revision of the ABM Treaty or that might be so perceived (i.e., ATM). Neither homeland BMD options nor the long-dormant ATM question have been raised explicitly by the United States with West European governments. It is possible that, when and if a BMD question is raised officially, political and technical circumstances will have changed significantly from the current situation. Highly convincing BMD technology (especially for NNK), perceptions of an increased Soviet threat, heightened feelings of dependence on U.S. military power, favorable (i.e., U.S.) financing arrangements, and/or other factors could combine to persuade West Europeans to accept, however begrudgingly, a new U.S. and NATO strategy of damage-limiting for defense and deterrence—i.e., an ability to deny the Soviets victory by defending selected military targets.

No reassessment of NATO's general strategic outlook could be more fundamental. At present, NATO's deterrent strategy is based on the assumption

---

60. Some observers have suggested that an ATM could also be employed as a surface-to-surface missile, in which case an arms control problem might be posed in terms of "changing the numbers" of systems that might target the Soviet Union. The Patriot used in a surface-to-surface role would not, however, have the range to threaten the Soviet Union. Nor does it seem probable that any ATM system now likely to be developed would have a range encompassing Soviet territory.

that any East–West conflict could rapidly lead to an escalation process ultimately including U.S. strategic nuclear strikes against Soviet society. The U.S., Western Europe, and NATO military forces are almost completely vulnerable to Soviet nuclear threats, except for limited air and civil defenses and, above all, the threat to severely punish Soviet society in retaliation.

Both aspects of the posture—the ultimate reliance on a threat to kill millions of Soviet citizens, and the virtually complete absence of effective defenses—could prove most unsatisfactory guarantees of security in war. If technically feasible and cost-effective, damage-limiting capabilities could deny the Soviets part of their ability to threaten the United States and Western Europe, and make it less necessary for the Alliance to threaten harm to Soviet society. An ability to physically deny the Soviets their plausible military objectives could become a yardstick for Alliance strategy superior to the ambiguities of "flexible response." Since the most plausible Soviet military objectives are not cities but military targets that could be relatively (though imperfectly) well defended, Soviet strategies for nuclear war-fighting and victory could be thwarted, and deterrence strengthened, by damage-limiting capabilities that increase Soviet uncertainties about prospects for successful pre-emptive attack.

The lack of damage-limiting capabilities in the West tends to drive Alliance strategy into embracing politically convenient ambiguities. The ambiguities about what "flexible response" might mean operationally are partly intended to conceal the scarcity of militarily sensible retaliatory options (given the lack of damage-limiting means) from the Soviets and Western publics. "Flexible response" thus ultimately rests heavily on the threat to unleash a conflict that could lead to the destruction of much of North America and Eurasia. Damage-limiting could assist NATO in becoming less dependent on this threat. All that Alliance strategists can hope for at present is that mutual restraint in a "crisis management" process will be able to control nuclear conflict, for the West has virtually no ability to enforce damage limitations through active (non-counterforce) defenses.[61]

Reassessment of the merits of damage-limiting is long overdue, and change

---

61. Anti-submarine warfare (ASW) capabilities are here assumed to constitute a form of counterforce, as opposed to active defenses that could intercept ballistic missile RVs or air-breathing systems in flight. Even if one hypothesizes that Western ASW could neutralize Soviet SSBNs pre-emptively (a highly improbable feat for several reasons, including the fact that land-based ASW communications and detection means would be at risk in war), the threat to Western society from Soviet air-breathing systems and land-based ballistic missiles would remain dire.

may be imminent. BMD, perhaps in conjunction with a deceptive-basing system, is one of three concepts for long-term MX basing under consideration by the Reagan administration. Congress, in the Fiscal Year 1982 Appropriations Act, has mandated reporting a final selection of an MX-basing mode by July 1983, and decisions may be made by April 1983.[62] While the scheduled October 1982 second five-year review of the ABM Treaty may therefore pass without either U.S. or Soviet proposals for revision, the United States may elect to propose amendments later, perhaps as a result of the MX-basing decision. Amendments to the ABM Treaty may be proposed at any time; and either party may withdraw, with six months' notice, if it judges that "extraordinary events" related to the Treaty's subject matter have "jeopardized its supreme interests."

On the other hand, doubts about the maturity of BMD technology persist in some political and technical circles. The possibility that the Soviets have a superior ABM Treaty "break-out" potential owing to their greater investments in BMD (and air defense) research may be another argument for U.S. caution in proposing major changes in the ABM Treaty. Extensive research (particularly on NNK) seems likely before ATM will become an immediate option. The BMD technology and costs will be key factors in determining whether a shift away from the West's prevailing theory of deterrence (societal punishment) to one partaking of greater elements of damage-limiting is feasible. One might then consider whether, given possible Soviet countermeasures, a damage-limiting strategy—or at least improved damage-limiting capabilities, especially for defense of selected strategic and theater military targets—would be sensible.

Technological opportunity may not, however, determine the rejection or choice of damage-limiting measures as much as established convictions regarding BMD and its political, diplomatic, and financial costs. Less costly BMD technology might reduce the financial burden, but the question of political will to pursue damage-limiting programs over the long term would remain. Without domestic or Alliance consensus on the strategic merits of damage-limiting, the United States would find it hard to sustain BMD deployment decisions. Moreover, U.S. political will to pursue analysis of BMD options seriously and to reassess Alliance strategy in the light of potential damage-limiting opportunities is not likely to be stiffened by encouragement from the Allies in Western Europe.

---

62. *Aviation Week and Space Technology*, April 5, 1982, p. 23.

# OFFICIAL DOCUMENTS

# Directed Energy Missile Defense in Space

## Background Paper

### April 1984

Prepared under contract for the
Office of Technology Assessment

by Ashton B. Carter
Massachusetts Institute of Technology

**This is an OTA Background Paper that has neither been reviewed nor approved by the Technology Assessment Board**

# Directed Energy Missile Defense in Space

**Background Paper**

**April 1984**

Prepared under contract for the
Office of Technology Assessment

by Ashton B. Carter
Massachusetts Institute of Technology

**OTA Background Papers** are documents containing information that supplements formal OTA assessments or are outcomes of internal exploratory planning and evaluation. The material is usually not of immediate policy interest such as is contained in an OTA Report or Technical Memorandum, nor does it present options for Congress to consider.

# Preface

This background paper was prepared by Dr. Ashton B. Carter under a contract with the Office of Technology Assessment. OTA commissions and publishes such background papers from time to time in order to bring OTA up to date on technologies that are the subject of frequent congressional inquiry. After Dr. Carter's work was under way, Senators Larry Pressler and Paul Tsongas of the Senate Foreign Relations Committee requested that the resulting paper be made available to that Committee as soon as possible. OTA is issuing the paper in the belief that others in Congress and members of the public will find it of interest and importance.

An OTA background paper differs from a full-fledged technology assessment. Background papers generally support an ongoing assessment of broader scope or explore emerging technological issues to determine if they merit a fuller, more detailed assessment. On March 22, 1984, the Technology Assessment Board directed OTA to carry out a full-fledged assessment of "New Ballistic Missile Defense Technologies," for which this background paper will serve as one point of departure.

This paper was prepared for OTA's International Security and Commerce Program, under the direction of Lionel S. Johns (Assistant Director, Energy, Materials, and International Security Division) and Peter Sharfman (Program Manager).

# Contents

Section 1
# INTRODUCTION

# INTRODUCTION

This Background Paper describes and assesses current concepts for directed-energy ballistic missile defense in space. Its purpose is to provide Members of Congress, their staffs, and the public with a readable introduction to the so-called "Star Wars" technologies that some suggest might form the basis of a future nationwide defense against Soviet nuclear ballistic missiles. Since these technologies are a relatively new focus for U.S. missile defense efforts, little information about them has been readily available outside the expert community.

Directed-energy or "beam" weapons comprise chemical lasers, excimer and free electron lasers, nuclear bomb-powered x-ray lasers, neutral and charged particle beams, kinetic energy weapons, and microwave weapons. In addition to describing these devices, this Background Paper assesses the prospects for fashioning from such weapons a robust and reliable wartime defense system resistant to Soviet countermeasures. The assessment distinguishes the prospects for perfect or near-perfect protection of U.S. cities and population from the prospects that technology will achieve a modest, less-than-perfect level of performance that will nonetheless be seen by some experts as having strategic value. Though the focus is technical, the Paper also discusses, but does not assess in detail, the strategic and arms control implications of a major U.S. move to develop and deploy ballistic missile defense (BMD).[1]

This Background Paper grows indirectly out of President Reagan's celebrated television speech of March 23, 1983, in which he called for a "long-term research and development program to begin to achieve our ultimate goal of eliminating the threat posed by strategic nuclear missiles."[2] Pursuant to the President's speech, the Department of Defense established a Defensive Technologies Study Team under James C. Fletcher (of the University of Pittsburgh) to prescribe a plan for the R&D program. A parallel effort, called the Future Security Strategy Study and headed by Fred S. Hoffman (of Research and Development Associates), addressed the implications for nuclear policy of renewed emphasis on BMD. This Paper covers the same technologies and issues as these Defense Department studies. The ABM Treaty reached at SALT I[3] severely restricts the development, testing, and deployment of BMD systems. Though this Background Paper treats the strategic roles of missile defenses, including many of their arms control implications, it does not treat the vital international political implications of a major U.S. move to BMD.

Focused on directed-energy intercept of missiles in their boost phase, i.e., on "Star Wars" proper, this Paper does not analyze midcourse and reentry BMD systems or non-BMD applications of directed-energy weapons.[4] "Star Wars" efforts generally further concentrate on intercept of intercontinental ballistic missiles (ICBMs) rather than the related but somewhat different problems of intercept of submarine launched ballistic missiles (SLBMs) or intermediate-range ballistic missiles (IRBMs). This Paper is therefore not a substitute for a more complete treatment of the entire subject of BMD.[5] Moreover, BMD itself is only part of the larger subject of strategic defense, comprising defense against bombers and cruise missiles, civil defense, passive defense of military targets, anti-submarine warfare (ASW), and pre-emptive counterforce attack in addition to BMD.

---

[1]BMD is the most common of four roughly equivalent acronyms covering defense against nuclear ballistic missiles. Such defenses were formerly called anti-ballistic missile (ABM) systems, but this designation fell out of favor after the debate over and eventual demise of the Sentinel and Safeguard ABM systems in the late 1960's and early 1970's. BMD largely replaced ABM as the term of choice, but recently the more self-explanatory Defense Against Ballistic Missiles (DABM) has gained popularity. Within the Executive Branch, BMD efforts pursuant to President Reagan's so-called "Star Wars" speech are referred to as the Strategic Defense Initiative (SDI). Normally the term strategic defense comprehends other methods for limiting damage from nuclear attack besides BMD.

[2]The relevant portions of President Reagan's speech are reproduced in Appendix A.
[3]The ABM Treaty and related documents are reproduced in Appendix B.
[4]Appendix C describes briefly, but does not assess, other proposed military applications of directed-energy weapons.
[5]For a more complete treatment, see *Ballistic Missile Defense*, ed. Ashton B. Carter and David N. Schwartz (Brookings, 1984).

It is unusual for the President to express himself on, and for the Congress and public consequently to concern themselves with, long-term research and development. "Star Wars" is thus a somewhat unusual subject for a technology assessment intended for a public policy audience. It is in the nature of this subject that unknown or unspecified factors outweigh what is known or can be presented in concrete detail. Many of the technologies discussed in this Paper, and most certainly all of the schemes for fashioning a defense system from these technologies, are today only paper concepts. In the debate over the Safeguard ABM system a decade ago, or over basing modes for the MX missile in recent years, one could analyze in detail the technical properties of well-defined systems in engineering development. So vague and tentative are today's concepts for "Star Wars" BMD that a comparable level of analysis is impossible. Fashions and "front runners" are likely to change. Nonetheless, one is faced with assessing the concepts receiving attention today within the Executive Branch and which underlie the President's Strategic Defense Initiative. Fortunately, judgments deduced from generic properties of these concepts, which are unlikely to change, are sometimes telling.

This Paper is based on full access to classified information and studies performed for the Executive Branch. But it turns out that a fully adequate picture of this subject can be presented in unclassified form. One reason is that the important features of the directed-energy BMD concepts are based on well-known physics, and many have in fact been discussed for 20 years. The second reason is that at this early stage of conceptualization there is simply no point in (and little basis for) discussion at the detailed level where classified particulars make a difference. The properties of actual weapon systems in engineering development, by contrast, are normally and understandably classified.

The author and OTA wish to thank officials of the Army's Ballistic Missile Defense Systems Command (BMDSCOM), the Lawrence Livermore National Laboratory, the Los Alamos National Laboratory, the Defense Advanced Research Projects Agency (DARPA), the Sandia National Laboratory, the Air Force Weapons Laboratory (AFWL), the Office of the Secretary of Defense (OSD), and the Central Intelligence Agency (CIA) for their hospitality and cooperation. Many individuals aided the research for this Background Paper, though none shares the author's sole responsibility for errors of fact or judgment. The author and OTA wish especially to thank Hans Bethe, Richard Briggs, Al Carmichael, Albert Carnesale, Paul Chrzanowski, Robert Clem, Sidney D. Drell, Dick Fisher, John Gardner, Richard L. Garwin, Ed Gerry, Jack Kalish, Glenn A. Kent, Louis Marquet, Michael M. May, Tom Perdue, Theodore A. Postol, George Rathjens, Victor Reis, Jack Ruina, George Schneiter, David N. Schwartz, Robert Selden, Leon Sloss, Daryl Spreen, Robin Staffin, John Steinbruner, Sayre Stevens, Thomas Weaver, Stephen Weiner, and Wayne Winton.

This Background Paper contains information current as of January 1, 1984.

# BOOSTER CHARACTERISTICS

# BOOSTER CHARACTERISTICS

Intercept of ICBMs in their boost phase offers advantages and disadvantages relative to intercept of reentry vehicles (RVs) later in the trajectory. The boosters are fewer and generally more easily disrupted or destroyed than the RVs. Decoy boosters would have to match an ICBM's huge heat output, making this offensive tactic attractive only in certain circumstances. The disadvantages of boost phase intercept are that boost phase is only a few minutes long and comprises the earliest stage of an attack, and that sensing and intercept must be accomplished from outer space and over enemy territory.

Figure 2.1 shows an ordinary (minimum energy) trajectory of a hypothetical future Soviet ICBM that has been given, for illustration, the boost profile of the U.S. MX Peacekeeper. Pressure from a steam generator expels the missile from its stor-

age cannister. Once clear of the cannister, the missile ignites its first stage motor. The first stage burns for about 55 seconds, burning out at an altitude of about 22 kilometers. The second stage also burns for 55 sec, burning out at 82 km. The third stage burns for 60 sec and carries the remainder of the missile to about 200 km, the altitude of the lowest earth orbiting satellites.

When the third stage is jettisoned at the end of the 3-minute boost phase, the remainder of the missile consists of the post-boost vehicle (PBV) or "bus" and its cargo of 10 reentry vehicles. At this point the bus and RVs are in ballistic free-fall flight to the United States. Even if they are disrupted in some way or destroyed, these objects or their debris will reenter the atmosphere over the United States. The last few seconds of third stage burn are crucial for giving the payload

**Figure 2.1.—The Flight of a Hypothetical Future Soviet ICBM With the Booster Characteristics of the U.S. MX Peacekeeper, Drawn to Scale**

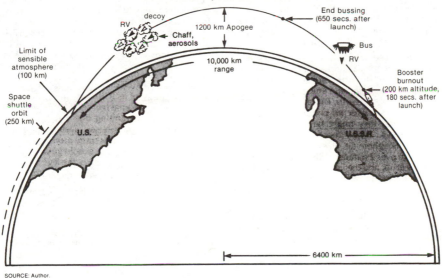

SOURCE: Author.

179

enough speed to reach the United States, so disruption of boost phase any time right up to burnout will cause the warheads to fall far short.

For the next 500 seconds or so after burnout—almost until it reaches apogee—the bus uses its thrusters to make small adjustments to its trajectory. After each adjustment, it releases an RV. RVs released on different trajectories continue on to different targets as multiple independently targeted reentry vehicles (MIRVs). Decoys and other penetration aids for helping the RVs escape defenses later in the trajectory are deployed during busing.

The bus itself is a target of declining value as it dispenses its RVs. Destroying it early in the deployment process would obviously be useful: the RVs not yet deployed from the bus would still arrive at the United States, but perhaps nowhere near their intended targets. If cities are the targets, relatively small aiming errors might be inconsequential. In any event, tracking the bus to allow some form of intercept requires a different type of sensor from that which tracks the booster for boost phase intercept, since the bus's thrusters are small and operate intermittently. Because of its small size, the bus (or at least critical elements of it) might be more easily hardened against directed-energy weapons than the booster. For all these reasons, the value of attempting bus intercept is very unclear, and it usually does not figure prominently in BMD discussions.

From apogee, the slowest point in their free-fall trajectory, the RVs and empty bus gain speed as they fall back to earth. RVs are more resistant to damage from directed-energy weapons than boosters, and they might be accompanied by many decoys. When these objects enter the upper atmosphere at about 100 km altitude somewhat over 2 minutes before impact, they begin to heat up, and the lighter objects slow down. Still lower, below 50 km altitude and less than a minute before impact, the objects undergo violent deceleration and the bus breaks up. The RVs, now glowing with heat, streak toward their targets at an angle of about 23 degrees to the horizontal.

The trajectory shape can be altered at the expense of payload (see Figure 2.2). A lofted trajectory takes longer but reenters faster, and a depressed trajectory can offer unfavorable viewing angles to defensive sensors late in the trajectory.

The most important trajectory variations from the point of view of boost phase intercept are variations in boost profile. Boosters like MX were designed with no regard for boost phase BMD, and optimizing their design gave rise to rather long boost times. But boost phase can be shortened—giving less time for boost phase weapons to act—and accomplished within the atmosphere—where certain directed-energy weapons cannot penetrate—with relatively little reduction in payload or increase in missile size. Fast burn is accomplished most easily with solid-fueled rockets. Liquid-fueled boosters like the Soviet SS-18s and SS-19s burn more slowly and burn out at higher altitudes. Thus while MX burns out at 200 km after 3 minutes of boost, the SS-18 burns out at 300-400 km after 5 minutes. The next generation of Soviet ICBMs will reportedly employ solid propellants.

Studies performed for the Defense Department[1] showed that with a 25 percent reduction in payload, a booster about the same size as MX could be built which would burn out in less than 1 minute at only 80 to 90 km, well within the sensible atmosphere. At 90 km the atmosphere is still too dense for extremely accurate RV deployment or for deployment of lightweight RV decoys and other penetration aids aimed at later defensive layers: these functions require an additional 10 to 15 seconds of precision deployment between 90 and 110 km. If the offense needs precision accuracy for some of its ICBMs but fears intercept during these additional few seconds of high-altitude operation, mounting one or two RVs on each of several "microbuses" instead of all the RVs on a single bus affords some protection. Each microbus would contain a simple guidance system only good enough to carry the RVs from upper stage burnout to 110 km. Instead of presenting one target above 90 km, therefore, such a booster would present several targets.

The United States is studying a "Midgetman" missile endorsed by the President's Commission on Strategic Forces (the Scowcroft Commission)

---

[1]"Short Burn Time ICBM Characteristic and Considerations," Martin Marietta Denver Aerospace, July 20, 1983 (UNCLASSIFIED).

Figure 2.2.—Normal (Minimum-Energy), Lofted, and Depressed ICBM Trajectories, Drawn to Scale

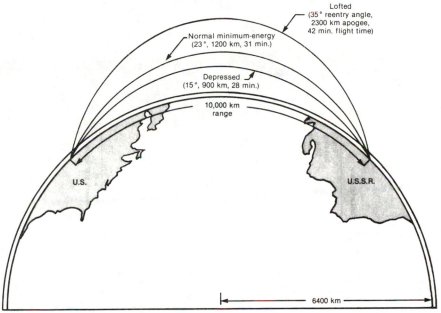

SOURCE: Author.

with weight 15 to 25 percent that of MX and carrying one warhead. Midgetman's warhead and bus are combined in one hardened structure. Table 2.1 shows the characteristics of Midgetman variants designed to face a boost-phase intercept system. The fast-burn version burns out at 80 km after 50 secs of boost. With a 10 percent increase in weight, the fast-burn version can carry a substantial payload of penetration aids. A low-flight-profile version is intended to stay within the atmosphere until burnout, protecting it from some types of directed-energy weapon. In the hardened version, one gram of ablative or other shielding material has been applied to each square centimeter of the entire booster body (if the boost-phase intercept system did not begin operation until a minute or so after launch, the first stage might not have to be hardened). These small boosters are all estimated to cost $10 to $15 million per copy, assuming a buy of 1,000 boosters. Costs for the second and subsequent thousand would of course be substantially smaller. These costs are two to three times higher per RV than MX.

The Soviet ICBM arsenal today comprises about 1,400 boosters, more than two-thirds of them MIRVd. Most are slow-burning liquid-fueled boosters. The U.S. arsenal contains about 1,000 faster-burning solid-fueled Minuteman boosters, about half of them MIRVd. Both sides are adding solid boosters to their arsenals in the 1980's.

The geographic distribution of offensive boosters can also be important to space-based boost-phase defenses. The number of satellites required in a defensive constellation usually increases if all opposing ICBM silos are concentrated in one region and decreases if the silos are spread over

**Table 2.1.—ICBM Booster Characteristics**

| | Gross weight (kg) | Length (m) | Width (m) | Type | Booster burnout time (seconds after launch) | Booster burnout altitude (km) | End bussing time (seconds after launch) | End bussing altitude (km) | Comments |
|---|---|---|---|---|---|---|---|---|---|
| SS-18 | 220,000[a] | 35.0 | 3.0 | 2-stage liquid | ~300 | ~400 | — | — | — |
| MX Peacekeeper | 69,000 | 21.3 | 2.3 | 3-stage solid | 180 | 200 | 650 | 1,100 | Carries 10 RVs on a single bus |
| MIRVd fast-burn booster | 87,000 | 22.9 | 2.1 | 2-stage solid | 50 | 90 | 60 | 110 | Deploys several "microbuses" carrying RVs and decoys |
| Midgetman | 19,000 | 12.1 | 1.5 | 2-stage solid | 220 | 340 | — | — | Carries one accurate RV |
| Midgetman fast-burn | 20,000 | 13.8 | 1.5 | 2-stage solid | 50 | 80 | — | — | Carries one accurate RV |
| Midgetman fast-burn with midcourse penetration aids | 22,000 | 14.3 | 1.5 | 2-stage solid | 50 | 80 | — | — | Carries one accurate RV plus decoys |
| Midgetman with low flight profile | 25,000 | 13.4 | 1.5 | 1st stage solid, 2nd liquid | 220 | 100 | — | — | Carries one accurate RV |
| Hardened Midgetman | 30,000 | 15.1 | 1.5 | 1st stage solid, 2nd liquid | 220 | 320 | — | — | Carries one accurate RV: entire booster covered with 1gm/cm$^2$ shielding |
| Pershing II | 7,500 | 10.5 | 1.0 | 2-stage solid | 100 | — | — | — | — |

[a]*Soviet Space Programs 1976-1980* (Committee on Commerce, Science, and Transportation, U.S. Senate, December 1982), Part 1, p. 63.

SOURCE: Except where indicated, "Short Burn Time ICBM Characteristics and Considerations," and accompanying backup, presented to DTST, July 20, 1983, by Martin Marietta Denver Aerospace (UNCLASSIFIED).

wide land areas. (On the other hand, too much concentration allows defensive satellites to be focused on one region by choosing the orbits judiciously.) Soviet SS-18 ICBMs, their largest MIRVd missiles, are organized into 6 wings of about 50 missiles each, spread out over a large region of the U.S.S.R. U.S. Minuteman missiles are organized into 6 wings of about 150 missiles each. Figures 2.3 and 2.4 show the geographic distributions.

The capabilities of a hypothetical future U.S. BMD should be measured against the *future* and *potential* Soviet ICBM arsenal, not against today's arsenal. The future arsenal will differ due to the natural retirement of old ICBMs and introduction of new ones, and because the Soviets might well react with new and different deployments when they learn of any U.S. plans to deploy defenses. It is impossible to project Soviet deployments far into the future. A reasonable "baseline" estimate for the Soviet ICBM arsenal 15 to 20 years from now might assign them the same number of boosters they have today, but with burn characteristics similar to the U.S. solid-propellant MX. This Background Paper indicates where and how the effectiveness of a hypothetical U.S. defense depends on the nature of the offensive arsenal it faces. In addition to having shorter average burn times, future Soviet ICBMs could be more numerous, deployed less widely geographically, less highly MIRVd, hardened against intercept, and so on.

**Figure 2.3.—Present U.S. ICBM Deployment Areas**

SOURCE: OTA, *MX Missile Basing*, p. 274.

**Figure 2.4.—Present U.S.S.R. ICBM Deployment Areas**

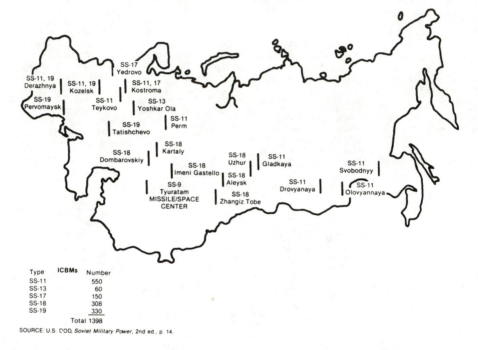

| Type | ICBMs | Number |
|------|-------|--------|
| SS-11 | | 550 |
| SS-13 | | 60 |
| SS-17 | | 150 |
| SS-18 | | 308 |
| SS-19 | | 330 |
| | Total | 1398 |

SOURCE: U.S. DOD, *Soviet Military Power*, 2nd ed., p. 14.

Section 3

# DIRECTED ENERGY WEAPONS FOR BOOST-PHASE INTERCEPT

# DIRECTED ENERGY WEAPONS FOR BOOST-PHASE INTERCEPT

This section describes the entire set of "beam weapons" being considered in the United States today for boost-phase ICBM intercept. Though these weapons receive the most attention, the "kill mechanism" that destroys the booster is not necessarily the most important or technically challenging part of an overall defense system. The next section describes other essential elements of a boost-phase defense.

A revisit to this subject several years from now might well find a new family of directed energy concepts receiving attention. But for now the devices described in this section are the basis for assessments of the prospects for efficient boost-phase defense, in the Defense Department and elsewhere (fig. 3.1). Though some of these concepts are new, many have in fact existed in one form or another for more than twenty years.

**Figure 3.1**

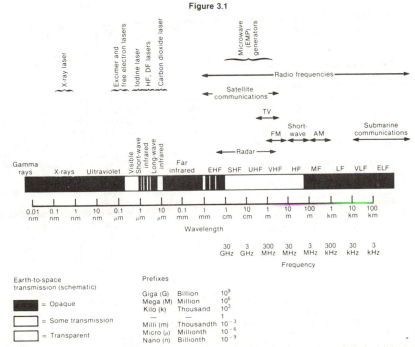

Figure 3.1 The electromagnetic spectrum, showing spectral regions of interest for directed energy BMD. Particle beams and kinetic energy weapons are not shown because their energy does not consist of electromagnetic radiation, but of atomic and macroscopic matter, respectively. Source: Author

For each concept this section attempts to work through, with some concreteness, the design of a hypothetical defensive system based on the concept. The resulting designs are *illustrative only*; no significance should be attached to precise numbers. Precision is simply not possible in the current state of technology and study of these concepts.

In all cases, the "current state of technology" (however this is defined in each case) is far from meeting the needs of truly efficient boost-phase intercept. The systems designed in this section illustrate the level to which technology would have to progress to be "in the ballpark." Much attention fastens on the gulf between the current state of technology and the ballpark requirements. This section does not emphasize such comparisons for several reasons. First, in some cases details of the precise status of U.S. research is classified. Second, and more importantly, quantitative comparisons (e.g., "A millionfold increase in brightness is required to fashion a weapon from today's laboratory device") can mislead unless accompanied by a deeper explanation of the technology; and the same quantitative measures are not appropriate for all technologies. Third, and most importantly, such comparisons imply that learning how to build the right device is tantamount to developing an efficient missile defense, which is far from true: equally crucial are design of a sensible system architecture, cost, survivability, resilience to countermeasures, and the myriad detailed limitations that do not turn up until later in development.

## 3.1 SPACE-BASED CHEMICAL LASERS: A FIRST EXAMPLE

This concept of directed energy weapon has been the one most frequently discussed in recent years for boost-phase ICBM intercept. For this reason (and not necessarily because it is the most plausible of all the concepts), it will be used to introduce certain features common to all the schemes that follow.

### Making and Directing Laser Beams

A molecule stores energy in vibrations of its constituent atoms with respect to one another, in rotation of the molecule, and in the motions of the atomic electrons. The molecule sheds energy in the form of emitted light when it makes transitions from a higher-energy state to a lower-energy state. Lasing takes place when many molecules are in an upper state and few are in a lower state: one downward transition then stimulates others, which in turn stimulate yet more, and a cascade begins. The result is a powerful beam of light.

Energy must be supplied to the molecules to raise most of them to the upper state. This process is called pumping. In the case of the chemical lasers considered in this section, the pumping energy comes from the chemical reaction that makes the lasant molecules: hydrogen and fluorine react to form hydrogen fluoride (HF) molecules in an upper state. The other requirement for lasing—few molecules in the lower state—is satisfied simply by removing the molecules from the reaction chamber after they have made their transitions to the lower state and replacing them with freshly made upper state molecules. The pumping process is not perfect: not all the pumping energy ends up as laser light. The ratio of pumping energy in to laser energy out is called the efficiency of the laser.

Laser light is special in two respects: its frequency is precise, since all the light comes from the same transition in all the molecules; and the light waves from all the molecules emerge with crests and troughs aligned, since the waves are produced cooperatively. These special features make it possible to focus the laser energy with mirrors into narrow beams characterized by small divergence angles (see fig. 3.2). Nonetheless, there is a limit to the divergence angle that even a perfect laser with perfect mirrors can produce. The divergence angle (in radians) can be no smaller than about 1.2 times the wavelength of the light divided by the diameter of the mirror. Thus a laser with 1 micrometer (=1 micron)

**Figure 3.2**

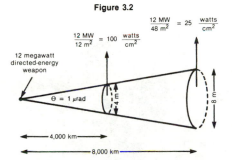

$$\frac{12 \text{ MW}}{48 \text{ m}^2} = 25 \; \frac{\text{watts}}{\text{cm}^2}$$

$$\frac{12 \text{ MW}}{12 \text{ m}^2} = 100 \; \frac{\text{watts}}{\text{cm}^2}$$

12 megawatt directed-energy weapon

$\Theta = 1 \; \mu\text{rad}$

4,000 km

8,000 km

Figure 3.2 Basic power relationships for directed energy weapons. If the directed energy weapon has a divergence angle of 1 microradian, the spot size at a range of 4000 kilometers is 4 meters (12 feet). In this figure, the divergence angle is exaggerated about 1 million times. (For comparison of scale, the Earth's radius is about 6,400 km.) If the directed energy weapon emits 12 megawatts of power, a target within the spot at 4,000 km receives 100 watts on each square centimeter of its surface. (For comparison, 100 watts is the power of a lightbulb, and a typical commercial powerplant produces 1,000 megawatts). Since a watt of power equals one joule of energy per second this weapon would take 10 seconds to apply a kilojoule per square centimeter (1 KJ/cm²) to the target at 4,000 km range. Source: Author

wavelength projected with a 1 meter mirror could have at best a 1.2 microradian divergence angle, making a spot 1.2 meters wide at a range of 1,000 kilometers (refer to fig. 3.2).[1] This perfect performance is called the diffraction limit. Dividing the laser power output by the size of the cone into which it is directed (cone size is measured in units called steradians; a divergence angle of x radians results in a cone of size $\pi x^2/4$ steradians) yields the laser's "brightness," the basic measure of a weapon's lethality.

### Destroying Boosters with Lasers

Assuming a high-energy laser with small divergence angle can be formed, stabilized so it does not wave about (jitter), and aimed accurately, what effect will it have on an ICBM booster? No

[1] The spot from a perfect laser with perfect mirrors would actually be brighter at the center than at the edges. The full angle subtended by this spot (the Airy disk from null to null in the diffraction pattern), is 2.4 times the wavelength divided by the mirror diameter, but most of the energy is in the central fourth of this area: hence the use in the text of the multiplier 1.2.

clear answer to this question can be given without more study and testing. Estimates of the hardness achievable with future boosters are probably reliable within a factor of two or three, though estimates of the hardness of current Soviet boosters are probably reliable only to a factor of 10 or so.

Roughly speaking, laser light can damage boosters in two distinct ways. With moderate intensities and relatively long dwell times, the laser simply burns through the missile skin. This first mechanism is the relevant one for the chemical lasers described in this section. The second mechanism requires very high intensities but perhaps only one short pulse: the high intensity causes an explosion on and near the missile skin, and the shock from the explosion injures the booster. This mechanism, called impulse kill, is more complex than thermal kill and is less well understood. It will be discussed in the next section.

Bearing in mind the uncertainties in these estimates, especially the complex interaction of heating with the mechanical strains of boost, the following estimates are probably reliable: A solid-fueled booster can probably absorb without disruption up to about 10 kilojoules per square centimeter (kJ/cm²) on its skin if a modicum of care is taken in the booster's design to eliminate "Achilles' heels." This energy fluence would result from 1 second of illumination at 10 kilowatts per square centimeter (kw/cm²), since one watt equals one joule per second.[2] Applying ablative (heatshield) material to the skin can probably double or triple the lethal fluence required. Applying a mirrored reflective coating to the booster is probably not a good idea, since abrasion during boost could cause it to lose its lustre. Spinning the booster triples its hardness, since a given spot on the side of the booster is then only illuminated about a third of the time.[3] On the other

[2] The lethal fluence (in kJ/cm²) must accumulate over a relatively short time, so that the booster wall suffers a high rate of heating. Thus a flux of 30 watts/cm² would deposit 10 kJ/cm² in 330 sec of dwell time, but such a slow rate of heating would probably not damage the booster.

[3] It is possible that uniform heating around the circumference of the booster introduces lethal mechanisms distinct from those that apply to heating a single spot on the side of the booster. In that case, spinning the booster might not lengthen the required dwell time by the full amount dictated by geometry.

hand, currently deployed boosters, especially the large liquid Soviet SS-18s and SS-19s, might be vulnerable to I kJ/cm² or even less. These too could be hardened by applying heatshield material.

## An Orbiting Chemical Laser Defense System

Consider a space-based BMD system comprised of 20-megawatt HF chemical lasers with 10 meter mirrors. The HF laser wavelength of 2.7 microns is attenuated as it propagates down into the atmosphere, but most of the light gets down to 10 km or so altitude. Deeper penetration is not really needed, since the laser would probably not be ready to attack ICBMs until after they had climbed to this altitude, and in any event clouds could obscure the booster below about 10 km. (Substituting the heavier and more expensive deuterium, an isotope of hydrogen, to make a DF laser at 3.8 micron wavelength would alleviate attenuation, but the longer wavelength would require larger mirrors.)

A perfect 10 meter mirror with a perfect HF laser beam yields 0.32 microradian divergence angle. The spot from the laser would be 1.3 meters (4.0 ft) in diameter at 4 megameters (4,000 kilometers) range. 20 megawatts distributed evenly over this spot would be an energy flux of 1.5 kw/cm². The spot would need to dwell on the target for 6.6 seconds to deposit the nominal lethal fluence of 10 kJ/cm². At 2 megameters (Mm) range, booster destruction would require only a fourth of this time, or 1.7 seconds of illumination. Since light takes about a hundredth of a second to travel 4 megameters and the booster is traveling a few kilometers per second, the booster moves about 50 meters in the time it takes the laser light to reach it. The laser beam must therefore lead the target by this distance.

The next step is to choose orbits for the satellites so that the U.S.S.R.'s ICBM silos are covered at all times and so that there are enough satellites overhead to handle all 1,400 of the present Soviet booster population. Equatorial orbits (fig. 3.3) give no coverage of the northern latitudes where Soviet ICBMs are deployed. Polar orbits give good coverage of northern latitudes but concentrate satellites wastefully at the poles where there are no ICBMs. The optimum constellation consists

**Figure 3.3**

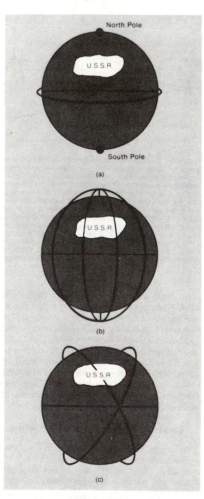

Figure 3.3 Designing a constellation of directed energy weapon satellites for optimum coverage of Soviet ICBM fields. Equatorial orbits (a) give no coverage of northern latitudes. Polar orbits (b) concentrate coverage at the north pole. Inclined orbits (c) are more economical. Slight additional economies are possible in some cases with further elaboration of the constellation design. Source: Author

of a number of orbital planes inclined about 70° to the equator, each containing several satellites.

The shorter the lethal range of the directed energy weapon, the lower and more numerous the satellites must be. For instance, with a lethal range of 3 Mm, 5 planes containing 8 satellites each, or a total of 40 satellites, are needed to ensure that Soviet boosters exiting Soviet airspace would be within lethal range of one satellite. If the lethal range is increased to 6 Mm, only 3 planes of 5 satellites each are needed. This dependence of constellation size on weapon range is displayed in figure 3.4. (It is possible to adjust these numbers a bit by using slightly elliptical orbits with apogees over the northern hemisphere, adjusting inclinations and phasing, etc.). In the present example, requiring that at least one HF laser be no further than 4 Mm from each So-

viet ICBM site at all times (corresponding to no longer than 6.6 seconds dwell time per booster) results in the illustrative constellation of 32 orbital positions shown in figure 3.5.

Since the 1,400 Soviet boosters currently deployed are spread out over most of the Soviet Union, perhaps 3 of the 32 orbital positions would be over or near the Soviet Union at a time, able to make efficient intercepts. That is, only one in 11 deployed U.S. battle stations would participate in a defensive engagement. The ratio of the total number of battle stations on orbit to the number in position to participate in a defensive engagement is called the absentee ratio. The inevitable waste reflected in the absentee ratio—

**Figure 3.5**

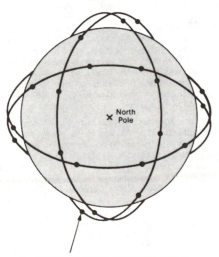

**Figure 3.4**

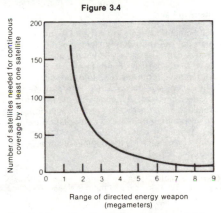

Range of directed energy weapon
(megameters)

Figure 3.4 The number of satellites needed in a constellation to ensure that at least one satellite is over each Soviet ICBM field at all times depends on the effective range of the directed energy weapon. For every one defensive weapon required overhead a Soviet ICBM field to defend against a rapid Soviet attack, an entire constellation must be maintained on orbit. Since there are many Soviet ICBM fields distributed over much of the Soviet landmass, more than one satellite in each constellation would be in position to participate in a defensive engagement. The ratio of the number of satellites in the constellation to the number over or within range of Soviet ICBM fields is called the absentee ratio. If all Soviet ICBMs were deployed in one relatively small region of the U.S.S.R., the absentee ratio would be the same as the number of satellites in the constellation. Source: Author

0 = Cluster of 5 chemical laser battle stations
Total of 32 × 5 = 100 battle stations

Figure 3.5 Constellation of hypothetical directed energy weapon satellites with 4,000 km range. The orbits are circular with 1000 km altitude. Each of the four orbital planes consists of eight positions spaced 45° apart around the circle. In the example given in the text, five chemical laser battle stations are clustered at each point shown in this figure, for a total of 32 x 5 = 160 battle stations. Source: Author

usually on the order of 10—offsets an oft-cited theoretical advantage of boost-phase intercept, namely, that intercepting one booster saves buying 10 interceptors for the booster's 10 RVs. On the other hand, coverage of the U.S.S.R.'s ICBM fields automatically gives good coverage of essentially all submarine deployment areas. Obviously the absentee ratio would be 32—the full constellation size—and not 32/3 = 10.7 if Soviet ICBM silos were not spread out so widely over Soviet territory but were deployed over a third or less of the Soviet landmass, so that only one of the 32 U.S. satellites was within range.

Three of the earlier described laser satellites in position over the Soviet ICBM fields are not enough to intercept 1,400 boosters if all or most of the boosters are launched simultaneously. Each satellite can only handle a few boosters because it must dwell for a time on each one. The time a chemical laser must devote to each booster depends on the satellite's position at the moment of attack—6.6 seconds for 4 Mm range, 1.7 seconds for 2 Mm range, etc. Taking 2 Mm as an average range for the 32-satellite constellation (hoping the Soviets do not choose a moment when most of the U.S. satellites are farther than 2 Mm from the ICBM flyout corridors to launch all their boosters simultaneously), a laser must devote an average of 1.7 seconds to each booster.

If the boosters in the future Soviet arsenal resemble the U.S. MX, and the defense waits 30 seconds or so to confirm warning and to wait for the boosters to climb to an altitude where the HF laser can reach, each booster is accessible for 150 seconds of its 180 second burn time. Each laser can therefore handle no more than 90 boosters, even with instant slewing of the beam from target to target. If 1,400 Soviet boosters were launched simultaneously, (1,400)/(90) ≅ 15 lasers would be needed in position, for a worldwide total (multiplying by the absentee ratio) of (10.7) × (15) = 160 satellites.

If the Soviets doubled their arsenal to 2,800 boosters, the United States would need to deploy another 160 satellites, possibly an uncomfortable cost trade for the United States.

What is worse, if the Soviets deployed 1,400 missiles in a *single region* of the U.S.S.R. (at a U.S.-estimated cost of $21 billion for Midgetman-like ICBMs; see section 2 above), the US would have to build, launch, and maintain on orbit an additional (32)×(1400)/(90) ≅ 500 lasers plus their fuel and support equipment.

If Soviet boosters were covered with shielding material and spun during flight to achieve an effective hardness of, say, 60 kJ/cm², a laser would have to devote 10 seconds to each booster at 2 Mm range, requiring a sixfold increase in the number of satellites, to 960. Alternatively, the average range of each engagement could be reduced to keep the dwell time at 1.7 seconds, with corresponding increase in constellation density (fig. 3.4). Either way, the number of U.S. satellites would grow to nearly the number of Soviet boosters intercepted.

Table 3.1.—Variation of the Number of Chemical Laser Battle Stations Needed to Handle a Simultaneous Launch of Soviet ICBMs, Depending on Characteristics of the Soviet Arsenal and the U.S. Laser Defense

| Departure from baseline | Number of Soviet boosters | Booster characteristics | Geographic distribution | Hardness (kJ/cm²) | Laser power (MW) and aperture diameter (m) | Approximate number of battle stations needed |
|---|---|---|---|---|---|---|
| Baseline .................. | 1,400 | MX-like | Current Soviet | 10 | 20/10 | 160 |
| Booster number ............ | 2,800 | MX-like | Current Soviet | 10 | 20/10 | 320 |
| Deployment geography ...... | 1,400 | MX-like | One region | 10 | 20/10 | 500 |
| Booster hardness .......... | 1,400 | MX-like | Current Soviet | 60 | 20/10 | 960 |
| Laser brightness ........... | 1,400 | MX-like | Current Soviet | 10 | 80/50 (100 times brighter) | 20-30 |
| Booster burn time ......... | 1,400 | Fast-burn | Current Soviet | 10 | 20/10 | 800-1,600 |
| Booster burn time ......... | 1,400 | SS-18-like | Current Soviet | 10 | 20/10 | 90 |

SOURCE: Author.

If the United States developed a battle station 100 times brighter (using, say, a 80 MW laser with an effective mirror diameter of 50 meters), a few lasers overhead (20 to 30 total worldwide) could easily handle an attack of 1400 boosters hardened to 10 kJ/cm². If the boosters were hardened to 60 kJ/cm², over 100 such lasers would be needed.

Deployment by the Soviets of 1400 fast-burn boosters would give the U.S. lasers just 20 to 40 seconds, rather than 200 seconds, to destroy all the boosters. The U.S. constellation would consequently need to grow by a factor 5 to 10, to 800 to 1600 satellites!

Table 3.1 summarizes how the size of the defensive deployment varies with the parameters assumed.

### Requirements for a Chemical Laser Defense

Figure 3.6 displays the performance of various hypothetical HF lasers. Keeping the size of the battle station constellation down to a hundred rather than several hundred satellites means lethal ranges of at least 4 Mm with illumination times less than about 1 second, assuming the defense must be capable of intercepting 1,000 to 2,000 Soviet boosters with launches timed to keep the boosters as far from the U.S. lasers as possible. Further assuming Soviet booster hardening to at least 10 kJ/cm² results in a requirement for chemical lasers considerably brighter than the 20 MW, 10-meter laser described above. A hundredfold increase in brightness would be achieved by a laser with power 80 MW and effective mirror diameter 50 meters.

Such a laser would be about 10 million times brighter than the carbon dioxide laser on the Air Force's Airborne Laser Laboratory. The current Alpha laser program of the Defense Advanced Research Projects Agency (DARPA) aims at a construction of an HF laser of just a few megawatts and built only for ground operation. Nonetheless, there is no fundamental technical reason why extremely bright chemical lasers cannot be built. In theory, several lasers can be operated together so that the brightness of the resulting beam increases with the square of the number

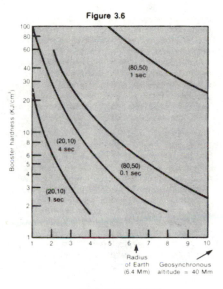

**Figure 3.6**

Radius of Earth (6.4 Mm)   Geosynchronous altitude = 40 Mm

Lethal range (megameters)

Figure 3.6 Lethal range versus booster hardness for HF chemical lasers of various sizes and dwell times. The labels on the curves have the format (laser power in megawatts, mirror diameter in meters) followed by the amount of time the laser devotes to destroying each booster. Source: Author

of lasers: 10 lasers combined in this way would produce a beam 100 times brighter than each individual laser. The trick is to arrange for the troughs and crests of the light waves from all the lasers to coincide. This theoretical prospect is unlikely to be realized with HF lasers, since their light is actually emitted at several wavelengths and with shifting patterns of crests and troughs.

To yield diffraction-limited divergence, the mirror surface must be machined to within a fraction of a wavelength of its ideal design shape over its entire surface. Since the mirror is over a million wavelengths across, avoiding small figure errors is a severe requirement. A number of small mirrors can obviously combine to produce one large optical surface if their positions are all aligned to within a fraction of a wavelength. The mirrors must maintain perfect surface shape in the face of heating from the laser beam, vibration from the chemical reaction powering the laser, and vibrations set up in the mirror as it is slewed. Substantial hardening of mirrors to radiation from nuclear bursts in space and to the x-ray laser (described below) would be a challenging task. The 2.5-meter diameter mirror on NASA's Space Telescope was produced without these constraints.

An extremely optimistic outcome of HF laser technology—near the theoretical limit for converting the energy of the chemical reactants to laser energy—would require more than a kilogram of chemicals on board the satellite for every megajoule radiated. A spot diameter of 2 meters at the target and a lethal fluence of 10 kJ/cm² over this area results in an energy expenditure of 300 MJ per booster. Destroying 1,000 Soviet boosters therefore requires, reckoning very crudely, 300,000 kg of chemicals in position over the Soviet ICBM field, or perhaps 10 million kg on or-

bit worldwide. The space shuttle can carry a payload of about 15,000 kg to the orbits where the satellite battle stations would be deployed. About 670 shuttle loads would therefore be needed for chemicals, with perhaps another half as many for the spacecraft structures, the lasers and mirrors, construction and deployment equipment, and sensors. 1,000 shuttle missions for every 1,000 Soviet boosters (perhaps Midgetmen) deployed in reaction to the U.S. defense is an impractical competition for the United States. Use of HF chemical lasers for BMD therefore requires remarkably cheap heavy-lift space launch capability in the United States.

The remaining components of the chemical laser defense system—sensors, aiming and pointing technology, and communications— are for the most part generic to all directed energy weapons and are discussed in section 4. Section 5 presents countermeasures the Soviets might take to offset or nullify a chemical laser defense.

## 3.2 GROUND-BASED LASERS WITH SPACE-BASED MIRRORS

A slight variant of the previous concept puts the laser on the ground and mirrors in space, reflecting the light back down toward Earth to attack ascending boosters. This scheme avoids placing the laser and its power supply in space, though mirrors, aiming equipment, and sensors remain. The excimer and free-electron lasers considered for this scheme are in fact likely to be rather cumbersome, so ground basing them might be the only practical way to use them for BMD. The lasers would emit at visible or ultraviolet wavelengths about ten times shorter than the near-infrared wavelengths of the HF and DF chemical lasers in the space-based concept. Shorter wavelengths permit use of smaller (though more finely machined) mirrors. The high power available with ground basing suggests at least the possibility of impulse rather than thermal kill of boosters.

The term excimer is a contraction of "excited dimer." A dimer is a molecule consisting of two atoms. The dimers considered for these lasers

contain an atom of noble gas and a halogen atom, making dimers like xenon fluoride (XeF), xenon chloride (XeCl), and krypton fluoride (KrF). The laser light comes from dimers in an excited upper state decaying to a lower state, just like in the HF laser. Excimer lasers tend to emit light in pulses rather than in a continuous wave. The population of upper-state molecules is provided by pumping with electric discharges in a rather complicated process. The population of lower-state molecules remains small because the lower-state dimer is unstable and quickly breaks up into its two constituent atoms. The pumping process for excimer lasers is inefficient, so only a small fraction of the energy put into the laser in the electric discharge emerges as laser light. Powerful excimer lasers would therefore be large and would need to vent large amounts of wasted energy; these characteristics make them unsuitable for space basing. Development of excimer lasers is at an early stage, and no excimer lasers exist with anything remotely approaching the characteristics needed for this boost phase intercept concept.

**Figure 3.7**

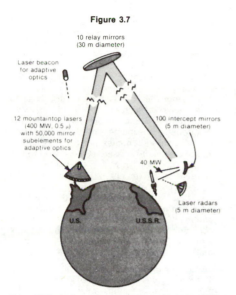

10 relay mirrors
(30 m diameter)

Laser beacon
for adaptive
optics

12 mountaintop lasers
(400 MW, 0.5 μ)
with 50,000 mirror
subelements for
adaptive optics

100 intercept mirrors
(5 m diameter)

40 MW

Laser radars
(5 m diameter)

U.S.          U.S.S.R.

Figure 3.7 Illustrative configuration of ground-based excimer or free-electron laser and space-based mirrors for thermal kill of Soviet ICBM boosters. Source: Author

Power outputs achieved in the laboratory are still several orders of magnitude less than the average power needed for thermal kill, and the energy achieved in a single pulse is much smaller than the single-pulse energies needed for impulse kill.

The working of a free-electron laser (FEL) is more complicated.[4] As the name suggests, the light-emitter (lasant) is free electrons emitted from a particle accelerator. Pumping therefore originates in the electrical source powering the accelerator. The free electrons from the accelerator are directed into a tube called the wiggler that has magnets positioned along its length. The magnets cause the electrons to wiggle back and forth as they transit the tube. As they wiggle, the

[4]See Charles A. Brau, "Free Electron Laser: A Review," *Laser Focus*, March 1981; *Physics Today*, December 1983, p.17.

electrons emit some of their energy as light. The presence of light from one electron causes others to emit in the usual cooperative manner of a laser, and a cascade begins. By adjusting the positions of the magnets and the energy of the electrons, the wavelength of the light can be tuned to any value desired. The only advantage of the FEL over excimer lasers is the high efficiency that can (theoretically) be obtained with the former. It has been suggested that it might even be possible to position FELs in space like HF chemical lasers. FEL operation at visible wavelengths is in its infancy, and the experimental devices used are many millions of times less powerful than those required in this BMD.

The BMD scheme calls for a large ground based excimer or free electron laser, relay mirrors at high altitude to carry the laser beam around the curve of the Earth, and intercept mirrors to focus the beams on individual boosters (fig. 3.7). The characteristics of a nominal system for thermal booster kill are easily ascertained. Suppose first that there are enough intercept mirrors so that the average range from mirror to booster is 4 Mm, and suppose the Soviet boosters are destroyed with 10 kJ/cm² deposited on a spot as small as several centimeters wide. Assume the excimer or free electron laser operates at about 0.5 microns, in the visible band. Then a 5 m intercept mirror will produce a spot 50 cm wide at 4 Mm range. If a half second of the main laser beam is devoted to each booster, then the required 10 kJ/cm² will be accumulated if the power reflected from each intercept mirror is 40 MW.

Only about a tenth of the power emitted by the ground based laser in the United States would be focused on the booster over the U.S.S.R. The remainder would be lost in transit through the atmosphere and in reflection from the two mirrors. Thus a 400 MW laser is required.

Passage through the atmosphere poses a number of problems for the primary laser beam. The most important source of interference is turbulence in the air, causing different parts of the laser beam to pass through different optical environments when exiting the atmosphere. Each part of the beam suffers a slightly different disruption, and the beam that emerges does not have the orderly arrangement of crests and troughs needed for diffraction-limited focusing from the intercept mirror. Without compensation for atmospheric turbulence, the ground-based laser scheme is completely impractical. Fortunately, the pattern of turbulence within the laser beam, though constantly changing, remains the same for periods of a few milliseconds. Since it takes only 0.1 millisecond for light to make a round trip through the atmosphere, the effect of turbulence on the laser beam can be compensated for with the following technique, called adaptive optics: A low-power laser beacon is positioned near the relay mirror. A sensor on the ground observes the distortion of the beacon beam as it passes through the atmosphere. The beam from the ground-based laser is then predistorted in just such a way that its passage through the same column of air transited by the beacon beam re-forms it into an undistorted beam.

Figure 3.7 shows the many components required by the ground-based laser concept. Since each Soviet booster requires 0.5 sec of beam at some time during its 200 sec. boost phase, four beams would be needed to handle 1,400 Soviet boosters launched simultaneously (assuming no retargeting delays). The lasers should be deployed on mountain tops to make atmospheric effects manageable, and enough should be deployed that at least four sites are always clear of cloud cover. The mirror on the ground would need to be tens of meters across and divided into tens of thousands of individually adjustable segments for predistortion of the wavefront. Each relay mirror would need to be accompanied by a beacon. Four large interception mirrors would be needed within 4 Mm of each Soviet ICBM flyout corridor, giving a worldwide constellation of a hundred or so.

The small laser wavelength means that all mirrors must be more finely machined than the mirrors for the chemical laser and can tolerate smaller vibrations and stresses due to heating from the laser beam. The small wavelength also results in a spot 10 times smaller at the target than the spot from a chemical laser beam at the same range. This small spot requires pointing accuracy ten times finer. Perhaps most important of all, the plume from the booster motor is too large to serve as target for such a narrow beam. Some way of seeing the actual missile body against the background of the plume is needed for the short-wavelength laser schemes (and for some configurations of chemical lasers). One answer to this problem, described in section 4 below, is to position near each intercept mirror a low-power laser and a telescope (a laser radar or ladar): the laser illuminates the booster and the telescope observes the reflected laser light, directing the pointing of the intercept mirror. The ladar telescope must have a mirror as large as the intercept mirror, since it must be able to "see" a spot as small as that made by the beam.

A single immense laser pulse that deposits 10 kJ/cm² in a very short time—millionths of a second rather than a second— might cause impulse kill rather than thermal kill. In impulse kill, the laser pulse vaporizes a small layer of the booster skin and surrounding air. The superheated gases then expand explosively, sending an impulsive shockwave into the booster. A strong enough shockwave might cause the booster skin to tear. The advantage of this kill mechanism is that it would be very difficult to protect boosters from it. The disadvantages are that impulse kill requires prodigious laser pulses and mirrors that can withstand them, and that the mechanism is poorly understood and depends on myriad factors like the altitude of the booster at the moment it is attacked.

## 3.3 NUCLEAR BOMB-PUMPED X-RAY LASERS: ORBITAL AND POP-UP SYSTEMS

The U.S. Government has revealed efforts at its weapon laboratories to use the energy of a nuclear weapon to power a directed beam of x-rays. Such devices are said to constitute a "third generation" of nuclear weapons, the first two generations being the atomic (fission) and hydrogen (fusion) bombs. Each succeeding generation represented a thousandfold increase in destructive energy, from a ton of high explosive to a kiloton fission weapon to a megaton fusion weapon. The third generation weapon uses the same amount of energy as the fusion weapon, but directs much of that energy toward the target rather than allowing it to escape in all directions. At the target, therefore, the energy received is much greater than the energy that would be received from a hydrogen bomb at the same range.

X rays lie just beyond ultraviolet light on the electromagnetic spectrum and have wavelengths about a thousand times smaller than visible light (see fig. 3.1). Compared to the infrared, visible, and ultraviolet lasers in the previous sections, the x-ray laser produces much more energy from its bomb pump, but the energy is spread out over a larger cone. The lethal ranges for boosters turn out to be roughly comparable for all these types of directed-energy device. Obviously the x-ray laser delivers all its energy in one pulse, so there is no question of dwell time on the target. Very short-wavelength x-rays penetrate some distance into matter (witness dental and medical x-rays), but the longer-wavelength x-rays produced by a laser device do not penetrate very far into matter or into the atmosphere.

Orbiting and ground-based "pop-up" systems have been proposed as ways to make use of the x-ray laser for boost phase BMD. Both of these schemes have attractive features but also serious drawbacks. It could well be that the x-ray laser device, if a powerful one can eventually be built, will be more useful in other strategic roles than boost-phase BMD.

## X-Ray Lasers

Little has been revealed about the characteristics of the bomb-pumped x-ray laser being studied by the United States (the so-called Excalibur device), but some general information can be deduced from the laws of physics and, to a lesser extent, from the scientific literature here and in the Soviet Union.[5]

[5]F. V. Bunkin, V. I. Derzhiev, and S. I. Yakovlenko, *Soviet Journal of Quantum Electronics* 11 (8), August 1981, p. 981. R. C. Elton, R. H. Dixon, and J. F. Seely, *Physics of Quantum Electronics 6* (1978), p. 243. Michael A. Duguay, *Ibid.*, p. 557. G. Chapline and L. Wood, *Physics Today*, June 1975, p. 40.

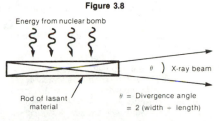

**Figure 3.8**

Energy from nuclear bomb

$\theta$ ) X-ray beam

Rod of lasant material

$\theta$ = Divergence angle
= 2 (width ÷ length)

Figure 3.8 In an x-ray laser, a rod of lasant material is pumped to upper energy states by a nuclear bomb. Those cascades of downward transitions that travel lengthwise build up more energy than sideways-going cascades. As a result, most of the energy emerges from the ends of the rod into a cone with divergence angle equal to twice the rod width divided by its length. Source: Author

The pumping source for the x-ray laser is a nuclear bomb. The radiant heat of the bomb raises electrons to upper energy levels in atoms of lasant material positioned near the detonation (the chemical nature of the lasant material has not been revealed). As the electrons fall back again to lower levels, it can happen that for a moment many atoms are in a given upper level and few in a lower level; this is the necessary condition for lasing from the upper level to the lower level. The wavelength of the emitted x-ray is determined by the energy levels involved. The wavelength of the laser under study in the United States is classified. We will use a round number of 1 nm.

Since x-rays are not back-reflected by any kind of mirror, there is no way to direct the x-rays into a beam with optics like the visible and infrared lasers. Nonetheless, some direction can be given to the laser energy by forming the lasant material into a long rod. Recall that a laser beam builds up when light from one lasant atom stimulates the upper-to-lower-level transition in another atom, which stimulates a third, and so on. The result is a cascade of light heading in same direction as the light from the original atom. The light pulse gets stronger and stronger as it traverses the lasant medium stimulating more and more transitions. In a long rod of lasant material, cascades that get started heading lengthwise down the rod are highly amplified by the time they leave the rod, whereas sideways-going cascades remain small. The result is that most of the laser energy emerges as a beam aligned along the rod axis (fig. 3.8).

The projected capabilities of the x-ray lasers being studied in the U.S. are classified; but it is fairly easy to determine the upper limit to how powerful such a laser could possibly be. Whether R&D will succeed in making such a perfect laser cannot be said. But it will become clear that something very close to the perfect laser is required for boost phase intercept, though a less successful development would still yield a potent antisatellite weapon.

A 1-megaton nuclear weapon releases about 4 billion megajoules of energy. By surrounding the bomb with lasant rods, most of this energy can be harnessed to pump the laser. Since the pumping mechanism for the x-ray laser is rather disorganized and wasteful, like the pumping mechanism for excimer lasers, at most a few percent of the bomb energy can be expected to end up in the laser beam.

The resulting 100 million megajoules or less of laser energy emerges from the rods into cones with relatively large divergence angle. It is easy to see why this divergence angle is much larger than the divergence angle obtained with the mirror-directed lasers treated in the previous section. The divergence angle is determined by the ratio of the width of the rod to its length, as in figure 3.8. A practical length for a rod is no more than about 5 meters. Making the rod thinner decreases the divergence angle, but beyond a certain point no further narrowing of the beam cone is possible. The limit arises from diffraction, just as with the infrared and visible lasers: the divergence angle of light emitted from an aperture (mirror, rod tip, or anywhere else) cannot be less than about 1.2 times the wavelength of the light divided by the diameter of the aperture. A very narrow rod therefore actually aggravates diffraction and produces a wide cone. Making the rod thinner results in no further narrowing of the beam when (1.2) (wavelength)/(rod width) $\cong$ (2) (rod width)/(rod length). For an x-ray wavelength of 1 nm and a rod length of 5 meters, this equation yields an optimum rod width of 0.06 mm and a minimum achievable (diffraction-limited) divergence angle of 20 microradians.

A 1-megaton bomb-pumped x-ray laser can therefore deposit no more than about 100 million megajoules into a cone no narrower than about 20 microradians. The x-ray pulse from detonating such a perfect laser would deposit about 300 kJ/cm² over a spot 200 meters wide at 10 Mm range.

## Interaction of X-rays with Matter

X-rays of 1 nm wavelength do not penetrate very far into matter: all the energy from such a laser would be absorbed in the first fraction of a millimeter of the aluminum skin of a missile. This paper-thin layer would explode, sending a shockwave through the missile. Thus the x-ray laser works by impulse kill.

Another consequence of the opacity of matter to x-rays is that the laser beam would not propagate very far into the atmosphere. The altitude to which the beam would penetrate depends on the precise wavelength, which is classified. For the nominal 1 nm wavelength described above, boosters below about 100 km would be quite safe from attack. If the wavelength were much shorter, the x-rays would penetrate lower, reaching perhaps 60 km altitude or so. In what follows, it will be assumed that boosters are safe from x-ray laser attack below about 80 km.

One last consequence of the physics of x-ray interaction with matter is noteworthy. When an atom of matter absorbs an x-ray, it emits an electron. As x-rays are absorbed, it becomes harder and harder to remove successive electrons. Finally further x-rays cannot remove further electrons, and the matter becomes transparent. This phenomenon, called bleaching, means that a strong x-ray laser beam can force its way through a column of air by bleaching the column, but a weak laser beam is completely absorbed. An x-ray laser in the atmosphere might therefore be able to attack an object in space because the beam is intense enough in the vicinity of the laser to bleach the air, whereas an x-ray laser in space could not attack objects within the atmosphere. This fact bodes ill for defensive space-based x-ray lasers attacked by similar lasers (or even weaker ones) launched from the ground by the offense.

As with visible and infrared lasers, the lethality of an x-ray laser is subject to large uncertainties. The proper order of magnitude for the amount of x-ray energy per square centimeter that needs to be deposited on the side of a booster to damage it can be estimated fairly easily. But the actual hardness of a booster would depend on many design details in a way that is not fully understood at this time. A simple calculation[6] indicates that 20 kJ/cm² is a reasonable number to take for the hardness of a booster. This is about the same as for impulse kill by visible laser. An RV would be harder, and a satellite softer.

### Orbital Defense Concept

The "perfect" x-ray laser whose characteristics were deduced above would be capable of intercepting a booster from geosynchronous orbit 40,000 km above the Earth. One laser would be needed for each Soviet booster. At lower altitudes, the rods surrounding the bomb could be gathered into several bundles and each bundle aimed at a different booster. At these lower altitudes, though, the absentee problem means that roughly one x-ray laser device would still need to be placed in orbit for each Soviet booster. Though the x-ray lasers are small and light compared to a chemical laser, the cost tradeoff involved in launching a new laser every time the Soviets deploy a new ICBM is obviously not a tolerable one for the United States.

The x-ray laser can attack the boosters after they have left the protective atmosphere but before burnout. Simultaneous launch of all Soviet boosters is not a problem for x-ray lasers in

the way it is a problem for chemical lasers that must dwell on each target before passing on to the next. Fast-burn boosters are likewise not a crippling problem for an orbiting x-ray laser system *unless* they burn out before they leave the atmosphere. Other countermeasures, most notably the vulnerability of U.S. orbital x-ray lasers to Soviet x-ray lasers, are treated in section 5.

### Pop-Up Defense Concept

The pop-up concept represents an attempt to avoid the one-laser-per-booster cost exchange and the vulnerability associated with basing the lasers in space (though crucial sensors remain space-based even in the pop-up scheme). The small size and light weight of the bomb-pumped lasers makes it possible to consider basing them on the ground and launching them into space upon warning of Soviet booster launch.

Figure 3.9 shows why basing the pop-up lasers in the United States is not practical. During the 200 seconds or so of burn time of a Soviet MX-

**Figure 3.9**

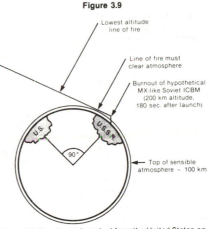

Figure 3.9 X-ray lasers launched from the United States on warning of Soviet ICBM launch would have to climb at least as high as the line of fire shown in the figure within three minutes to intercept an MX-like Soviet ICBM. Such a huge fast-acceleration defensive booster would be many times larger than the Saturn V that took astronauts to the moon.
Source: Author

[6]A mallet or soft hammer blow applies an impulse per unit area of about 10 ktaps (0.5 kg hammer head, 5 m/sec striking velocity, 3 cm radius contact area; 1 tap = 1 dyne-sec/cm²). To apply an impulse of this strength to the entire side of an ICBM booster requires a fluence F, whose order of magnitude can be estimated as follows: The cold mass absorption length (a) for 1nm x-rays is about 0.5 milligrams/cm². If all the energy absorbed by the paper-thin absorbing layer were converted to kinetic energy, the boil-off velocity would be $(F/a)^{1/2}$, meaning an impulse per unit area of order $(Fa)^{1/2}$. 10 ktaps is therefore produced if F = 20 kJ/cm².

In reality, not all the deposited energy couples to the booster in this way. A more careful calculation of this lessened coupling has been performed by Hans Bethe (private communication).

like ICBM, the U.S.-based pop-up lasers would have to climb high enough to see the Soviet boosters over the Earth's horizon and have a line-of-fire unobstructed by the absorbing atmosphere. Climbing so high so fast requires a booster for the x-ray lasers that is many thousands of times larger than the Saturn V rocket that carried U.S. astronauts to the Moon.

If the British Government allowed the U.S. Government to base x-ray lasers in the United Kingdom, the lasers would be separated from Soviet ICBM silos by only 45 degrees of arc rather than 90 degrees as with U.S. basing. Even so, popping up to attack an MX-like Soviet booster would require an enormous fast-burn booster for the x-ray laser and would put it into position to attack the Soviet booster only seconds before burnout. If the Soviets depressed the trajectory or shortened the burn time of the offensive booster very slightly, or if the United States suffered any delay whatsoever in launching the defensive boosters after Soviet launch (instantaneous warning), this hypothetical U.K.-based system would be useless.

A final possibility would be launch of defensive lasers from submarines stationed immediately off Soviet coasts—in the Kara Sea or Sea of Okhotsk, separated from Soviet silos by about 30 degrees of arc—on SLBM-sized fast-burn boosters. With instantaneous warning, a sea-based laser might be able to climb to firing position a few seconds before burnout of a Soviet MX-like ICBM and would enjoy almost an entire minute of visibility to a slow-burning, high-burnout-altitude booster like the SS-18. Because of the short range, each bomb-pumped laser of the perfect design described above could attack many (over 100) boosters using many individual lasing rods. Such efficiency could well be essential, since a submarine cannot launch all its missiles simultaneously and might only be able to fire one defensive missile in the required few seconds. If the MX-like Soviet boosters were flown on slightly depressed trajectories, if warning were not communicated to the submerged submarine promptly, if a human decision to launch defensive missiles were required, or if the Soviets deployed boosters that burned faster than MX, the sub-launched system would be nullified. Last, submarine patrol very near to Soviet shores suggests the possibility of attacking the submarine with shore-based nuclear missiles as soon as its position has been revealed by the first defensive launch. Other countermeasures are discussed in Section 5.

## 3.4 SPACE-BASED PARTICLE BEAMS

Beams of atomic particles would deposit their energy within the first few centimeters of the target rather than at the very surface as with lasers. The effects of irradiation with the particle beam could be rather complex and subtle and would probably depend on design details of the attacking Soviet booster. The result is uncertainty of several factors of ten in the effective hardness of an ICBM booster to beam weapon irradiation.

Only charged particles can be accelerated to form high-energy beams, but a charged beam would bend uncontrollably in the Earth's magnetic field. (There is one theoretical exception to this statement, described below.) For this reason

neutral particle beams, consisting of atomic hydrogen (one electron bound to one proton), deuterium (one electron, one proton, one neutron), tritium (electron, proton, two neutrons) or other neutral atoms are considered. To produce a neutral hydrogen ($H^o$) beam, negative hydrogen atoms ($H-$) with an extra electron are accelerated; the extra electron is removed as the beam emerges from the accelerator.

Two features of neutral particles beams dominate their promise as boost phase intercept weapons (leaving aside entirely the issue of countermeasures). The first is the uncertain lethality of the beam. The second is the fact that the beam cannot propagate stably through even the thinnest atmosphere and must wait for an attacking booster to reach very high altitude.

## Generating Neutral Particle Beams

The accelerator that accelerates the negative hydrogen ion is characterized by its current in amperes, measuring the number of hydrogen ions per second emerging from the accelerator; and by the energy of each accelerated ion in electron volts (eV; 1 eV = 1 watt per ampere). Multiplying the current by the energy gives the power of the beam, so that a l-amp beam of 100 MeV particles carries 100 MW of power. Ground-based high-current accelerators and ground-based high-energy accelerators have been built and are operated daily in laboratories. One of the challenges for neutral particle beams as weapons is that they require *both* high current *and* high energy. Another challenge is to provide multi-megawatt power sources and accelerators in a size and weight suitable for space basing.

Magnets focus and steer the beam as it emerges from the accelerator. The last step is to neutralize the beam by passing it through a thin gas where the extra electron is stripped off in glancing collisions with the gas molecules, forming H° from H-. The divergence angle of the beam is determined by three factors. First, the acceleration process can give the ions a slight transverse motion as well as propelling them forward. Second, the focusing magnets bend low-energy ions more than high-energy ions, so slight differences in energy among the accelerated ions lead to divergence (unless compensated by more complicated bending systems). Third, the glancing collisions that strip off the extra electron give the H atom a sideways motion. This last source of divergence

is unavoidable and, by the Heisenberg uncertainty principle, cannot be controlled or compensated. It sets a lower limit on the divergence angle achievable with this method of producing neutral particle beams. Table 3.2 shows the divergence angle resulting from this third source, assuming perfect control of the first two sources. The divergence cone from a neutral particle beam is therefore about 10 times larger than the beam from the chemical laser of section 3.1 and 10 times smaller than from the x-ray laser of section 3.3.

A 100 MeV, 0.5 amp neutral tritium (T°) beam thus directs 50 MW of power into a cone of divergence angle 2 microradians, producing a spot 10 meters across at 5 Mm range. A target within this spot receives only 65 watts/cm², requiring 1.5 seconds of dwell time to deposit only 100 J/cm².

## Booster Vulnerability to Particle Beams

As soon as the neutral particle beam hits the target, the remaining electron is stripped off, leaving the energetic proton (or deuteron or triton) penetrating deeply into the target. The proton scatters electrons in its path, giving up a small amount of its energy to the electron in each collision. When it has given up all its energy, it stops. For most of its path, it deposits energy uniformly. Thus if a 100 MeV T° beam penetrates 4 cm into the propellant in a missile, it deposits about 25 MeV along each cm. Protons penetrate more deeply than tritons of the same energy, and all particles penetrate more deeply as they are given more energy (table 3.3).

**Table 3.2.—Neutral Particle Beam Divergence Angle**

|  | H° | T° |
|---|---|---|
| 100 MeV | 3.6 microradians | 2.0 |
| 500 MeV | 1.4 | 1.0 |

Divergence angle introduced by stripping electrons from a beam of negative hydrogen or tritium ions to produce a neutral beam. This source of divergence is an unavoidable consequence of the Heisenberg uncertainty principle applied to the sudden stripping of the electron. If the satellite-based accelerator of the negative ions were absolutely perfect, this amount of divergence would remain.

SOURCE: Author.

**Table 3.3.—Penetration Range of Neutral Particle Beams Into Matter (in centimeters)**

|  | H° | | T° | |
|---|---|---|---|---|
|  | 100 MeV | 250 MeV | 100 MeV | 250 MeV |
| Solid propellant or high explosive (density 1.0 gm/cm³) | 9.5 | 46.6 | 4.2 | 20.2 |
| Aluminum .......... | 3.5 | 17.2 | 1.6 | 7.6 |
| Lead .............. | 0.8 | 3.9 | 0.4 | 1.7 |

SOURCE: Author.

## Figure 3.10

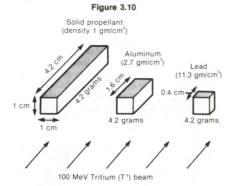

Solid propellant
(density 1 gm/cm³)

Aluminum
(2.7 gm/cm³)

Lead
(11.3 gm/cm³)

4.2 cm

4.2 grams

1.6 cm

4.2 grams

0.4 cm

4.2 grams

1 cm

1 cm

100 MeV Tritium (T°) beam

Figure 3.10 A neutral particle beam penetrates farther into an aluminum target than into a lead target but deposits the same energy per gram. Though the energy per gram needed to melt aluminum is well known, the utility of particle beam BMD concepts rests on the less certain destructive effects at lower levels of irradiation. Source: Author

### Table 3.4.—Effects of Particle Beam Irradiation

| Harmful effect | Energy deposition (Joules per gram) |
|---|---|
| Disruption of electronics .......... | 0.01—1.0 |
| Destruction of electronics ......... | 10 |
| Detonation of propellants, high explosive ................. | 200 |
| Softening of uranium and plutonium ................. | hundreds |
| Melting of aluminum.............. | 1,000 |

Approximate energy deposition (radiation dose) required to produce various harmful effects in components of a missile booster and its payload. Many other effects, such as melting of glue and plastic and rate-dependent effects, might also be important.

SOURCE: Author.

The target electrons that recoil from collisions with beam particles eventualy stop, and their energy appears as heat. The 100 MeV T° beam described above, depositing 100 J/cm² on an aluminum target, penetrates to a depth of 1.6 cm. The 1.6 cubic centimeter volume of aluminum that absorbs this 100 joules of energy weighs about 4 grams. The effect of the beam is therefore to deposit about 25 joules per gram throughout the first 1.6 cm of the target. The penetration depth is inversely proportional to the density of the absorbing material, so the same beam on a lead target would not penetrate as far but would

deposit the same energy *per gram* as it did in aluminum (fig. 3.10).

The destructive effects of penetrating particle beams are therefore expressed in joules/gram deposited within the target rather than in joules/cm² on the surface of the target as with lasers. Table 3.4 shows the energy deposition needed to produce certain harmful effects. Melting the target is straightforward, but for the other effects at lower levels of irradiation the criteria are less clear. Heat effects in solid booster propellants and in the high explosive and special nuclear materials in warheads depend on the design of the target. Effects on electronics, particularly transient disruption of computer circuits when electrons are scattered by a passing proton, are poorly known and doubtless quite complicated and specific to the target. Other components not shown in table 3.4 —plastics, glue, guidance sensors— make for a very complicated analysis. What is more, the particle beam might have to suffer the attenuation of passage through, say, two layers of aluminum and a layer of plastic before reaching a sensitive component.

Uncertainties in the destructive or disruptive effects of small amounts of radiation from a particle beam weapon is the principal obstacle to stating what energy, current, and divergence angle would make this concept a candidate for boost-phase intercept.

Shielding to protect components from a neutral particle beam would necessarily be heavy but could still be an attractive countermeasure. It is discussed in Section 5.

### An Orbiting Neutral Particle Beam System

A critical limitation of neutral particle beams is that they cannot be aimed through even the thinnest atmosphere—air so thin that even the x-ray laser beam could pass through easily. A neutral beam could not attack a Soviet booster until the booster reached at least 160 km altitude (versus about 80 km for the x-ray laser)[7]. Collisions

---

[7]The stripping cross section on oxygen is about 1.5 megabarns. Elastic scattering can also be important for beam loss, since the RMS scattering angle can be larger than the beam divergence. The author is indebted to Dr. George Guillespe of Physical Dynamics, Inc., in La Jolla for results of his Born approximation cross section calculations.

between air molecules and $H^o$ strip the electron from the $H^0$, and gradually all the remaining protons spiral off the beam axis into 200 km wide circles under the action of the earth's magnetic field.

An MX-like Soviet booster could be attacked between 160 km altitude and burnout at 200 km, a period of about 10 seconds. This short attack window means that the neutral beam cannot afford to dwell for long on each booster.

It is impossible to state with confidence the resilience of an ICBM booster to irradiation with a neutral particle beam. But it is likely that faith would have to be placed in degradation of electronics and other subtle effects, rather than in gross structural damage, for the beam weapon to stand a chance as an economical defense system (ignoring the issue of countermeasures entirely).

Consider again a battle station producing a 0.5 ampere beam of 100 MeV tritium ($T^o$) atoms with 2 microradian divergence. This beam carries 400 watts/cm² at 2 megameters range. To do structural damage to the outer few centimeters of a missile's body might take some 2 kJ/cm² (depositing 500 J/gm in 1.6 cm depth of aluminum, for instance), requiring 5 seconds of dwell time at this range. Since the available dwell time is only about 10 seconds, each beam could handle only two boosters. With a constellation size of almost 100 for 2 Mm range (fig. 3.4), this kill criterion results in a preposterous system where the U.S. deploys 50 space-based accelerators for every one Soviet booster deployed in one region of the U.S.S.R.

If the assumed Soviet booster hardness is reduced by 100 times, corresponding perhaps to transient upset of unshielded electronics, each satellite can destroy 200 boosters at 2 Mm range, meaning an overall tradeoff of one U.S. accelerator deployed for each two Soviet boosters deployed. Alternatively, the constellation can be thinned out to an effective range of 5 Mm, where each satellite at this range can destroy only 32 boosters but the constellation size is only about 16—still a one-to-two trade of battle stations for

Soviet boosters. Such a system scarcely seems promising in terms of cost exchange.

Obviously the neutral particle beam would stand no chance of intercepting a fast-burn booster that burns out well within the protective atmosphere. Even an MX-like booster that flew a slightly depressed trajectory would be invulnerable.

## A Theoretical Electron Beam System

Physical theory[8] holds out the prospect of one other type of beam besides the neutral particle beam. Under certain circumstances, an electron beam might be able to propagate through the extremely thin air of near-earth space without bending. In this scheme, a laser beam would first remove electrons from air molecules in a thin channel stretching from the battle station to the target, leaving a tube of free electrons and positive ions. The high-energy, high-current electron beam would then be injected into the channel. The beam electrons would quickly repel the free electrons from the channel, leaving the beam propagating down a positively charged tube. The attractive positive charge would prevent the electrons from bending off the beam path under the influence of the geomagnetic field and would also prevent the mutual repulsion of electrons within the beam from causing the beam to diverge. The result would be straight-line propagation to the target, where their effect would be similar in most respects to the neutral particle beam. This scheme will not work for a proton beam.

The physics of intense beam propagation through thin gases is so complex that experiments will be needed to determine whether this concept is even feasible in principle. If so, the concept would resemble the neutral particle beam, with the added requirement for the channel-boring laser and perhaps the ability to intercept boosters at slightly lower altitudes than the neutral counterpart.

[8]R. B. Miller, *The Physics of Intense Charged Particle Beams*, New York, 1982, ch. 5.

# 3.5 SPACE-BASED KINETIC ENERGY WEAPONS

Kinetic energy is the name given to the energy of a moving projectile. Use of this term makes ordinary weapons using aimed projectiles into "directed kinetic energy" weapons.

The phenomenology of high-velocity collisions between a projectile and a structure like a booster is surprisingly complex, but in general lethality is not an issue for kinetic energy boost phase intercept concepts. Rather, the problem is getting the projectile from its satellite base to the ascending booster in time to make an intercept. Schemes where the projectile is carried by a small rocket launched from the satellite suffer most directly (leaving aside countermeasures) from a combination of the large number and large size of the rockets needed for adequate coverage. In particular, the most conspicuous public example of the kinetic energy approach, the High Frontier Project's Global Ballistic Missile Defense (GBMD) concept,[9] has extremely limited capability for boost phase intercept of current Soviet ICBMs and would have no capability at all against a future generation of MX-like boosters.

## Kinetic Energy Concepts

Rocket attack of ICBM boosters is obviously not as novel as beam attack, but it entails rather more complexity than appears at first blush. The rocket needs radio or other guidance by long-range sensors on its carrier satellite (or other satellites) to direct it to the vicinity of its target, since it is impractical to put a long-range sensor on each rocket. Once in the vicinity of the target booster, the interceptor needs some form of terminal homing sensor and rather sizeable divert rocket motors. Homing on the plume of the ICBM booster is not straightforward, since attacking the plume will obviously not harm the booster: the booster body must be located in relation to the plume. These complications introduce opportunities for offensive countermeasures.

An alternative to rocket propulsion would be to expel the homing vehicles at high velocity from a gun. So-called rail guns use a clever scheme

[9]General Daniel O. Graham, *The Non-Nuclear Defense of Cities: The High Frontier Space-Based Defense Against ICBM Attack* (Cambridge: Abt Books, 1983).

to convert electrical energy to projectile kinetic energy. Since a 10 kilogram projectile ejected with 5 km/sec velocity carries 125 megajoules of energy (the amount of energy expended by a 25 megawatt chemical laser in 5 seconds of dwell on a booster), the power requirements of the gun schemes are imposing. Providing chemical fuel or explosives to power a gun therefore involves the same magnitude of on-orbit weight as the chemical laser.

Doing away with the homing sensor and replacing the guided projectile with many small fragments is not an attractive alternative, since the needed fragments end up weighing far more than the guided projectile.

## The Importance of Projectile Velocity

In the 300 or so seconds from launch to burnout of a slow-burning booster like the SS-18, the defensive rocket or other projectile must fly from its satellite to the path of the booster. Such a booster burns out at about 400 km altitude, so if the projectile wishes to use the entire 300 seconds of boost phase to travel to its quarry, it must make its intercept at 400 km altitude.

Suppose now that the projectile's rocket or gun launcher can give it a maximum velocity of 5 km/sec with respect to the carrier satellite. In the 300 seconds of available travel time to its target, the projectile cannot fly more than (5 km/sec) × (300 sec) = 1.5 Mm from its carrier. Each carrier therefore has an effective range of 1.5 Mm (fig. 3.11a).

Referring to figure 3.4, a constellation of about 240 carrier satellites are needed for continuous coverage of the Soviet Union. Since Soviet ICBMs are spread over much of the country, 10 or so of the carrier satellites might be able to participate in a defensive engagement. The absentee ratio is therefore about 24. The 10 satellites over the U.S.S.R. at the moment of a massive Soviet attack need to be able to handle all 1,400 boosters, meaning each satellite needs to carry 140 projectiles.

An idealized rocket accelerating a 15 kg guided projectile to 5 km/sec velocity would need to weigh about 80 kg (a real rocket with this capa-

Figure 3.11(a)

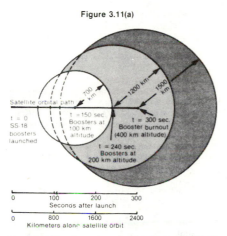

Figure 3.11(b)

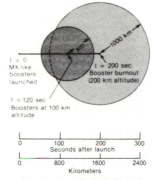

(b) Same as (a), except the target is the faster-burning MX.

Figure 3.11 View from above (looking down on earth) of coverage by a satellite carrying kinetic energy boost-phase intercept vehicles. The satellite is deployed in a 400 km orbit. At time t=0, offensive boosters are launched. The satellite can make intercepts by shooting downward or wait until the boosters rise to their burnout altitude and fire nearer to the horizontal. The longer after launch the intercept is made, the farther the rocket intercept vehicles can travel from the satellite to make the intercept. Smaller circles thus correspond to downward firing, larger circles to horizontal firing. The satellite moves from left to right in accordance with its 8 km/sec orbital velocity. The area enclosed by all the circles taken together gives the total coverage of the satellite and determines how many satellites are needed in the worldwide constellation for continuous coverage of opposing ICBM fields. All dimensions are to same scale.

(a) The satellite-based kinetic energy interceptors are capable of 5 km/sec velocity relative to the satellite. The attack is on a slow-burning Soviet SS-18.

bility would weigh more like 200 kg). Each carrier satellite must therefore weigh (140) × (80) = 11,000 kg. Less than twice this weight can be carried by the space shuttle into the appropriate orbits, so establishing the total 240 satellite constellation requires over 120 shuttle launches in this highly idealized model with idealized rockets and weightless carrier satellites. (A more careful estimation of interceptor design would more than double this load.)

From this baseline, we can consider five excursions:

1. Suppose the velocity capability of the interceptor is doubled to 10 km/sec, doubling the effective flyout range to 3 Mm. At this range, 48 satellites complete the constellation, with perhaps as many as eight of them in position to participate in the engagement. Each of the eight satellites must handle 175 Soviet boosters.

   Doubling the velocity capability more than doubles the weight of the rocket required. The reason is simple: to increase the velocity requires more propellant, and the extra propellant must itself be accelerated, requiring yet more propellant. The rocket weight thus grows exponentially with velocity capability. The idealized 10 km/sec rocket weighs 420 kg. Each satellite carrying 175 rockets then weighs 75,000 kg and requires some five shuttle launches to orbit. The result is that over 200 shuttle launches are required to orbit the entire (idealized) defense system. Increasing the velocity capability is therefore no escape from large on-orbit weights.

2. The current Soviet ICBM force consists largely of slow-burning liquid-fueled boosters distributed widely over the Soviet Union. Consider the consequences for the U.S. kinetic energy defense system if the Soviets de-

**Figure 3.11(c)**

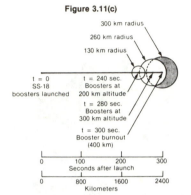

(c) The High Frontier Global Ballistic Missile Defense (GBMD) concept, with intercept vehicles capable of only 1 km/sec velocity relative to the satellite. Intercept of SS-18. In the actual High Frontier proposal, the satellites are in 600 km orbits, giving them even less coverage than shown here.

**Figure 3.11(d)**

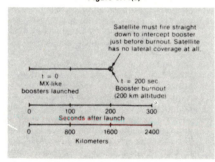

(d) The High Frontier concept has no capability whatsoever for boost-phase intercept of MX.

ploy 100 faster-burning MX-like boosters in one region of the country, so that defensive interceptors have less time to fly to their targets and only one satellite overhead participates in the engagement. MX burns out 200 seconds after launch, so each satellite has an effective range of 1 Mm, requiring a constellation of 400 satellites (fig. 3.11b).

Each satellite must carry the 100 80-kg rockets needed to handle the attack. An exchange ratio of four 8,000 kg U.S. defensive satellites for every Soviet offensive booster deployed is surely an economic advantage for the U.S.S.R.

3. Soviet deployment of 1,000 Midgetman-like boosters would require a compensating deployment of 400 U.S. satellites, each weighing at least 80,000 kg. A system that forces the United States to such a response is clearly absurd.

4. Soviet fast-burn boosters would be totally immune to the kinetic energy defense system. An interceptor on a satellite in 400 km orbit (lower orbits shorten satellite lifetimes because of atmospheric drag) could not even descend straight down to the fast-burn booster's 100 km burnout altitude in the required 50 seconds, much less have any lateral radius of action.

5. Intercepting SS-18 or MX post-boost vehicles is clearly easier, from the point of view of flyout velocities, than boost-phase intercept. Satellites at 700 km or so altitude would have 500 seconds to fly out to meet the bus when it ascended to their altitude, giving a 2.5 Mm lethal radius.

In conclusion, a rocket-propelled kinetic energy system acting against today's Soviet ICBM arsenal (with no Soviet countermeasures) would require many heavy satellites and would be a dubious investment for the U.S. Soviet deployment of MX-like or Midgetman-like boosters would nullify the United States defense or force the U.S. to large investments in new satellites.

### Analysis of the High Frontier Concept

The High Frontier Program[10] proposes a Global Ballistic Missile Defense (GBMD) using rocket propelled interceptors for boost phase intercept. This concept claims to have some utility, at least against the present Soviet ICBM arsenal.

The concept consists of 432 satellites (24 planes of 18 satellites in circular orbits inclined 65 degrees) at an altitude of 600 km. A velocity

[10] Ibid.

capability of I km/sec relative to the satellite ("truck") is attributed to the interceptor. The interceptors are apparently command guided to the vicinity of the target. The homing sensor is not specified, but short wave infrared homing on the hot rocket plume is implied.

Consider this concept defending against the SS-18 in its boost phase. Since the SS-18 burns out at 400 km altitude 300 seconds after launch, each GBMD satellite has a 0.3 Mm radius of action. Since the satellites are deployed at 600 km altitude, the interceptor must descend 200 km to make an intercept just before burnout, resulting in a lateral radius of action of 0.22 Mm (compare fig. 3.11c, where the satellites are assumed deployed at 400 km altitude). With a range this small, thousands of satellites would be needed worldwide for continuous coverage of Soviet ICBM fields. The High Frontier concept with only 432 satellites would therefore have meager coverage of Soviet ICBM fields.

The GBMD concept would have no capability whatsoever against an MX-like booster. Such a booster would burn out before the interceptor could reach it, even if the interceptor were fired straight down (fig. 3.11d).

It is possible that the High Frontier concept is designed for post-boost intercept rather than boost phase intercept. Its coverage for post-boost intercept, though greater than for boost-phase intercept, would still be only partial. The only example given in the description[11] of the system is of *boost phase* intercept of an SS-18, however. In this example the interceptor is launched 53 seconds *before* launch of its target booster, though no explanation is given of how the U.S. defense knows *in advance* the precise moment at which the Soviets would launch a given booster. This early launch allows the interceptor to reach its target seconds before burnout. Plume homing, a technique inappropriate for bus intercept, is also implied for the High Frontier concept. Post-boost intercept permits some RVs to be deployed on trajectories carrying them to the United States before intercept; and the entire bus, with its warheads, would continue on to the U.S. after the interceptor collision, with uncertain consequences.

It would therefore appear that the technical characteristics of the High Frontier scheme result in a defensive system of extremely limited capability for boost phase intercept of present Soviet ICBMs and no capability against future MX-like Soviet boosters, even with no Soviet effort to overcome the defense.

---

[11] Ibid., p. 103.

## 3.6 MICROWAVE GENERATORS

Microwaves are short-wavelength radio waves used in radar, satellite communications, and terrestrial communications relays. A number of ideas have been conceived for generating microwaves in space and directing them towards ascending ICBM boosters. The principal technical problem with this type of BMD, generator technology aside, is the uncertain effect the microwaves would have on their target.

The microwaves would propagate through the atmosphere unattenuated at all but the highest power levels. The weapon divergence angle would be very large, producing a spot many km wide at a few hundred km range. From these considerations the following concept emerges: As Soviet ICBMs lift off from their silos, a few microwave generators in space bathe the silo fields with microwaves.

At high power levels, as in a microwave oven, microwaves cause heating in many materials. But in the BMD scheme, the divergence cone is so large that even a prodigious amount of energy emitted from the generator would lead to very small energy deposition per square centimeter on the target (millions of times less than lasers). The microwave pulse received at the booster would resemble the high frequency component of the electromagnetic pulse (EMP) from a high-altitude nuclear detonation. However, even weak microwaves can upset sensitive circuitry if they can reach it.

A metal skin on the booster would stop the microwave pulse altogether from reaching internal electronics. The microwave defense must therefore hope that some aperture or conduit is available into the booster, whether by design (as in an antenna), inadvertence, or poor mainte-

nance. If so, and if the electronic circuitry is not or cannot be made resistant to disruption or burn-out, the part of the booster's performance dependent on those electronics (perhaps accurate guidance) would be affected.

Because of the very uncertain lethality of microwaves, deployment of space-based generators (if they can ever be built) would be a harassing tactic rather than a confident-kill ballistic missile defense.

## 3.7 OTHER CONCEPTS

Other directed energy concepts suitable in theory for ballistic missile defense have been broached from time to time. Some of them are listed below. It is quite possible that in a few years time a revisit to this subject will find a new panoply of concepts enjoying the front rank of discussion.

1. **Short-wavelength chemical lasers** would combine the simplicity and efficiency of the HF chemical laser with the small mirrors of the short-wavelength excimer and free-electron lasers. Though some ideas have been advanced along these lines, no laser exists which can be said to be a candidate to fulfill this theoretical promise.
2. **Explosive-pumped lasers and particle beams** are said to be under study in the Soviet Union.[12] Such devices might possibly be quite compact, each bomb generating a

single huge pulse for impulse kill of a booster. All these schemes are at a very early conceptual stage.

3. **Antimatter beams** would penetrate into a target just like ordinary particle beams, except that when the antiparticle reached the end of its range it would annihilate a particle in the target, freeing a large extra amount of harmful energy. Acceleration of antimatter beams is accomplished exactly as with particle beams, and laboratory beams of antimatter have been used routinely in pure research. One important difference is that antimatter is not freely available in the universe as is matter; the antimatter for the accelerator would have to be produced by the defense system, a formidable and complex undertaking. It is not clear that the extra energy released in the target by an antimatter beam would justify the trouble of producing the beam.

---
[12]*Aviation Week and Space Technology,* July 28, 1980, p. 47.

# OTHER ESSENTIAL ELEMENTS OF A BOOST-PHASE INTERCEPT SYSTEM

# OTHER ESSENTIAL ELEMENTS OF A BOOST-PHASE INTERCEPT SYSTEM

The previous section treated only the defensive weapon itself, the so-called "kill mechanism." But if beam weapons ever evolve to the point where deployment is a serious possibility, other elements of the overall defensive system will emerge as equally important determinants of cost and level of protection. After all, the interceptor missile in traditional BMDs has not been the central focus of attention or technical debate since the 1950's, when it became clear that a "bullet could hit a bullet." Discussion of BMD at that point passed to the difficult issues of radar performance, data processing capability, and vulnerable basing of defensive components—issues that had nothing to do with the kill mechanism. In a similar manner, the other essential elements of a boost-phase intercept system will figure more prominently in discussion of boost-phase BMD if and when the kill mechanisms—lasers, mirrors, accelerators—are in hand. These other essential elements introduce their own technological problems and opportunities for offensive countermeasures. If traditional BMDs are any guide, provision of a kill mechanism will be just the beginning of making an efficient, robust defensive system.

## 4.1 TARGET SENSING

Locating and tracking an ICBM booster with enough precision to aim a directed-energy weapon is not as straightforward as is sometimes supposed. It is true that booster motors emit hundreds of kilowatts of power at short- and medium-wave infrared (SWIR and MWIR) wavelengths of a few microns. Sensors can detect these plumes at great distances from the earth. Plume sensing is used today for early warning of missile attack to support launch of bombers and airborne command posts and launch under attack of ICBMs.

To be useful for directed-energy BMD, however, the sensor must localize the booster within an area as small as the beam spot. Otherwise the beam would have to sweep wastefully back and forth over the area of uncertainty. Small divergence beams must therefore be accompanied by sensors with small angular resolution.

Diffraction limits the angular resolution of a sensor in the same way it limits the divergence angle of a laser. A large infrared telescope with 5 m diameter mirror observing MWIR booster emission at 4 micron wavelength would have angular resolution no more precise than a microradian. Such a sensor affixed to each battle station in a defensive constellation would localize ascending boosters to within a spot 5 m wide at 5 Mm range. At this range, the (illustrative) systems described in Section 3 have spot sizes: 1.5 m for the HF laser, 0.6 m for the ground based laser, 10 m for the neutral particle beam, and 100 m for the x-ray laser. Even a large infrared sensor on each battle station would therefore be inadequate for directing the laser beams at a point source of MWIR light, marginal for directing the neutral particle beam, and adequate for directing the x-ray laser. The actual situation would be worse still, since the booster is not a point source. The booster plume would be larger than the laser or particle beam spots, and the booster body would need to be located in relation to the plume to avoid wasting beam time attacking the plume.

For directed-energy weapons with small divergence angles, therefore, sensing the conspicuous rocket plume is inadequate. Another kind of sensor must be introduced into the BMD system. For finer angular resolution one looks to shorter wavelengths, in the visible or ultraviolet. At these wavelengths the sensor must provide its own illumination. A so-called laser radar or ladar is the only practical solution. In a ladar, a low-power visible or ultraviolet laser shines on the booster body, and a telescope on board the battle station senses the reflected light.

Besides the annoyance of a new laser and new sensor, the necessary introduction of ladar into the boost-phase system creates opportunities for the defense to spoof and blind the offensive sensor.

Kinetic energy systems do not need precision long-range sensing, since the rocket or guided projectile homes on the target when it comes within short range. The terminal homing might involve deducing the location of the booster body in relation to its MWIR plume, homing on low-power laser light shined from a defensive satel-lite and reflected from the target, or some other method. These homing methods are susceptible to countermeasures.

Though this Background Paper treats only intercept of the booster proper, it is worthwhile pausing to consider tracking of the post boost vehicle or bus. The low thrust levels of the post boost vehicles's rocket motors, their intermittent operation, the possibility of dimming them with propellant additives, and the possibility of building decoys with small rocket motors all suggest that MWIR plume sensing is not practical for post boost intercept. The alternatives are ladar or radar, suggesting again many opportunities for countermeasures.

## 4.2 AIMING AND POINTING

The directed-energy beam must be aimed and stabilized as accurately as it is collimated. If the beam waves around too much, the effective divergence increases, and the beam wastes energy missing the target. The mirrors or other mechanism steering the beam must be stabilized despite vibrations in the battle station caused by the beam's large power source.

In the 15 milliseconds the beam takes to travel from the battle station to a booster 5 Mm away, the booster moves about 50 m. A narrow beam must therefore lead the target. In one second of dwell time, the target moves several km; the beam must remain on the target, sweeping through the sky at the necessary angular rate while still maintaining its aim and jitter control.

## 4.3 INTERCEPT CONFIRMATION

A desirable, though perhaps not essential, function of BMD systems is confirmation that an attempted intercept succeeded. This function is sometimes called "kill assessment." Intercept confirmation would allow the beam to move onto subsequent boosters with more than a statistical estimate that its previous task was accomplished. Structural damage to the booster would presumably be revealed by an erratic course or burn pattern, though it might be difficult to say in advance exactly what the sensor's view of the wounded booster would be. Subtle damage inflicted by a particle beam or microwave generator might not be visible. Damage to a bus would be difficult to assess and interpret if the debris, including RVs (perhaps arranged by the offense to separate from the bus under extreme circumstances), continue on their ballistic course to the continental United States.

Related to intercept confirmation, and ultimately more serious, is the question of determining whether the beam is missing the target (perhaps by slight misalignment of sensor and beam boresights, miscalibration of aiming mechanisms, etc.) and, if so, by how much and in what direction. It might be possible to observe a glowing column of air where a laser beam passes through the atmosphere. Some clever but elaborate schemes have been devised to track a neutral particle beam. Obviously each new complication added to the defensive system potentially creates new opportunities for offensive countermeasures.

## 4.4 COMMAND AND CONTROL

The crucial infrastructure of command and control of a complex system is always the last to take shape, since it integrates the workings of all the separate components. It is easy to ignore the difficulty of accomplishing this last step at this early stage when the other components of a boost phase system are not yet remotely in hand. The command and control system of a boost-phase intercept system would comprise communications links among its far-flung components, data processing to support sensors and battle station operations, and "battle management" software incorporating all the instructions and decisions needed to run the defensive engagement and to coordinate the defense with U.S. offensive forces.

Communications and data processing are two areas of technology where there is the least pessimism—looking two or so decades into the future when boost-phase systems could presumably be deployed—that technology will be able to meet the needs of directed-energy defenses. Compact, lightweight, and rapid data processing hardware is virtually assured, though interesting questions attend on hardening, reliability, and lifetime in space. Software would be expensive and would introduce issues of reliability and security from programmer sabotage. Satellite-to-satellite communication via extremely high frequency radio and laser offers high data rates and virtual immunity to jamming from earth or from space.

Command and control for BMD does introduce two interesting issues to which technology cannot provide an answer. The first is the impossibility of testing the whole defense system from end to end in a realistic wartime setting. Unlike the air defense systems of World War II, which learned through attack after attack to exact kill rates of several percent, the BMD system would have to work near perfectly the very first time it was used. The second issue is the likely need for the defense to activate itself autonomously, since there would be no more than a minute for human decision.

## 4.5 SELF-DEFENSE

Consideration of anti-satellite (ASAT) attack (see Section 5.1), and analogy with traditional BMD systems (where vulnerability of key radars, data processors, and other components is usually the chief limitation on defense performance) suggest that self-defense mechanisms could well end up being a large part of the defense system. These mechanisms could include shields, escort weapons, and countermeasures to ASAT sensors. Unless and until a credible overall approach to satellite survivability is found, one cannot specify the needed hardware.

Ground-based BMD lasers and pop-up x-ray lasers would obviously need to be protected from precursor attack by cruise missiles and other delivery systems.

## 4.6 POWER SOURCES

Chemical lasers, x-ray lasers, and rocket-propelled kinetic-energy interceptors have power sources integral to the weapon, but excimer lasers, free electron lasers, neutral particle beams, and rail guns would need sources of electrical power and the equally important means to convert electricity into a form usable by the weapon ("power conditioning"). Space basing obviously complicates the task. Large commercial power plants on the ground produce about 1,000 MW of power, and directed energy weapons might require hundreds of MW. On the other hand, the power plants on defensive satellites need not work reliably for many years but only once for a short time, and they need not be very highly efficient. The three alternatives for space power are fuel burning, explosives, and nuclear power. Starting up a large power source in seconds from a condition of dormancy poses some interesting design issues.

# COUNTERMEASURES TO BOOST-PHASE INTERCEPT

# COUNTERMEASURES TO BOOST-PHASE INTERCEPT

Countermeasures that limit the effectiveness of traditional ballistic missile defenses—decoys, radar blackout, defense suppression, etc.—are well known. A comparable set of countermeasures, no less daunting for being less familiar, faces the designer of boost-phase defenses.

The need to resort to countermeasures imposes a cost on the offense. This cost is measured in money to build more or specialized offensive hardware, but also in the time needed to do so, in constraints upon the type of attack the offense can incorporate in its nuclear planning, and in the confidence with which it can predict a "successful" outcome of the strike.

Every BMD system actually proposed for deployment would be accompanied, at least ideally, by, first, an analysis of its degradation in the face of an improving Soviet offense and, second, by an analysis of how much it would cost for the United States to improve its defense in such a way as to avoid being overcome.[1]

---

[1]See *Ballistic Missile Defense*, ed. Ashton B. Carter and David N. Schwartz (The Brookings Institution, 1984), ch. 4.

**Figure 5.1**

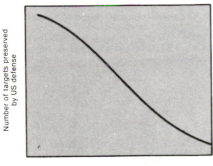

Fig. 5.1. Schematic drawdown curve, showing how the performance of a BMD system degrades as the size and sophistication of the attacking force increase.

**Figure 5.2**

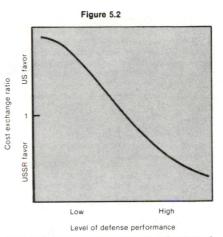

Fig. 5.2. The marginal cost exchange ratio measures the outcome of a race between the Soviet offense to enhance its penetration and a U.S. defense to maintain its level of protection. In general, modest defense goals (e.g., "preserve 40 percent of the targets") are easier to sustain than high goals ("preserve 95 percent of the targets") against improvements in Soviet offensive forces, including deployment of countermeasures.

The first analysis would be expressed in a drawdown curve such as that shown in figure 5.1. The Soviets can overcome the defense and destroy a large number of U.S. targets, but to do so the Soviets must "pay" an "attack price."

The second analysis would be encapsulated in the cost exchange ratio. The marginal cost exchange might be defined as follows: "Assume that in the year 2000 the U.S. defense and Soviet offense have evolved so that each has a certain level of effectiveness. Suppose the Soviets wish to improve their position and the U.S. resolves to maintain the status quo. Which side spends more in the competition?" For example, suppose every time the Soviets add 100 ICBMs to their arsenal, the United States has to add 20 satellites to its defensive constellation to intercept them: Which costs more, 100 ICBMs or 20 satellites?

In general, high levels of defense performance are harder to enforce in the face of offensive improvements than low levels: this important fact is shown schematically in figure 5.2 (see also Section 8.2).

All of the boost-phase intercept schemes discussed in this report are in such an early stage of conceptualization that nothing remotely like the analyses represented by Figures 5.1 and 5.2 can be done for them. Nonetheless, countermeasures are known for every boost-phase system devised, and in many cases simple heuristic estimates of the cost tradeoffs are suggestive.

Technical experts disagree not so much about the facts and calculations underlying these countermeasures as about the interpretation to be given to them. Should an apparently fatal flaw uncovered at this early stage of study of a defensive concept be decisive, or should work (and the inevitable expectations that accompany it) continue on the chance that a new idea will turn up to rescue the concept? Would the Soviets really resort to a subtle tactic or exotic piece of hardware as a confident basis for their nuclear policy? Some analysts see BMD as a way of "forcing" the Soviets to take a certain direction in their pursuit of the arms race, e.g., away from large, slow-burning MIRVd boosters to single-warhead Midgetman-like boosters. In this view, defeat of the BMD is purchased at the price of a theoretically more stable and desirable Soviet offensive posture. All these questions of judgment loom large in making a final assessment of a given countermeasure.

## 5.1 ANTI-SATELLITE (ASAT) ATTACK, INCLUDING DIRECTED-ENERGY OFFENSE

All boost-phase intercept BMD concepts have crucial components based in space. Even a pop-up-defense would need warning and very probably target acquisition sensors on satellites over the Soviet Union. Ground-based laser defenses would have mirrors and sensors—their most fragile components—in space. Vulnerability of these satellites is a cardinal concern because their orbits are completely predictable (they are in effect fixed targets), they are impractical to harden, conceal, or proliferate to any significant degree, and because successful development of effective directed-energy BMD weapons virtually presupposes development of potent anti-satellite (ASAT) weapons. ASAT is the clear boost-phase analogue of familiar defense suppression tactics against traditional BMDs, where attack is first made upon the defensive deployment (especially fixed radars) and then upon the defended targets.

The interplay of ASAT techniques—missiles (nuclear or conventional), space mines, directed energy—and satellite defense (DSAT) techniques is a complex one. It is difficult to generalize, but in the specific case of large battle stations in low-earth orbit it would seem that the advantage is very likely to lie with ASAT, not DSAT. For one thing, the offense need not destroy a large number of defensive satellites, but only "cut a hole" in the defensive constellation. Second, the traditional military refuges all offer complications: concealment from radar, optical, infrared, and electronic detection, while possibly successful for small payloads in supersynchronous orbits, is impractical for large, complex spacecraft at most a few thousand km from the earth's surface; decoy satellites must generate heat, stationkeep, and give status reports, and they are in any event only useful if the ASAT designer is somehow restrained (perhaps by cost) from shooting at all suspicious objects; hardening imposes weight penalties, and massive shields could interfere with the constant surveillance and instant response required of the defense; proliferation is useless for expensive satellites facing inexpensive ASAT methods. As a consequence, discussions of DSAT for BMD battle stations usually emphasize large keep-out zones around the satellites and active self-defense. A third reason why ASAT is likely to prevail over DSAT is that possession by the offense of the same type of directed energy satellites used by the BMD probably assures successful first strike. Fourth, the Soviets would pick the time and sequence of their attack, and it would occur over Soviet territory.

Two rather novel ASAT threats are worthy of note. The first is the x-ray laser itself. The x-ray laser, if it could be developed, would constitute a powerful space mine. Because of its long range, it could lurk thousands of km from its quarry. The Soviets might also launch x-ray lasers a few seconds before launch of their main attack. Recall that the well-known phenomenon of bleaching (see Section 3.3) would probably allow such x-ray lasers to shoot out of the atmosphere at a U.S. x-ray laser defense, but the U.S. x-ray lasers could not shoot down into the atmosphere at the ascending lasers.

A second ASAT tactic, discussed for many years, imagines the Soviet Union exploding nuclear weapons at high altitudes in peacetime with the intent of shortening the orbital lifetimes of the U.S. defensive satellites. The nuclear bursts inject further radiation into the van Allen belts that circle the earth's equator from about 1,500 to 10,000 km altitude. Satellites (more likely carrying sensors than weapons at these altitudes) passing through the belts accumulate a radiation dose that gradually degrades electronics, sensors, and optical surfaces. This possibility, if taken seriously, would require defensive satellites designed to withstand rather substantial accumulated radiation doses.

A detailed treatment of the ASAT problem is beyond the scope of this Background Paper. The following "parable" illustrates some of the problems encountered in trying to ensure the survivability of a defensive constellation, taking the 20 MW HF lasers of Section 3.1 as an example.

The United States deploys the HF lasers in this hypothetical system in low orbits at 1,000 km altitude. Higher altitude would place them too far from their targets. This is unfortunate: higher altitude (say, between 2,000 km and semisynchronous orbit at 20,000 km) would move the satellites further from ground-based ASAT weapons and put them into lesser-used orbits where staking out a sanctuary would involve less interference with foreign spacecraft.

Suppose the battle station designers have succeeded in the considerable task of making the satellites resilient to multi-megaton nuclear space mines (bombs, not x-ray lasers) as little as 100 km away. To keep all Soviet spacecraft (i.e., all potential mines) at least 100 km away, the United States claims for itself the orbital band between 900 and 1,100 km altitude. Perhaps the Soviets are awarded some other orbital zone for their own military purposes. The United States establishes the following rules in its zone: 1) No foreign spacecraft may transit the zone without prearrangement; 2) All transiting vehicles must remain at least 100 km from all U.S. battle stations, passing through a "hole" in the constellation; 3) Foreign spacecraft failing to obey these rules may be destroyed by the U.S. lasers.

Consider first a Soviet kinetic energy ASAT deployed at 1,100 km altitude, just outside the U.S. keepout zone. Suppose the rocket interceptors on the Soviet satellites have the same propulsive capacity—one km/sec—as the proposed High Frontier Global BMD system. The Soviet ASATs are then just 100 seconds away from the U.S. lasers. The U.S. lasers must therefore be very vigilant to avoid surprise attack. Fortunately, at 100 km range the 20 MW laser with 10 m mirror would burn up even a heavily hardened ASAT rocket in short order. Since starting up the main laser for self-defense might be awkward, wasteful of fuel, or time consuming, each U.S. battle station might be escorted by a satellite carrying a smaller laser or rockets for self-defense.

A constellation of Soviet 20 MW, 10 m HF lasers (the same technology as the U.S. lasers) at 1,100 km is another matter. These lasers could attack the U.S. lasers seconds before launch of a Soviet ICBM attack. The United States would have to keep these Soviet spacecraft *thousands* of km away from the U.S. constellation. That is, the United States would have to dominate near-earth space. Suppose the United States does so.

Now the Soviets build a fleet of pop-up x-ray lasers. These lasers climb to 100 km or so altitude, where information radioed to them from the ground allows them to point their rods at the U.S. lasers and detonate. The Soviets have had poor success at building an x-ray laser; theirs are 100 times less bright than the ideal x-ray laser described in Section 3.3. Nonetheless, by pointing all its lasing rods at the same target, a Soviet x-ray laser can destroy a U.S. laser battle station at 10 Mm range. The U.S. chemical lasers attack the Soviet x-ray lasers as they ascend, but at this range long dwell times are required to destroy the Soviet lasers. By launching enough x-ray lasers simultaneously, the Soviets succeed in getting some to 100 km altitude, where they can

shoot out through the thin atmosphere, before the U.S. lasers can destroy them. In this way, the Soviets "punch a hole" in the U.S. defensive constellation. (At a minimum, the Soviet ASAT attack consumes precious laser fuel aboard the U.S. battle stations.)

Just to make sure, the Soviets also deploy some powerful ground-based excimer or free electron lasers to destroy the U.S. battle stations as they orbit helplessly through space.

## 5.2 FAST-BURN BOOSTERS

Shortening the boost time and lowering the burnout altitude is easily accomplished at little sacrifice in useable ICBM payload (see Section 2). Shorter boost time increases the number of lasers needed for space-based laser or ground-based laser systems to handle simultaneously launched boosters. Short burn time makes rocket-propelled kinetic energy systems impractical, since the radius of action of each satellite becomes too small. Short burn time, together

with low burnout altitude, would severely compromise the effectiveness of x-ray lasers popped up even from subs near Soviet shores. Low burnout altitude nullifies the neutral particle beam, which cannot penetrate very far into the atmosphere.

Fast-burn boosters would therefore be a potent, even decisive, countermeasure against almost all concepts for boost-phase intercept.

## 5.3 COUNTER C[3]I TACTICS

Countermeasures to the crucial functions of target sensing and command and control are a relatively unexplored, but probably key, problem area for directed energy BMD. In the case of terminal and midcourse defenses, the issues of decoy discrimination, confusion caused by chaff and aerosols, radar blackout and infrared redout, radar jamming, and traffic handling have always been and remain central limitations. It is likely that analogues will be found for boost-phase systems. Devising countermeasures requires a degree of specificity about the nature of the defense system which cannot be provided in the present conceptual stage. There follow a few examples of C[3]I countermeasures, by no means an exhaustive list.

A first point to note is that sensors are likely to be the most vulnerable part of a defensive sat-

ellite. A laser shined into an optical sensor can daze or injure the focal plane elements, though viewing in frequency bands absorbed by the atmosphere offers protection from ground-based lasers. Mirrors would be very susceptible to damage from a Soviet x-ray laser. A Soviet neutral particle beam could disrupt electronic circuits on U.S. satellites. Radiation pumped into the van Allen belts by nuclear bursts would affect sensors and electronics.

A single nuclear burst causes the upper atmosphere to glow brightly over areas 100 km in radius for over a minute. Calculated radiances[2] are large enough to cause background problems for MWIR tracking sensors.

---

[2]S. D. Drell and M. A. Ruderman, *Infrared Physics*, Vol. 1, p. 189 (1962).

Some directed-energy weapons produce spots only meters wide at the target, requiring target sensing to commensurate precision via laser radar (see Section 4.1). Laser radars sense laser light reflected from the target. A small corner reflector affixed to the target would produce a bright glint of reflected light, as would other corner reflectors launched on sounding rockets, ejected from the target, or attached to the target by extendable booms. These proliferated corner reflectors might force the beam weapon to attack them all.

The homing sensor of a kinetic energy interceptor could be susceptible to spoofing, depending on its type.

Jamming satellite-to-satellite communications crosslinks is probably *not* an effective offensive tactic, since the links would have narrow beamwidths, requiring the jammer to locate itself directly between the two satellites; and wide bandwidths, requiring high jammer power.

## 5.4 SHIELDING

A degree of shielding from lethal effects is practical for all but the kinetic energy weapons but involves in each case different methods suited to the different physical principles at work. At the same time, large uncertainties plague all lethality estimates, and further testing and study will be needed before firm answers can be given for any of the systems. For thermal kill with a laser, a solid booster designed with some attention to a laser threat can probably easily be made to withstand 10 kJ/cm². Application of a gram or so of heatshield material on each square centimeter of booster skin can probably triple this hardness, and spinning the booster enhances hardness by another factor of three. Heatshield material is ablative, meaning that when heated it burns off, carrying away the heat in the combustion gases rather than conducting it through to the missile skin underneath. A factor of nine increase in hardness requires the defensive laser to dwell on the booster nine times as long or to approach within a third of the range. Though hardening a new booster from scratch is clearly easiest, there is no serious impediment to retrofitting ablative coatings on existing boosters. Applying a gram per square centimeter of ablative material to the entire body of the MX missile would require removing several RVs from the payload, since the coating would weigh well over 1,000 kg.

An interesting possibility, requiring further study, would involve injecting into the atmosphere or producing from atmospheric gases, either throughout the ICBM flyout corridors or in the vicinity of individual boosters, smoke or laser absorbing molecules. Likewise, dust clouds raised by groundburst weapons (delivered by cruise missiles or by ICBMs that "leak" through the defense) might cause serious propagation problems for the ground-based laser scheme.

Hardening to an x-ray laser involves quite different physical principles. Recall that the x-ray energy is deposited in a paper-thin layer of the booster skin. The superheated layer explodes, applying an impulsive shock to the booster. Obviously a paper-thin shield between the booster and the laser will stop x-rays from reaching the booster wall. But the problem then becomes the debris from the exploding shield. One can easily show by calculation that the debris applies virtually the same impulse to the wall of the booster as would result from direct impinging of the x-rays! A number of schemes can be devised to divert the debris from striking the booster, but these require more study to implement in practice. One factor acting in favor of the shield designer is that the booster is not vulnerable to x-ray attack until it leaves the atmosphere. The lightweight shields therefore do not have to be designed to suffer large drag forces.

The neutral particle beam presents a third distinct type of hardening problem. The energetic beam particles penetrate into the target, and several centimeters of lead would be required to stop them. Since the beam cannot penetrate very far into the atmosphere, only the upper booster stages need be hardened. But if the third stage, say, of the MX were covered with a few grams per square centimeters of lead, the shielding alone would weigh as much as several RVs. On the other hand, if the neutral particle beam is only designed to disrupt or damage sensitive electronics, but is not powerful enough to do damage to other parts of the missile, only the sensitive components need be shielded. The weight penalty then becomes small.

It is possible that the offense can extend the protection of the upper atmosphere against the neutral particle beam by exploding a few nuclear weapons at moderate altitudes before the beam can reach them. The detonations heat the air, which rises, effectively elevating the altitude at which the neutral beam is stripped of its remaining electron and bent in the geomagnetic field. This phenomenon is called atmospheric heave. It is as yet unresolved whether atmospheric heave will loft enough air to make a difference to the engagement altitude of the x-ray laser.

## 5.5 DECOYS

There is no way for a decoy booster to mimic closely the hot exhaust plume of an ICBM booster except by burning a similar rocket stage. One can add chemicals to the propellants to brighten a small booster's plume and dim the ICBM's, but as a first approximation a faithful decoy must be another booster.

Decoy tactics are therefore not as attractive for boost-phase intercept systems that use plume sensing as they are for midcourse and reentry defenses, where large numbers of cheap, lightweight decoys can be carried with negligible offload of RVs. Still, the usefulness of a decoy depends not on how expensive the decoy is, but on how the cost of the decoy compares to the cost of the defense that intercepts it. Booster decoys would not be nearly as expensive as true ICBMs, since they carry no warheads or precision guidance system, they need not be highly reliable, and they might not need to be based in underground silos but can be deployed above ground next to the ICBM silos.

Some of the boost phase intercept systems must grow in the number of their deployed battle stations in direct proportion to the number of Soviet boosters. Deploying one decoy (with a dummy payload) next to each of the 1,400 Soviet ICBM silos might cause the United States to have to double the number of battle stations overhead (and thus worldwide, multiplying by the absentee ratio) to handle the extra traffic. If the defensive battle stations were at all expensive, this would be an unpleasant prospect for the United States.

Many directed energy schemes would not rely on plume sensing alone (see Section 4.1). Decoy tactics against laser radars (including corner reflectors; see Section 5.3) might be much easier for the offense to implement than mimicking the booster plume.

## 5.6 SALVO RATE

The worst-case attack for all the boost-phase intercept schemes is massive, simultaneous launch of all Soviet ICBMs. The defensive satellites over the Soviet silo fields at the moment of launch then have to handle the entire attack.

A more leisurely salvo rate would allow laser and particle beam defenses that have to dwell on their targets more time to handle more targets. Slow attack also allows pop-up defenses to climb to intercept position. An attack drawn out 10 minutes or longer allows fresh defensive satellites to move along their orbits into position overhead, replacing depleted satellites. The orbital period of satellites in low earth orbit is about 90 minutes. If there are 8 to 10 satellites in each ring of the defensive constellation, satellites replace one another every 10 minutes or so.

An exception to this simultaneous-launch worst-case analysis is the x-ray laser, which delivers all its energy in an instant.

There would seem to be few military penalties for the Soviet Union to adopting plans to launch all their ICBMs in large attack within a few minutes. Indeed, if one of their objectives is to destroy U.S. ICBMs in their silos, rapid attack would be the best Soviet choice. Some, though not all, of the successful attacks that could be mounted on Closely Spaced Basing (Densepack) for MX involve very slow or intermittent salvo rates, however.

More importantly, in many circumstances the Soviets might wish to launch only a fraction of their ICBM force. A U.S. defense deployment too small to intercept all boosters in a massive attack would still be able to handle a small attack. A light defense might therefore establish a "threshold" of attack intensity below which Soviet boosters would face intercept. This prospect is discussed further in Section 9.

## 5.7 OFFENSIVE BUILDUP

The most straightforward way for the offense to compete with the defense is to grow in size. If for every new ICBM added to the Soviet arsenal, the U.S. defensive satellite constellation must be augmented at comparable or greater cost, the Soviets could challenge the United States to a spending race to their net advantage. On the other hand, if the defensive buildup is cheaper than the offensive buildup, the defense forces the offense either to accept limitations on its penetration or to resort to qualitative changes in its arsenal.

As an illustrative example of such a cost trade-off, consider the hypothetical HF chemical laser system described in Section 3.1. Each laser in that system requires 1.7 seconds of dwell time to destroy a booster. During the 200 seconds of boost, each laser overhead can therefore destroy 120 boosters. But for each laser overhead, 32 are needed worldwide. Suppose now that the Soviets deploy, in one region of the U.S.S.R., 1,000 Midgetman missiles at a cost of 10 to 20 billion U.S. dollars (see Section 2). The U.S. defense now needs to be "beefed up" with addition of $(1,000) \times (32)/120 = 270$ laser battle stations. A tradeoff of more than one complex U.S. satellite, launched and maintained on orbit, for every 4 Soviet boosters (or decoy boosters) deployed on the ground would certainly appear to be a losing proposition for the United States. This is true even though the hypothetical HF laser system represents a very favorable outcome of laser technology.

Note that Soviet deployment of new ICBMs in one region of the Soviet Union, within coverage of only a single U.S. satellite, gives them the best leverage in the cost exchange.

## 5.8 NEW TARGETING PLANS

Truly efficient ICBM defenses would presumably force upon the superpowers a stricter attention to targeting priorities. With thousands of warheads in today's arsenals able to be literally lobbed into any target area unimpeded, the superpowers have less need to be discriminating or parsimonious in their nuclear targeting. Such a shift might have both desirable and undesirable consequences. For example, the offense might decide that in view of the cost of countermeasures it could no longer afford to threaten the other side's ICBM silos. The warheads "freed up" from the countersilo mission might then be dedicated to heavier targeting of other aimpoints (perhaps cities) as a hedge against poor penetration. How the superpowers would greet these hypothetical defenses is not clear, but it is probably quite wrong to imagine future defenses acting against offensive forces targeted according to the war plans of today.

## 5.9 OTHER MEANS OF DELIVERING NUCLEAR WEAPONS

One additional Soviet response to an efficient defense against their ICBMs would be increased emphasis on submarine-launched ballistic missiles (SLBMs), bombers, cruise missiles, and whatever novel methods time and ingenuity might in the future devise for introducing nuclear weapons to the United States. As defenses forced up the cost-per-delivered-warhead of ICBM forces, these other methods would become relatively more attractive. Though they would sidestep the BMD, these delivery means have higher pre-launch survivability than ICBMs, and bombers and cruise missiles have longer times of flight. These attributes are usually seen as "stabilizing." Shifting the emphasis of the arms competition away from ICBMs is therefore sometimes viewed as adequate payoff for the BMD effort.

SLBMs would obviously be vulnerable to the same boost-phase weapons as ICBMs. The same worldwide coverage, reflected in the absentee ratio, that plagues the anti-ICBM cost exchange means that orbiting boost-phase defenses threaten SLBMs the world over. However, midcourse and terminal tiers of a layered defense would in general have much less capability against SLBMs, because of the latter's short time of flight, possibly depressed trajectory, and uncertain direction of attack. Thus SLBMs could conceivably enjoy greater penetration of a layered defense than ICBMs.

If one takes an optimistic view of emerging defensive technologies, or if one contemplates technological "breakthroughs," it is at least conceivable that such developments will spawn new ways of delivering or aiding the delivery of nuclear weapons as well as new ways of interdicting them.

Section 6

# A WORD ON "OLD" BMD AND "NEW" BMD

# A WORD ON "OLD" BMD AND "NEW" BMD

No one knows whether directed-energy weapons can be built with the characteristics assigned to the hypothetical systems of Section 3. Even if such systems can be built, it is not clear that their performance will match, much less exceed, the performance of terminal and midcourse BMD systems in level of protection (attack price) and in cost relative to offsetting offensive improvements. The boost-phase BMD systems receiving so much attention today were a year ago at the periphery, to say the least, of technical discussion of missile defense. It is important not to lose sight of the status of traditional reentry and "advanced" (as they were called a year ago) "overlay" midcourse BMDs.

Naturally, the promise of the better-understood terminal and midcourse systems does not seem so grandiose, nor the flaws so clear-cut, as they do for the conceptual boost-phase defenses discussed in this Background Paper. Sounder technical assessments can be made of the "old" BMDs than of the "new" concepts. Rough concepts gloss over all the difficult design problems that inevitably limit achievable performance and turn up serious problems; nonetheless, identifying potentially unsolvable problems at this early stage of study does not mean they will remain insurmountable. BMD architectures incorporating boost-phase intercept are not known to be able to perform better, dollar-for-dollar, than BMD architectures incorporating only midcourse and reentry intercept. They are just not known to be worse. Terminal defense systems have been studied, designed, and tested for years, and it is generally agreed that such systems, acting alone, can enforce a modest attack price of between 2 and 8 RVs (perhaps equivalent to 20 to 80 percent of a booster) per defended aimpoint. Though their capabilities are modest, reentry and midcourse defenses suffice for modest defensive goals. There is no need to incur the technical risk of "new" boost-phase intercept schemes unless one aspires to levels of performance clearly beyond those possible with "old" concepts.

Many of the "new" concepts for boost-phase intercept are not new at all. They have been studied and discussed in one form or another for 20 years. Conversely, there are some new ideas for improving terminal and midcourse BMDs.[1] The spirit of technical optimism that accompanied the new emphasis on boost-phase intercept in the past year affected thinking about "old" BMD as well.

For terminal defense systems, the new features receiving attention are, first, non-nuclear warheads on interceptor missiles and, second, airplane-borne infrared sensors as supplements to ground-based radars. The principal benefit of non-nuclear intercept is that interceptors can be deployed nationwide without public concern about the safety of defensive nuclear warheads. Non-nuclear kill does not permit the defense to avoid all the disruptive effects of nuclear bursts, however, since the offense can still arrange for RVs to detonate when they sense interceptor impact ("salvage fuzing"). The miss-distance/weight relationship of the non-nuclear warhead requires the interceptor to approach more closely to the RV, and this in turn requires a homing seeker on the interceptor. Terminal homing obviously creates new opportunities for offensive countermeasures.

Airborne optical sensors obviously do not suffer radar blackout, but they can suffer the analogous problem of infrared redout. Decoy discrimination remains a problem, though it acquires some interesting new features. Details of these new aspects of terminal BMD are obviously classified. Though important, these aspects are fairly straightforward extensions of traditional techniques rather than revolutionary "breakthroughs."

New thought about midcourse defense focuses on alleviating the Achilles' heel of systems that use infrared sensing to support intercept in space: the ease with which the offense can accompany attacking RVs with clouds of decoys. One approach receiving attention is simply to cheapen the interceptor and shoot at everything, RVs and decoys alike. Another is to probe the attacking

---

[1] See Julian Davidson, "BMD: Star Wars in Perspective," *Aerospace America*, January 1984, p. 78.

objects with an active sensor, rather than relying on their thermal emissions, in the hope of discriminating RVs from decoys. Some of these "active discrimination" schemes are complex and expensive and might in turn be susceptible to offensive spoofing. A third aid to discrimination is the boost-phase defensive layer itself, which might constrain the number and type of penetration aids the offense could mount on each boost-er in addition to reducing the total number of objects approaching the midcourse tier. Fourth, extensive use of space-based sensors would allow the defense to observe penetration aids throughout their flight (including during deployment from the bus) rather than just as they approach the United States. It remains unclear whether these techniques will be worth the costs and new countermeasures they would bring to the defense.

# A HYPOTHETICAL SYSTEM ARCHITECTURE

# A HYPOTHETICAL SYSTEM ARCHITECTURE

Most analysts of boost-phase BMD assume that midcourse and terminal BMDs will augment the boost-phase layer. This section assembles a hypothetical layered defense system *in toto*. This system is *purely illustrative*, taking current BMD concepts at their face value and conveying a concrete image of the defensive architectures analysts apparently have in mind when they speak of nationwide defense. Obviously there are many choices for such a "strawman" system. The particular system described below was chosen for its illustrative value and not because it represents some "most plausible" alternative. It would be meaningless to suggest a "front runner" in the present state of study and technology development. Rather, the purpose of this example is to show how the layers interact and to indicate the overall scale of the deployments contemplated, without implying that anything remotely like it ever could or would be built.

A defense with several layers presents the offensive planner with some of the variety of problems that afflicts the BMD designer, who never knows in advance which attack tactic or countermeasure the offense will choose and must include responses to all of them in the system design. Layered defense forces the offense not only to develop responses to all the layers, but to develop responses that can be accomplished simultaneously. Thus, for example, the method chosen to avoid boost-phase intercept must not prevent deployment of lightweight midcourse decoys. The synergistic effect of the several layers obviously works strongly in the defense's favor.

Nonetheless, one must compare the performance of a three-tiered defense to the performance of a two-tiered defense of the same cost. Thus it should be no surprise if a $200 billion system with boost (and possibly even post-boost), midcourse, and terminal layers performs better than a $50 billion system with no boost phase layer.

The correct questions are whether the additional $150 billion is worth the extra performance, and whether spending the $150 billion on more terminal and midcourse defense would in fact be a better investment.

Occasionally one sees a simplified leakage calculus applied to layered defense. The calculus assigns a "leakage" of, say, 25 percent (0.25) to the boost phase layer, 15 percent to the midcourse layer, and 10 percent to the reentry layer, deducing an "overall leakage" of 0.4 percent on the basis of the equation $(0.25) \times (0.15) \times (0.10) = 0.004$. Though the term leakage can be defined so that this calculus holds, the result actually bears little relation to the number of targets preserved by the defense. For one thing, a given defensive layer does not have an associated leakage *fraction* independent of attack size: the leakage fraction for each layer usually increases with attack size, most obviously (but not only) when the defensive arsenal becomes saturated. Second, the performance levels of the individual layers are not independent. For example, if the midcourse layer's interceptor arsenal is sized to handle only 25 percent of the attack, and the boost-phase layer works poorly and in fact allows 50 percent of the attack through, the midcourse layer obviously cannot display the same fractional efficiency against the attack of double the expected intensity. Conversely, improvements in one layer might improve performance of another: effective boost-phase or midcourse layers might force the offense to abandon the highly structured "laydowns" of RVs in space and time that limit a terminal layer's effectiveness. Third, the raw number or fraction of leaking RVs does not indicate the number or fraction of targets destroyed because of the tactics of preferential offense and defense. For these reasons, the leakage calculus is not a helpful way to encapsulate layered defense performance.

## 7.1 SYSTEM DESCRIPTION

The system design described below takes literally the goal of comprehensive nationwide defense. It seeks the capability (at least on paper) to engage all attacking Soviet missiles, whether targeted at cities, U.S. silos, or other military installations. Clearly the precise numbers and kinds of components in this description can be adjusted to suit any set of assumptions. The point of this description is merely to convey the flavor of these massive architectures. Most assumptions are favorable to the defense.

Suppose that at some time in the future the Soviet ICBM arsenal still consists of 1,400 boosters, as it does today. For simplicity, suppose further that each booster is an MX-sized solid propellant missile carrying 10 RVs and that all silos are located in one large region of the U.S.S.R. The boosters are not specially shielded against lasers, but some care in their design has given them an effective hardness of 10 kJ/cm$^2$, and they are further spun during ascent. Each booster carries a small number of decoys, but their small number is offset by the decoys' high fidelity. Each RV is accompanied by 9 lightweight infrared replicas and 1 high altitude reentry decoy. (This is a very modest penetration aid loading. One can assume larger numbers with perhaps poorer fidelity, chaff and aerosols, etc.)

The hypothetical U.S. defense system comprises both HF chemical laser battle stations and x-ray laser battle stations for boost-phase intercept, land-based midcourse interceptors carrying LWIR homing vehicles (the so-called "Overlay"), and land-based high-endoatmospheric homing interceptors with non-nuclear warheads for reentry intercept.

The HF chemical laser system resembles that described in Section 3.1, except that it has only five 20 MW lasers at each position in the 32-position constellation, for a worldwide total of 160. At a range of 2 Mm, a laser must dwell on each spinning booster for 5 seconds, so each laser at this range can handle 30 simultaneously launched boosters if defense begins 30 sec into the 180 sec boost phase of an MX-like booster. The five lasers overhead the Soviet silos at any one time can therefore only handle 150 of the 1,400 Soviet ICBMs. For small Soviet attacks, however, this non-nuclear boost-phase layer suffices.

For a large-scale Soviet attack, the United States deploys in addition a nuclear boost-phase system of x-ray lasers. A "perfect" laser with characteristics such as those derived in Section 3.3 can intercept ideally about 50 boosters at 4 Mm range. Therefore 28 lasers need to be in position over the Soviet silos at any time to handle a massive launch, giving a worldwide total of 900 (absentee ratio 32).

Warning for the boost-phase system is provided by MWIR warning satellites in synchronous or supersynchronous orbits. Also, each of the 160 HF laser battle stations has an MWIR telescope with 4 m mirror and an ultraviolet or visible ladar with 2 m mirror for pointing. Each x-ray laser is accompanied by an MWIR telescope tracker with 1 m mirror.

The 1,400 Soviet boosters carry 14,000 RVs, 126,000 midcourse decoys, and 14,000 reentry decoys. The United States assumes that only 10 percent of the boosters will survive the boost phase defense, so the midcourse tier needs to face 1,400 RVs and 12,600 midcourse decoys. The midcourse interceptors are given extremely long range, so only two bases are needed to cover the entire United States. However, each base must be prepared to absorb the entire attack, since the Soviets could target one half of the country more heavily than the other. There-

**Table 7.1.—Hypothetical Future U.S. Defense Designed for Nationwide Protection Against Hypothetical Future Soviet Offense**

| U.S. Defense | Soviet Offense |
|---|---|
| 5 warning satellites | 1,400 MX-like ICBMs |
| 160 HF laser satellites | deployed in one |
| 160 laser radars | region |
| 900 x-ray laser satellites | 10 RVs per ICBM |
| 900 MWIR trackers | 9 midcourse decoys |
| 28,000 midcourse intercept | per RV |
| vehicles and boosters | 1 reentry decoy per RV |
| 20 LWIR satellites | |
| 75 radars | |
| 140,000 terminal non-nuclear | |
| interceptors | |
| 25 aircraft with LWIR sensors | |

SOURCE: Author.

fore the United States needs 14,000 midcourse interceptors at each base, for a total of 28,000 interceptors. A constellation of 20 satellites with large LWIR sensors provide long-range acquisition and target assignment to these interceptors.

The United States next estimates that 90 percent of the RVs that enter the midcourse layer will be successfully intercepted. The terminal defense must handle 140 RVs plus 140 reentry decoys.[1] The reentry decoys are, by assumption, completely faithful in mimicking the signatures of RVs as seen by ground-based radars and aircraft-borne infrared sensors during early reentry—large decoy numbers have been sacrificed for this high fidelity. If the U.S. defense takes literally its charge of nationwide defense, it must be prepared to make 280 intercepts anywhere in the country.

---

[1]The number of reentry decoys the terminal system must face actually depends on whether the decoys fool the midcourse as well as the terminal system's sensors and on whether the terminal and midcourse layers cooperate in discrimination. Suppose first that the reentry decoys look like RVs (and therefore also like midcourse decoys) to the midcourse layer's sensor. Then the midcourse layer will intercept 90 percent of them (more midcourse interceptors must be bought to do this). If, on the other hand, the midcourse layer correctly identifies the reentry decoys as non-lethal objects, it might (a) not intercept them, requiring the terminal layer to plan to face 140 RVs plus 1400 reentry decoys, increasing enormously the required arsenal of terminal interceptors; (b) intercept them, in which case a reentry decoy is a perfect midcourse decoy, making possible a new threat—large numbers of midcourse decoys that look like reentry decoys rather than RVs; (c) radio the information, object-by-object, to the terminal defense fields.

Each terminal defense site consists of a phased-array radar and a number of high altitude non-nuclear interceptors. Additional target acquisition support is provided by a fleet of aircraft patrolling the U.S. periphery, carrying LWIR sensors. Each radar has a radius of action of over 200 km, so 75 or so cover the entire United States. However, an interceptor only covers an area about 50 km in radius. Since the area of the United States is about 8 million square km, over 1,000 interceptor sites would be needed for nationwide coverage. Should the defense have to reckon with intensive Soviet attack on some regions and no attack on others? Clearly, yes. But equipping each interceptor site to handle all 280 objects passing through the first two layers would require buying 280,000 interceptors! The defense would need to buy this many interceptors if it wanted to claim the literal capability to engage *all* Soviet RVs, *no matter* where they landed. Suppose, then, that the United States hopes for a more evenly distributed Soviet attack and deploys just half of the arsenal needed for complete coverage—140 interceptors per region. One radar might not suffice to handle all the RV traffic in its sector if the Soviets attack some sectors preferentially, but the United States nonetheless buys just 75 radars. Five aircraft on patrol at all times requires a backup fleet totalling perhaps 25.

Table 7.1 summarizes the offensive and defensive deployments.

## 7.2 ASSESSMENT

It is obviously not possible to assess the performance of a system, such as the hypothetical layered defense described above, whose components are not (and in many cases cannot be) designed today, much less assembled in an overall architecture. It is nonetheless worth sketching, in the illustrative spirit of this section, the issues that would require analysis if anything resembling Table 7.1 were ever proposed for deployment.

The first issue concerns the cost of the improvements to the system needed to offset growth in the Soviet ICBM arsenal—that is, the cost exchange ratio. The number of defensive weapons is proportional to the number of Soviet ICBMs. If the Soviets were to double the size of their ICBM arsenal, the United States would need to double the number of its x-ray lasers and interceptor missiles. (The number of sensors would generally have to increase also, though perhaps not in proportion to the Soviet buildup. The number of HF lasers could remain the same if the United States continued to intend to use this non-nuclear boost-phase layer to engage only Soviet attacks of 150 boosters or less.) Comparison of the two columns of Table 7.1 indicates that an arms race of Soviet offense and U.S. defense seems certain to favor the Soviet side greatly.

A second issue concerns the huge inventories of midcourse and reentry interceptors needed for nationwide defense. The arsenal of reentry interceptors shown in Table 7.1 is in fact only half the size needed for literally complete nationwide coverage, as remarked above. The cause of these large interceptor inventories—besides the obvious presence of decoys—is twofold: first, the low leakage sought by the defense precludes preferential defense, the tactic that makes silo defense so much more economical; and second, the limited coverage of each interceptor battery makes preferential *offense* possible for the attacker. The goal of nationwide low-leakage defense therefore forces the BMD system to forfeit the two sources of leverage that have historically impelled BMD towards the technically modest mission of defending compact silo deployments to relatively low survival levels.

Soviet countermeasures—besides straightforward buildup of ICBMs—is a third issue. The particular boost-phase layers described above are susceptible in varying degrees to all of the countermeasures described in Section 5, and the midcourse and reentry layers to their respective sets of countermeasures.

Defensive coverage against submarine-launched ballistic missiles (SLBMs) is a fourth issue. The boost-phase layers can intercept SLBMs launched from all points on the globe. Though the average range from laser satellite to booster is larger at equatorial and polar latitudes than at the mid-latitudes where Soviet ICBMs are located, the number of SLBMs that could be launched in a short time from each ocean area is also much smaller than the huge number of ICBMs that could lift off simultaneously from the U.S.S.R. Even the chemical laser deployment alone might suffice for boost-phase coverage of SLBMs. The midcourse and reentry layers, however, would in general not perform as well against SLBMs as against ICBMs. SLBM trajectories present bad viewing angles to the midcourse layer's LWIR sensors, and the short timeline limits interceptor coverage. The reentry layer would need to be augmented with more airborne sensors and more radars (or radar faces) to cover attack from the ocean. In general, then, a layered system optimized for ICBM defense could not necessarily handle SLBMs as well.

Section 8

# DEFENSIVE GOALS I: THE PERFECT DEFENSE

# DEFENSIVE GOALS I: THE PERFECT DEFENSE

No assessment of whether a defensive system "works" or not is meaningful without a clear and direct statement of the goal of the deployment. Though there has been much discussion of the feasibility of boost-phase BMD, proponents and skeptics alike frequently leave unstated the standards against which they are judging the technical prospects. A "successful" BMD deployment could be defined as anything from a truly impenetrable shield, to a silo defense that merely costs less to build than it costs the Soviets to overcome, to a tangled deployment that just "creates uncertainty" for the attacker.

The most ambitious conceivable goal for BMD would be to take at literal face value the words of President Reagan in his so-called "Star Wars" speech of March 23, 1983, when he called for development of a defense capable of making nuclear weapons "impotent and obsolete."[1] It is not clear that the President intended his words to be taken literally, nor that the Administration or anyone else is suggesting the United States seek a truly perfect or near-perfect defense.[2] Nonethe-

less, so much writing and debate focuses on this prospect, and its importance is so great, that it is taken up in this section. Section 9 treats the many other possible goals for less-than-perfect defenses.

There is some confusion in the literature about the use of the term "mutual assured destruction" (MAD) in connection with the notion of perfect defense. In common strategic parlance, MAD refers to the *technological circumstance* of mutual vulnerability to catastrophic damage from nuclear weapons, not to a *chosen policy* to promote such vulnerability. There is a strategic school of thought that advocates a policy usually called "minimum deterrence," maintaining that the capability for assured destruction of Soviet society is the *only* requirement of U.S. strategic forces. However, many experts believe that effective deterrence and other national security objectives require nuclear forces capable of many other tasks than assured destruction. This section addresses itself to the question of whether MAD is an avoidable technological circumstance, *not* to whether minimum deterrence is a prudent strategic policy.

A sensible start at judging the prospect for near-perfect defense must involve two steps: first, an exact statement of what perfect defense means in the context of attack on society with nuclear weapons; second, some way of gauging the likelihood of success when the technological future cannot be accurately predicted.

---

[1]*Weekly Compilation of Presidential Documents*, Vol. 19, No. 12, March 28, 1983, pp. 447-448. The text of the relevant part of the speech is reprinted as Appendix A.

[2]Defense Secretary Caspar Weinberger appeared to confirm a literal interpretation in a March 27 interview on NBC's *Meet the Press*, when he said (as reported in the *Baltimore Sun*, March 28, 1983, p. 1):

> The defensive systems the President is talking about are not designed to be partial. What we want to try to get is a system which will develop a defense [sic] that is thoroughly reliable and total. I don't see any reason why that can't be done.

Later, the Defense Department stated that the purpose of the President's initiative was not to save lives, but to deter war. Responding to the "Congressional Findings" section of the proposed "People Protection Act" (H.R. 3073, 98th Congress, lst session) which stated, "The President has called for changes in United States strategic policy that seek to save lives in time of war," Defense Department General Counsel William H. Taft IV wrote:

> It is clear that portions of the "Congressional findings" section [of H.R. 3073] vary from the purpose of the President's initiative. First, and most importantly, the purpose of the President's initiative is to strengthen our ability to deter war by, as the President has said, ". . . rendering [ballistic missiles] impotent and obsolete." In short,

the purpose of the Administration's policy is to reduce the likelihood of war. The finding [of H.R. 3073] that the purpose is to "save lives in time of war" departs from our goal of deterring war.

Dr. Charles Townes, a frequent adviser to Secretary of Defense Weinberger and leader of two DOD task forces studying basing modes for the MX missile, said that a perfect defense proposal is "quite impractical. There is no technical solution to safeguarding mankind from nuclear explosives." (*New York Times*, April 11, 1983, p. 14).

# 8.1 NUCLEAR ATTACK ON SOCIETY

**Figure 8.1.—The Effect of Attack Size on the Extent of Prompt Fatalities in U.S. Urban Areas**

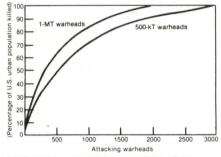

Note: Aimpoints chosen to maximize prompt human fatalities. U.S. urban population is estimated to be 131 million, as in the 1970 census.

SOURCE: Arms Control and Disarmament Agency, *U.S. Urban Population Vulnerability* (ACDA, 1979), quoted in Arthur M. Katz, *Life After Nuclear War: The Economic and Social Impacts of Nuclear Attacks on the U.S.* (Ballinger, 1982). Adapted from Ashton Carter and David N. Schwartz, eds., *Ballistic Missile Defense* (The Brookings Institution, 1984), p. 168.

Suppose one wants to take literally the goal of removing from the hands of the Soviet Union the ability to do socially mortal damage to the United States with nuclear weapons, so that the Soviet Union no longer possesses the elemental capability of assured destruction.) What does this mean?

Figure 8.1 shows how the percent of the U.S. urban population (about 130 million people in the 1970 census) killed promptly increases with the number of Soviet warheads detonating over U.S. cities. No such curve of the effects of nuclear attacks on cities should be taken as anything but suggestive: uncertainties are very great in such estimates, and no attempt has been made to reflect long-term and indirect effects of the detonations.[3] The curve also accounts only for

[3]For discussion of some of the effects of nuclear weapons on population and cities, see: U.S. Arms Control and Disarmament Agency, *An Analysis of Civil Defense in Nuclear War* (ACDA, 1978); OTA, *The Effects of Nuclear War* (GPO, 1979); ACDA, *The Effects of Nuclear War* (ACDA, 1979); "Economic and Social Consequences of Nuclear Attacks on the United States," prepared for the Joint Committee on Defense Production, 96 Cong. 1 sess., 1979;

fatalities, not for the many additional people injured. If one wishes to account for the possibility of civil defense evacuation, then the curve should be taken to represent not the number of people killed, but the number of people whose homes, businesses, historic monuments, schools, and places of worship have been destroyed.

No one supposes that the Soviet Union actually chooses aimpoints for its nuclear weapons with the goal of maximizing human fatalities, as has been done in preparing Figure 8.1. If the United States possessed a defense capable of intercepting all but a few of the 8,000 to 10,000 Soviet nuclear warheads, however, the Soviet Union might retarget its forces to wreak the most destruction possible with its few penetrating warheads. At any rate, any defense promising U.S. society genuine immunity from nuclear attack must reckon with Soviet determination to keep its arsenal from being "rendered impotent," and therefore with targeting plans contrived to do the most damage to the fabric of U.S. society.

Where on the curve of Figure 8.1—after how many detonations—does one locate the boundaries of "assured destruction," "assured survival," "impotent and obsolete," and similar phrases? Clearly there is no analytical prescription for these boundaries: they are the subject of a broader human judgment. 500 half-megaton warheads kill half the urban population, injure most of the rest, and totally destroy all American cities and large towns. Just 5 megatons, about one two-thousandth of the Soviet arsenal, detonated over the 10 largest U.S. cities could kill several million people and wound over 10 million more.

For the sake of discussion, we shall use 100 megatons—about 1 percent of the Soviet Union's arsenal and 1.5 percent of its ICBM force—as the level of penetration for which a defense would

J. Carson Mark, "Global Consequences of Nuclear Weaponry," *Annual Review of Nuclear Science,* Vol. 26 (1976), pp. 51-87; National Research Council, *Long-Term Worldwide Effects of Multiple Nuclear-Weapons Detonations* (National Academy of Sciences, 1975).

be judged "near-perfect." This definition is obviously very generous to the notion of perfect defense, since most people would presumably not regard 100 megatons of explosive force as "impotent and obsolete." Still, it is a definite reference for assessment.

## 8.2 THE PROSPECTS FOR A PERFECT DEFENSE

There is not and cannot be any "proof" that unknown future technologies will not provide near-perfect defensive protection of U.S. society against Soviet ICBMs. The question that needs to be answered is whether the prospects for near-perfect defense are so remote that such a notion has no place in reasonable public expectations or national policy. It is, after all, not provable that by the next century the United States and U.S.S.R. will not have patched up their political differences and have no need to target one another with nuclear weapons. The issue of the perfect defense is unavoidably one of technical judgment rather than of airtight proof.

Four misapprehensions seem common among non-technical people addressing the prospects for perfect defense.

The first misapprehension is to equate successful technology development of individual *devices*—lasers, power sources, mirrors, aiming and pointing mechanisms—with achievement of an efficient and robust defensive *system*. Millionfold increase in the brightness of some directed-energy device is a necessary, but is far from a sufficient, condition for successful defense. In the early 1960's, intercept of RVs with nuclear-tipped interceptor missiles was demonstrated—"a bullet could hit a bullet"—but 20 years later systems incorporating this "kill mechanism" are still considered relatively inefficient. In general, skeptics about the future of space-based directed-energy BMD do not confine their doubts to, or even emphasize, unforeseen problems in developing the individual components.

A second misappprehension arises in attempts to equate BMD development to past technological achievements, such as the Manhattan project's atomic bomb or the Apollo moon landings.[4] The technically minded will recognize a vital difference between working around the constraints imposed by nature, which are predictable and unchanging, and competing with a hostile intelligence bent on sabotaging the enterprise. A dynamic opponent makes of BMD, first, a more difficult design problem, since the offense constructs the worst possible barriers to successful defense; and second, not one problem but many problems that need to be sidestepped simultaneously in the design, since the designer cannot be certain which tactics the offense will use.

A third misapprehension concerns the prospect for a "technological breakthrough" that would dispel all difficulties. Such breakthroughs are not impossible, but their mere possibility does not help in judging the prospects for the perfect defense. For one thing, an isolated technological breakthrough creating a new defensive component would not necessarily alleviate the system issues—vulnerability, dependability, susceptibility to countermeasures, cost—that determine overall effectiveness. Second, one can just as easily imagine offensive "breakthroughs," sometimes involving the same technologies. Thus the x-ray laser, if it matures, might turn out to have been

---

[4]Defense Secretary Caspar Weinberger has written (*Air Force Magazine*, Nov. 1983):

The nay-sayers have already proclaimed that we will never have such technology, or that we should never try to acquire it. Their arguments are hardly new . . . In 1945 President Truman's Chief of Staff, Adm. William Leahy, said of the atomic bomb: "That's the biggest fool thing we've ever done. The bomb will never go off, and I speak as an expert in explosives." In 1946 Dr. Vannevar Bush, Director of the Office of Scientific Research and Development said of intercontinental ballistic missiles, "I say technically I don't think anybody in the world knows how to do such a thing, and I feel confident it will not be done for a long time to come." These critics were proved wrong; what is more, they were proved wrong quickly.

better termed a breakthrough in strategic offense than a breakthrough in strategic defense.

A fourth misapprehension concerns the confidence with which predictions could be made about the performance of a complex system once in place. The "performance" of a system, as quoted in analyses, is the most likely outcome of an engagement of offense versus defense. Other outcomes, though less likely, might still be possible. Computing the relative likelihoods of *all* possible outcomes would be difficult even if one could quantify all technical uncertainties and statistical variances. Still, there would remain a residue of uncertainty about the performance of a system that had never been tested once in realistic wartime conditions, much less in a statistically significant ensemble of all-out nuclear wars. The defense would also have no chance to learn and adapt. In World War II, by contrast, air defense crews learned in raid after raid to inflict losses of several percent in attacking bombers. The only reason these modest losses assumed strategic significance was that they accumulated over many raids. Of course, the same uncertainties plague the offense as plague the defense. In general, the offense would tend to overestimate the defense's capability. This natural tendency toward "offense conservatism" is probably vitally important to the psychological and deterrent value of BMD as it is applied to less-than-perfect goals. For the perfect defense goal (as defined above), however, it would seem that the uncertainty weighs heaviest on the defense. To the reckless, non-conservative defense, a wrong estimate of defense performance spells the difference between safety and socially mortal damage (or between deterrence and war). The reckless offense, on the other hand, is presumed desperate enough to try to inflict such damage on its enemy and willing to accept the consequences: it stands to lose little if its estimates are wrong and the defense *does* work perfectly after all.

With these misapprehensions out of the way, and recognizing clearly that there can be no question of "proof," it would seem that four major factors conspire to make extremely remote the prospect that directed-energy BMD (in concert with other layers if necessary) will succeed in re-

ducing the vulnerability of U.S. population and society to the neighborhood of 100 megatons or less.

1. **Near-perfect defense of society is much harder and more expensive than partial defense of military targets.** That is, the marginal cost exchange is much higher for near-perfect defense than for partial defense (see also Fig. 5.2). There are two reasons behind this well-known statement.

The first reason is illustrated schematically in Figure 8.2. In going from partial silo defense to perfect city defense, the BMD loses the leverage of preferential defense. Additionally, the offense gains the leverage of preferential offense against the terminal and midcourse layers with their limited geographic coverage, although *not* against the boost-phase layer. In Figure 8.2(a), the defense aims to save only 10 percent of the ICBM force, or one silo. Assuming perfect interceptors and adopting the tactic of adaptive preferential defense (using all its interceptors to save just one silo chosen randomly from the ten at the moment of attack), the defense concludes it needs to prepare to make one only intercept to counter the offense's 10 RVs. In Figure 8.2(b), the offense can focus all 10 of its RVs on any one of ten cities. The defense must prepare to make 10 intercepts for each city, buying a total of 100 interceptors, if it wants to try to save all the cities. If the Soviets double their RV arsenal, the United States must buy just one interceptor to satisfy the defense goal of Figure 8.2(a), but the United States must buy 100 interceptors to satisfy the defense goal of Figure 8.2(b). The cost exchange ratio is thus 100 times worse for the city defense, even though it uses the same technology as the silo defense.

The second reason why a near-perfect defensive goal shifts the cost burden in favor of the offense is that the offense can turn all its resources to improving or replacing just a portion of its ICBMs to sidestep the defense. The Soviet Union could therefore harden just 1 percent of its boosters, perhaps concealing exactly which ones were hardened. Moreover, it could deploy a few fast-burn boosters immune to x-ray lasers and neutral particle beams; build a few different ASAT de-

**Figure 8.2**

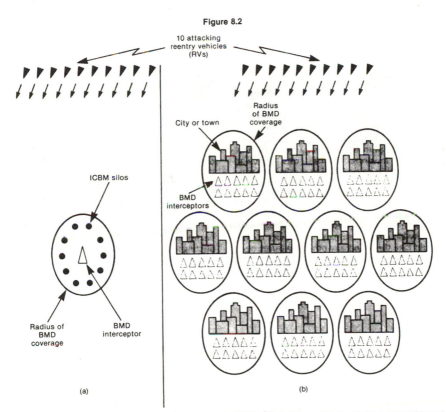

Figure 8.2. The cost to a U.S. reentry BMD to compete with increases in the size of the Soviet arsenal (the "marginal cost exchange") is much greater for near-perfect city defense (b) than for partial silo defense (a). In going from (a) to (b), the U.S. loses the leverage of preferential defense and the Soviet Union gains the leverage of preferential offense.

vices; and so on. More costly countermeasures are available to the offense if the countermeasures need only be implemented on a small scale rather than throughout the ICBM force. The offense can "experiment" with a number of different tactics with different portions of its force. The defense's costs also grow much larger if it must plan to face a *variety* of offensive countermeasures. In short, the defense must be able to stop all kinds of attack, but the offense only has to find one way to get through.

2. **For every defense concept proposed or imagined, including all of the so-called "Star Wars" concepts, a countermeasure has already been identified.** These countermeasures were enumerated in Section 5 and will not be repeated here. Three further generalizations about these countermeasures reinforce a poor prognosis for cost-effective near-perfect defense: 1) In general, the countermeasures could be implemented with today's technology, whereas the defense itself could not; 2) In general, the costs of the coun-

termeasures can be estimated and shown to be relatively low, whereas the costs of the defense are unknown but seem likely to be high; 3) In general, the future technologies presupposed as part of the defense concept would also be potent weapons for attacking the defense.

3. **The Soviet Union does not configure its nuclear missile forces today to maximize damage to U.S. society and population, but it could do so if faced with near-perfect U.S. defenses.** Targeting plans could focus exclusively on damaging cities. High missile accuracy would be unnecessary, lowering offense costs. Nuclear weapons could be designed to maximize harmful fallout. ICBM survivability measures—silos, racetracks, densepacks, etc.—would be unnecessary to a side striking first or possessing its own defense effective at saving many missiles (but not *all* cities); thus basing costs could be diverted to the city-kill goal. Presumably the Soviet Union would take these and any other measures necessary to prevent itself from being effectively disarmed by a U.S. defense, since otherwise it would be at its enemy's mercy.

4. **BMD by itself will not protect U.S. society from other methods of delivering nuclear weapons to U.S. soil or from other weapons of mass destruction.** Bombers and cruise missiles (and to a significant extent SLBMs and IRBMs) present very different defensive problems than ICBMs. Today the technical problems of air defense are no better resolved than the technical problems of BMD. Novel future offensive delivery vehicles can only be conjectured along with the future defense technologies discussed in this Paper. A desperate Soviet Union could introduce nuclear weapons into the United States on commercial airliners, ships, packing crates, diplomatic pouch, etc. Other methods of mass destruction or terrorism would be feasible for the U.S.S.R., including sabotage of dams or nuclear power plants, bacteriological attack, contaminating water, producting tidal waves with near-coastal underwater detonations, and so on.

Section 9

# DEFENSIVE GOALS II:
# LESS-THAN-PERFECT DEFENSE

# DEFENSIVE GOALS II:
# LESS-THAN-PERFECT DEFENSE

A host of less grandiose goals than perfect or near-perfect defense assume importance in certain theories about the workings of nuclear deterrence and the requirements of U.S. security. Thoughtful observers debate not just the feasibility of achieving these goals but the validity and importance of the goals as well. The urgency one attaches to these goals determines the costs, risks, and harmful side effects one is willing to incur to fulfill them. Assessing the wisdom of less-than-perfect defense thus involves a complex and subjective balancing of goals and risks, in which purely technical issues sometimes take a back seat. In discussion of perfect defense, by contrast, technical assessment is paramount. This section therefore calls up many issues of nuclear policy not subsumed under the title of this Background Paper, and no pretense is made here to complete treatment.[1]

Though various strategic goals for BMD can be distinguished in principle, in practice it might not be clear or agreed among all parties in the United States what the purpose of a proposed deployment actually was. Interpretations by the Soviet Union and other foreign nations of U.S. goals might be quite different yet.

Those familiar with BMD design and assessment will recognize that stating a general *strategic* goal is not enough: the *technical* specifications are essential. For instance, it makes an enormous difference in silo defense whether the defense seeks to charge the offense a price of five RVs (or half a booster) or 10 RVs (one booster) per silo.

For goals requiring very modest performance, terminal and midcourse defenses might suffice. Since no one knows whether boost phase defenses, when better defined, will surpass or even equal traditional defenses in terms of leakage and cost exchange, there is no point in turning to exotic technologies to satisfy modest goals. Virtually all observers agree, on the other hand, that terminal and midcourse systems are unequal to the more demanding goals; for these goals one is forced to direct one's hopes to the promise of future technologies.

This section sketches various goals for less-than-perfect defenses and the strategic thinking that attaches importance to them. It then points out a number of side effects against which fulfillment of these goals needs to be balanced. This short section is no substitute for a comprehensive assessment of the pros and cons of BMD.

---

[1]For a more complete treatment of the entire subject of BMD, see Ashton B. Carter and David N. Schwartz, ed., *Ballistic Missile Defense* (The Brookings Institution, 1984).

## 9.1 GOALS FOR LESS-THAN-PERFECT DEFENSE

1. **Strengthen deterrence by preventing preemptive destruction of retaliatory forces.** It is widely recognized that the Soviet Union will soon have, if it does not already, the combination of yields, numbers, and accuracy in its ICBM forces to destroy most U.S. Minuteman ICBMs in their silos. It is also widely agreed that vulnerable nuclear forces create unwanted temptations for both sides to strike first if war seems likely. The long and anguished search for survivable basing modes for the U.S. MX (Peacekeeper) ICBM has to date turned up no clear favorites when sur-

vivability is balanced against cost, technical risk, strategic effects, and environmental impacts.[2] BMD would substitute for or complement these other basing modes. By shooting down a fraction of the opponent's missiles, BMD would in effect "de-MIRV" ICBM forces.

Of course, turning to BMD to ease ICBM vulnerability is not without problems. One problem is the prospect of a compensating Soviet BMD.

---

[2]*MX Missile Basing*, Office of Technology Assessment, U.S. Congress, September 1981.

MX is presumably being bought and made survivable in the first place so that the U.S. can absorb a Soviet strike and retaliate with its ICBMs (in addition to its bombers and submarines) against Soviet targets. But modification or termination of the ABM Treaty to permit a U.S. defense would permit a Soviet defense as well. The surviving U.S. ICBMs guaranteed by the U.S. BMD might still not result in retaliatory damage to Soviet targets if these targets are defended by Soviet BMD. The U.S. BMD deployment, all bought and paid for, might therefore have been cancelled out by a Soviet counter-deployment.

Other elements of the U.S. retaliatory force comprise command and control links, bomber alert bases, and in-port submarines. Bomber bases, sub ports, and fixed command and control facilities are the worst type of target base for BMD to try to defend—a small number of high-value, soft, and interdependent targets. The important remaining category of mobile command and control facilities, on the other hand, does not easily lend itself to active defense with BMD.

2. **Strengthen deterrence by preventing the use of nuclear weapons as decisive military tools for high-confidence "limited" strikes on conventional forces.** This goal is associated with so-called "warfighting" strategies for nuclear weapons. According to analysts who hold this view, today's "offense dominated" world creates dangerous temptations to resort to nuclear weapons to accomplish militarily well-defined objectives. One can imagine warheads simply being lobbed unopposed into another country in any number or combination. Though surely the effects of these "limited" attacks on nearby communities would not be so well-defined, the effect on the opposing military machine might be truly dramatic, even decisive. This use of nuclear weapons in wartime is possible with today's unopposed offenses with considerable confidence and might therefore be tempting to the combatants. Such temptations threaten nuclear deterrence and should be eliminated. The goal of a comprehensive defense would be to make such limited attacks infeasible, or at least to complicate the offense's estimations of success to such a degree

that it would not attempt an "experiment."[3] Analysts who favor this approach usually maintain that Soviet military doctrine inclines the Soviets towards a view of nuclear weapons as military tools to a far greater degree than is common in U.S. thinking.[4]

To take an explicit example (in this case of Soviet failure to deter the U.S.) of a "warfighting" scenario (chosen randomly from a great many possibilities), suppose NATO were at war with the Warsaw Pact, and the Soviets were resupplying their ground offensive through just 10 or so rail trunks from the Soviet Union through eastern Europe. Just 10 well-placed nuclear weapons (according to a hypothetical analyst considering this type of scenario) would cut off a large fraction of supplies coming to the front, slowing the Pact offensive and giving NATO vitally needed time to marshall its defenses. Wouldn't the United States be sorely tempted to use just a few ICBMs for this decisive intervention in the course of the war?

Analysts who recommend attention to warfighting scenarios and doctrines are surely aware of the profound difference between conventional and nuclear weapons, but they maintain that the threat of punishment through retaliation upon cities is not an effective deterrent in such scenarios. Wouldn't it be preferable if these scenarios were simply closed off by defensive technology?

Critics of this BMD goal object both to the warfighters' emphasis upon the risk of this type of scenario and to the assumption that defense would materially diminish that risk. In their view, myriad detailed chinks in the armor of deterrence can always be found, with or without defense, and worrying about them represents a loss of

---

[3]Presidential Science Adviser George A. Keyworth, II has stated (interview with *U.S. News and World Report,* April 11, 1983, p. 24):

"The objective is to have a system that would convince an adversary that an offensive attack will not be successful. It has to be a very effective system, but it would not have to be perfect to convince a potential adversary that his attack would fail."

Dr. Robert Cooper, director of the Defense Advanced Research Projects Agency has also stated this view (*The New York Times,* Nov. 5, 1983, p. 32): "Even if only 50 percent of all incoming missiles were stopped, the Soviets could then have no confidence in the success of a first strike, and war would be more remote."

[4]*Ballistic Missile Defense,* op. cit., Chapter 5.

perspective on the basic difference between nuclear and conventional instruments of war. Besides, they say, suppose the effect of the Soviet BMD is to force the United States to attack each rail line with ten weapons instead of one to assure penetration: Is there truly a psychological divide between using 10 and 100 nuclear weapons, once the divide between 0 and 10 has been crossed? Third, would NATO not be adequately deterred by the prospect of Soviet retaliation with 10 of its nuclear weapons against 10 vital NATO targets? Last, suppose NATO used 10 cruise missiles, against which the BMD was powerless, instead of 10 ICBM RVs?

The persuasiveness of this second goal for less-than-perfect BMDs therefore depends on one's views of the roles and risks of nuclear weapons—views that are fundamental and deeply held. This goal is therefore one of the most controversial of all.

3. **Save lives.**[5] Another goal for BMD is purely humanitarian and seeks no military or strategic advantage. If the defense did not interfere too much with Soviet military targeting objectives (enough for the Soviets to try to overcome it), and assuming the Soviets have no explicit aim to inflict human casualties, the United States could expect some reduction in fatalities in a nuclear war even from a modest defense. This reduction would necessarily be limited, since Soviet military objectives include destruction of many targets collocated with population. BMD and civil defense measures would be mutually reinforcing.

Analogous discussion of civil defense has always revealed an inherent tension between the humanitarian objective of defense and a related strategic objective. The strategic objective seeks to reduce fatalities and damage in order to enhance U.S. "flexibility" in a crisis, to allow the United States to "coerce" the U.S.S.R. (or avoid coercion) from a position of reduced vulnerability, or to enhance U.S. ability to persist in its war effort despite receiving a Soviet nuclear strike. The supposed result of the BMD deployment is to allow the U.S. President, in dealing with the Soviet leadership in time of crisis, to be more willing

or appear to be more willing to resort to nuclear war because the consequences to the United States are presumed smaller.

The coexistence of the humanitarian and strategic objectives for the analogous case of civil defense is apparent in the literature on civil defense. The Defense Department[6] has argued that the United States should have the same crisis relocation options as the U.S.S.R. for two reasons, one strategic and one humanitarian: 1) "to be able to respond in kind if the Soviet Union attempts to intimidate us in time of crisis by evacuating the population from its cities"; and 2) "to reduce fatalities if an attack on our cities appears imminent." Prominent scientists arguing for civil defense have also maintained that, "A nation's civil defense preparedness may determine the balance of power in some future nuclear crisis. . . . In our opinion, we must strive for an approximately equal casualty rate".[7] More recently, the High Frontier Study urging strengthened U.S. strategic defenses stated: "The protection of our citizens must be prime, but civil defense . . . would reduce the possibility that the United States could be coerced in a time of crisis".[8]

In practice, therefore, the humanitarian and strategic objectives are likely to be difficult to disentangle. Unlike the humanitarian objective, the strategic objective might stimulate a Soviet effort to put the same number of American lives at risk regardless of the defense. In this way, the Soviet Union could retain the strategic advantage that, by hypothesis, the BMD deprives them of. The issue then becomes the usual one of the cost-exchange ratio measuring the price to the Soviet Union of retaining its "advantage" relative to the price of the U.S. defense.

The Defense Department has stated that saving lives in time of war is not the purpose of President Reagan's BMD initiative.[9]

---

[5]This discussion borrows from the author's previous work in *Ballistic Missile Defense*, op. cit., Chapter 4.

[6]*Annual Defense Department Report, FY 1976*, p. II-24.
[7]Arthur A. Broyles and Eugene P. Wigner, "Civil Defense in Limited War," *Physics Today*, vol. 29 (April 1976), pp. 45-46.
[8]Daniel O. Graham, *The Non-Nuclear Defense of Cities: The High Frontier Space-Based Defense Against ICBM Attack* (Abt Books, 1983), p. 122.
[9]See footnote 8.2.

4. **Shape the course of the arms competition and arms control.**[10] One version of this goal sees the Soviet tendency to upgrade and proliferate existing ICBM forces as the principal impediment to arms control. By introducing BMD (or even discussing it), according to this view, the United States makes the Soviets unsure about the next step in the arms competition and thus undercuts the momentum of Soviet strategic programs, especially ICBM modernization. Though fast-burning Midgetman boosters might defeat boost-phase defenses, this argument goes, the slow-burning SS-18s and SS-19s will not. BMD might not be able to make all nuclear weapons impotent and obsolete, but it can make large Soviet ICBMs impotent and obsolete—something the U.S. has been trying to do for a decade. Perhaps efficient defenses will ''force'' the Soviets to emphasize submarines, bombers, and cruise missiles in their strategic arsenal to the same degree the United States does. (One problem with this line of argument is that by the time the defense is in place, present-generation Soviet ICBMs might already be replaced.)

Another line of argument holds that a major BMD initiative strengthens the U.S. negotiating position at START. An aggressive BMD program demonstrates U.S. technological prowess and hints at what the Soviets could face if this prowess were unleashed. It would seem that new BMD initiatives might not coexist easily with the reductions in offensive arsenals proposed by the United States in START, however. Since U.S. BMD is equivalent to attrition of the Soviet ICBM arsenal, any anxieties the Soviets feel at reduc-

ing the size of their missile inventories would, logically at least, be enhanced by a simultaneous U.S. BMD buildup. Politically, it would seem unlikely, though certainly not impossible, that a climate favorable to far-reaching offensive arms control would also foster an amicable dismantling of the ABM Treaty.

5. **Respond to Soviet BMD efforts.** Many analysts view with alarm Soviet strategic defense activities, including upgrading of the Moscow ABM, development of a transportable terminal BMD system, construction of a radar in apparent violation of the ABM Treaty, development of defenses against tactical ballistic missiles, incorporation of limited BMD capability in air defenses, and continued attention to other damage-limiting methods (civil defense, air defense, antisubmarine warfare, and countersilo ICBMs). A strong U.S. BMD research and development program might deter the Soviets from breaking out of the ABM Treaty and from continued encroachments on the Treaty's provisions. It is frequently noted that aggressive U.S. research into penetration aids and other methods for countering defenses might be an even more effective way to demonstrate to the Soviets that they would be ill-advised to overturn the ABM Treaty's ''freeze'' on missile defenses.

6. **Protect against accidental missile launches and attack from other nuclear powers.** These goals have been put forward several times in the past, most notably in the late 1960's when the Johnson Administration proposed the Sentinel ABM system to counter Chinese ICBMs, believed at that time to be fast-emerging. Neither goal figures prominently in today's discussion of BMD in the United States, though defense against Chinese, British, and French missiles could well loom larger in Soviet thinking. Emerging nuclear powers or terrorists would be unlikely to use ICBMs to deliver their small nuclear arsenals to the United States. BMD is therefore of little importance in staving off the threat to U.S. security posed by nuclear proliferation.

---

[10]Presidential science adviser George A. Keyworth, speech before the Washington chapter of the Armed Forces Communications and Electronics Association, as reported in *Defense Week* (Oct. 17, 1983):

''Although the strategic defense program's goal would still be eventual deployment of a working system, we shouldn't overlook its potential beneficial impact on arms reduction as its progresses.''
Richard DeLauer, Undersecretary of Defense for Research and Engineering, has said that an arms control agreement is needed to prevent the Soviets from overcoming a defensive system: ''With unconstrained proliferation [of Soviet missiles], no defensive system will work.'' (Interview with *The New York Times*, May 18, 1983).

## 9.2 SIDE EFFECTS OF BMD DEPLOYMENT

The inevitable side effects of a major strategic initiative such as BMD might turn out to match, both in magnitude and in duration, the beneficial effects of satisfying the goal emphasized by the system's purveyors. The public and policymakers would therefore need to assess the net, long-term effect of adding BMD to the strategic equation, and not just the achievement of a certain discrete goal as if by surgical intervention. This section reviews the well-known list of BMD side effects. Many of these effects are detrimental to U.S. security and would need to be balanced against the benefits of fulfilling the modest goals of less-than-perfect defense. In making this assessment, it is impossible to ignore the many unknowns and uncertainties that make it impossible to compare today's world without BMD to a future world with BMD.

1. **First strike versus ragged retaliation.** It is frequently noted that BMD ends up being a better investment for the side that strikes first than for the side that retaliates. Weapon systems that create relative advantages to striking first in a crisis (rather than risking being struck while seeking a peaceful resolution) are defined to be "destabilizing." The side striking first uses its full arsenal in an organized penetration of the other side's defense; the retaliating side can only use its surviving arsenal in a possibly disorganized "ragged retaliation" against a forewarned and fully prepared defense.

Mitigating factors could in certain circumstances soften this classical statement of the destabilizing effect of BMD. First, truly effective defenses might prevent the first striker from destroying a substantial fraction of the other side's retaliatory forces. Second, with proper planning (involving post-attack retargeting and coordinated timing), the retaliating forces might still be able to mount a tailored, efficient strike. Third, there will seemingly always be a *relative* advantage to being the side that strikes first in a nuclear war, with BMD or without BMD; but this calculus of relative advantage is far from being the only factor in deterrence. Other stabilizing factors might be strengthened by BMD, offsetting this desta-

bilizing factor. Thus BMD might also *discourage* temptations to strike first, by threatening to disrupt the attack.

2. **Soviet BMD.** A U.S. BMD deployment would seem very likely to stimulate a Soviet deployment. Even if the Soviets saw no compelling military rationale for following suit, political appearances could prove decisive. A Soviet BMD counter-deployment would obviously blunt U.S. offensive striking power, which the U.S. has been spending a great deal to build up. If the U.S. deployment sought to protect its ICBMs from preemptive destruction in their silos (Goal 1 above), the Soviet BMD might nonetheless nullify the U.S. ICBMs—this time in flight to their targets. Soviet BMD would also introduce a threat to U.S. SLBMs, which are today thought to be virtually immune to Soviet disruption and to be significantly advanced relative to their Soviet counterparts. If the U.S. deployment sought to prevent "limited" strikes by the Soviets Union (Goal 2), the Soviet BMD might in turn preclude a U.S. option to use nuclear weapons selectively and flexibly in support of its NATO allies—an option sometimes seen as central to NATO strategy. Clearly the actual effect of the Soviet BMD counterdeployment would depend upon its technical characteristics and the targets it defended.

3. **Demise of the ABM Treaty.** An arms control treaty obviously cannot serve as its own justification, and presumably virtually everyone would agree to the abandonment of the ABM Treaty the moment it ceased genuinely to serve the national security. In addition to its concrete provisions limiting BMD deployment, however, the ABM Treaty has unavoidably assumed a symbolic political meaning in the United States and, in different forms perhaps, in Europe and the U.S.S.R. The Treaty stands for a decade of arms control and attempts at superpower understanding about nuclear weapons. As a practical matter, it is impossible to overturn the Treaty's technical provisions without calling into question U.S. commitment to the whole fabric of the SALT/START process. This side effect would have to be weighed against the purely military and strategic

benefit (if there were, in fact, a net long-term benefit) of a U.S. BMD deployment.

4. **Allied and Chinese missile forces.** The nuclear missile forces of Britain, France and China are obviously a greater threat to the Soviet Union than to the U.S. Most analysts agree that the existence of these forces enhances U.S. security. But a major BMD initiative sparking widespread Soviet defense would in effect disarm our allies (to a degree depending on the nature of the Soviet deployment).

5. **Accompanying strategic programs.** A number of new weapon systems and strategic programs would be natural, though perhaps not necessary, accompaniments to BMD. On the offensive side, the U.S. would need to develop and deploy penetration aids against the Soviet BMD and improve its bomber and cruise missile forces to reflect added reliance on non-missile delivery vehicles. On the defensive side, the overall category of "strategic defense" comprises, in addition to BMD: nationwide air defenses against Soviet bombers and cruise missiles, defensive coverage against SLBMs, civil defense shelters and evacuation plans, and passive "hardening" of military installations and industrial facilities.

6. **Opportunity Costs.** The initial investment in BMD deployment, the inevitable follow-ons, and any accompanying strategic programs would make a substantial, permanent demand on the defense budget, competing with other nuclear forces and with conventional forces, not to mention with nonmilitary expenditures.

In a more fundamental sense, the transition from a world with a near-total ban on BMD to a world with BMD deployments is probably an irreversible change. Reimposing a defensive "freeze" after a period of unrestrained deployment, much less dismantling defenses and returning to zero, would involve all of the problems that

attend upon arms control reductions at START today. Extra caution seems warranted where strategic actions cannot easily be reversed or recalled: the opportunity for a comprehensive ban on missile defense might not arise again.

7. **Bean counting.** Strategists, politicians, and diplomats place considerable emphasis on quantitative measures of the nuclear balance and on "proofs" that "parity" exists. Arms control negotiations also reduce themselves quickly to counting rules. It is unclear whether or how BMD should affect such "bean counting." For each U.S. battle station added to a defensive constellation, are the Soviets to be credited with fewer ICBMs, since the U.S. defense represents potential attrition of the Soviet force? How many Soviet interceptor missiles equals one U.S. laser? Whose estimate of the BMD's likely wartime performance—the defense's or the offense's—governs these counting rules? Experience indicates that these types of questions, however far-fetched and even preposterous they appear in prospect, in the end assume considerable perceived importance.

8. **Asymmetries.** The Soviet BMD deployment that could well follow U.S. deployment might not share the same defensive goal or the same technology, stimulating the usual anxieties about unequal intentions and capabilities. Defensive systems are also complex, leading different analysts to widely different conclusions about the likely wartime performance of the BMD systems on both sides. Moreoover the owner of the BMD, aware of all the system's hidden flaws, might credit it with little capability, whereas the offense planner will tend to give it the benefit of the doubt. Though some hypothetical future world with mutual BMD deployments might therefore appeal to one analyst or nation, another could easily have a completely different view of the technical and strategic "facts."

# PRINCIPAL JUDGMENTS AND OBSERVATIONS

# PRINCIPAL JUDGMENTS AND OBSERVATIONS

**1. The prospect that emerging "Star Wars" technologies, when further developed, will provide a perfect or near-perfect defense system, literally removing from the hands of the Soviet Union the ability to do socially mortal damage to the United States with nuclear weapons, is so remote that it should not serve as the basis of public expectation or national policy about ballistic missile defense (BMD).** This judgment appears to be the consensus among informed members of the defense technical community.

Technical prognosis for such a perfect or near-perfect defense is extremely pessimistic because of the concentration and fragility of society; because all concepts identified as candidates for a future defense of population are known to be susceptible to countermeasures that would permit the Soviet Union to retain a degree of penetration with their future missile arsenal despite costly attempts to improve the U.S. defense; because the Soviet Union would almost certainly make such a determined effort to avoid being disarmed by a U.S. defense; and because missile defense does not address other methods for delivering nuclear weapons to the United States.

Mutual assured destruction (MAD), if this term is applied to a state of technological existence rather than to a chosen national policy, is likely to persist for the foreseeable future.

**2. The wisdom of deploying less-than-perfect ballistic missile defenses remains controversial.** Less-than-perfect defenses would still allow the Soviet Union to destroy U.S. society in a massive attack but might call into question the effectiveness of smaller, specialized nuclear strikes.

Certain theories about nuclear war maintain that such defenses could lessen the chances of nuclear war and enhance U.S. security by protecting U.S. retaliatory forces; by interdicting "limited" nuclear strikes; by further confusing Soviet predictions of the outcome of a strike; by driving Soviet missile deployments in directions favored by the United States; by lessening the consequences of nuclear attack; and/or by fulfilling still other strategic goals.

Critics dispute the validity of some of these goals; dispute that technology can fulfill the truly useful goals; and/or argue that the many harmful side effects of introducing BMD to the strategic equation and altering the Anti-Ballistic Missile (ABM) Treaty regime are not worth satisfying these goals.

To address the wisdom of less-than-perfect defense, the public and policymakers would need a precise statement of the strategic goal of the deployment, an assessment of whether technology could satisfy that goal, and an analysis balancing fulfillment of the goal against the side effects and uncertainties of introducing a new ingredient into the strategic nuclear arena.

**3. The strategic goal of President Reagan's Strategic Defense Initiative calling for emphasized BMD research—perfect, near-perfect, or less-than-perfect defense against ballistic missiles—remains unclear.** No explicit technical standards or criteria are therefore available against which to measure the technological prospects and progress of this initiative.

**4. In all cases, directed-energy weapons and other devices with the specifications needed for boost-phase intercept of ICBMs have not yet been built in the laboratory, much less in a form suitable for incorporation in a complete defense system.** These devices include chemical lasers, excimer and free electron lasers, x-ray lasers, particle beams, lightweight high-velocity kinetic energy weapons, and microwave generators, together with tracking, aiming, and pointing mechanisms, power sources, and other essential accompaniments.

It is unknown whether or when devices with the required specifications can be built.

**5. Moreover, making the technological devices perform to the needed specifications in a controlled situation is not the crux of the technical challenge facing designers of an effective ballistic missile defense. A distinct challenge is to fashion from these devices a reliable defensive architecture, taking into account vulnerability**

of the defense components, susceptibility to future Soviet countermeasures, and cost relative to those countermeasures.

New intercept mechanisms—directed energy weapons and the like—therefore do not by themselves necessarily herald dramatically new BMD capabilities.

6. **It is clear that potent directed-energy weapons will be developed for other military purposes, even if such weapons are never incorporated into effective BMDs.** Such weapons might have a role in nuclear offense as well as defense, in anti-satellite (ASAT) attack, in anti-aircraft attack, and in other applications of concern to nuclear policy and arms control. Defense and arms control policy will thus need to face the advent of these new weapons, irrespective of their BMD dimension.

7. **For modest defensive goals requiring less-than-perfect performance, traditional reentry phase defenses and/or more advanced midcourse defenses might suffice.** Such defenses present less technical risk than systems that incorporate a boost-phase layer, and they could probably be deployed more quickly.

New ideas for improving such "old" BMD concepts have emerged in the atmosphere of technical optimism enjoyed by the boost-phase concepts.

8. **Deployment of missile defenses based on new technologies is forbidden by the Anti-Ballistic Missile (ABM) Treaty reached at SALT I.** The Treaty permits only restricted deployment of traditional BMDs using fixed, ground-based radars and interceptor missiles. Research into new technologies, and in selected cases development and testing of defense systems based on these technologies, are allowed within the Treaty.

9. **There is a close connection, not explored in detail in this Background Paper, between advanced BMD concepts and future anti-satellite (ASAT) systems.** This connection springs from four observations: 1) ASAT attack on space-based weapons and sensors is probably the most attractive countermeasure to boost-phase BMD; 2) directed-energy weapons are more likely to succeed in the easier mission of ASAT than in the more difficult mission of boost-phase BMD; 3) to a degree dependent on technical details, early stages of development of boost phase BMDs might be conducted in the guise of ASAT development, stimulating anxieties about the health of the ABM Treaty regime; 4) to a degree dependent on technical details, concluding a treaty with the Soviet Union limiting ASAT development would impede BMD research at an earlier stage than would occur under the terms of the ABM Treaty alone.

# APPENDIXES

# APPENDIX A: THE CONCLUSION OF PRESIDENT REAGAN'S MARCH 23, 1983, SPEECH ON DEFENSE SPENDING AND DEFENSIVE TECHNOLOGY

Weekly Compilation of

# Presidential Documents

Monday, March 28, 1983
Volume 19—Number 12
Pages 423–466

Now, thus far tonight I've shared with you my thoughts on the problems of national security we must face together. My predecessors in the Oval Office have appeared before you on other occasions to describe the threat posed by Soviet power and have proposed steps to address that threat. But since the advent of nuclear weapons, those steps have been increasingly directed toward deterrence of aggression through the promise of retaliation.

This approach to stability through offensive threat has worked. We and our allies have succeeded in preventing nuclear war for more than three decades. In recent months, however, my advisers, including in particular the Joint Chiefs of Staff, have underscored the necessity to break out of a future that relies solely on offensive retaliation for our security.

Over the course of these discussions, I've become more and more deeply convinced that the human spirit must be capable of rising above dealing with other nations and human beings by threatening their existence. Feeling this way, I believe we must thoroughly examine every opportunity for reducing tensions and for introducing greater stability into the strategic calculus on both sides.

One of the most important contributions we can make is, of course, to lower the level of all arms, and particularly nuclear arms. We're engaged right now in several negotiations with the Soviet Union to bring about a mutual reduction of weapons. I will report to you a week from tomorrow my thoughts on that score. But let me just say, I'm totally committed to this course.

If the Soviet Union will join with us in our effort to achieve major arms reduction, we will have succeeded in stabilizing the nuclear balance. Nevertheless, it will still be necessary to rely on the specter of retaliation, on mutual threat. And that's a sad commentary on the human condition. Wouldn't it be better to save lives than to avenge them? Are we not capable of demonstrating our peaceful intentions by applying all our abilities and our ingenuity to achieving a truly lasting stability? I think we are. Indeed, we must.

After careful consultation with my advisers, including the Joint Chiefs of Staff, I believe there is a way. Let me share with you a vision of the future which offers hope. It is that we embark on a program to counter the awesome Soviet missile threat with measures that are defensive. Let us turn to the very strengths in technology that spawned our great industrial base and that have given us the quality of life we enjoy today.

What if free people could live secure in the knowledge that their security did not rest upon the threat of instant U.S. retaliation to deter a Soviet attack, that we could intercept and destroy strategic ballistic missiles before they reached our own soil or that of our allies?

257

I know this is a formidable, technical task, one that may not be accomplished before the end of this century. Yet, current technology has attained a level of sophistication where it's reasonable for us to begin this effort. It will take years, probably decades of effort on many fronts. There will be failures and setbacks, just as there will be successes and breakthroughs. And as we proceed, we must remain constant in preserving the nuclear deterrent and maintaining a solid capability for flexible response. But isn't it worth every investment necessary to free the world from the threat of nuclear war? We know it is.

In the meantime, we will continue to pursue real reductions in nuclear arms, negotiating from a position of strength that can be ensured only by modernizing our strategic forces. At the same time, we must take steps to reduce the risk of a conventional military conflict escalating to nuclear war by improving our non-nuclear capabilities.

America does possess—now—the technologies to attain very significant improvements in the effectiveness of our conventional, non-nuclear forces. Proceeding boldly with these new technologies, we can significantly reduce any incentive that the Soviet Union may have to threaten attack against the United States or its allies.

As we pursue our goal of defensive technologies, we recognize that our allies rely upon our strategic offensive power to deter attacks against them. Their vital interests and ours are inextricably linked. Their safety and ours are one. And no change in technology can or will alter that reality. We must and shall continue to honor our commitments.

I clearly recognize that defensive systems have limitations and raise certain problems and ambiguities. If paired with offensive systems, they can be viewed as fostering an aggressive policy, and no one wants that. But with these considerations firmly in mind, I call upon the scientific community in our country, those who gave us nuclear weapons, to turn their great talents now to the cause of mankind and world peace, to give us the means of rendering these nuclear weapons impotent and obsolete.

Tonight, consistent with our obligations of the ABM treaty and recognizing the need for closer consultation with our allies, I'm taking an important first step. I am directing a comprehensive and intensive effort to define a long-term research and development program to begin to achieve our ultimate goal of eliminating the threat posed by strategic nuclear missiles. This could pave the way for arms control measures to eliminate the weapons themselves. We seek neither military superiority nor political advantage. Our only purpose—one all people share—is to search for ways to reduce the danger of nuclear war.

My fellow Americans, tonight we're launching an effort which holds the promise of changing the course of human history. There will be risks, and results take time. But I believe we can do it. As we cross this threshold, I ask for your prayers and your support.

Thank you, good night, and God bless you.

*Note: The President spoke at 8:02 p.m. from the Oval Office at the White House. The address was broadcast live on nationwide radio and television.*

*Following his remarks, the President met in the White House with a number of administration officials, including members of the Cabinet, the White House staff, and the Joint Chiefs of Staff, and former officials of past administrations to discuss the address.*

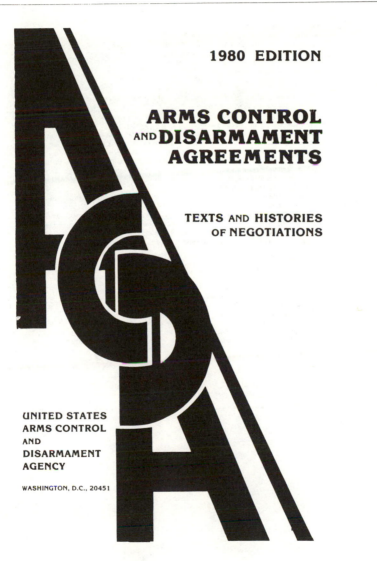

1980 EDITION

# ARMS CONTROL AND DISARMAMENT AGREEMENTS

TEXTS AND HISTORIES
OF NEGOTIATIONS

UNITED STATES
ARMS CONTROL
AND
DISARMAMENT
AGENCY

WASHINGTON, D.C., 20451

**Treaty Between the United States of America and the Union of Soviet Socialist Republics on the Limitation of Anti-Ballistic Missile Systems**

*Signed at Moscow May 26, 1972*
*Ratification advised by U.S. Senate August 3, 1972*
*Ratified by U.S. President September 30, 1972*
*Proclaimed by U.S. President October 3, 1972*
*Instruments of ratification exchanged October 3, 1972*
*Entered into force October 3, 1972*

The United States of America and the Union of Soviet Socialist Republics, hereinafter referred to as the Parties,

Proceeding from the premise that nuclear war would have devastating consequences for all mankind,

Considering that effective measures to limit anti-ballistic missile systems would be a substantial factor in curbing the race in strategic offensive arms and would lead to a decrease in the risk of outbreak of war involving nuclear weapons,

Proceeding from the premise that the limitation of anti-ballistic missile systems, as well as certain agreed measures with respect to the limitation of strategic offensive arms, would contribute to the creation of more favorable conditions for further negotiations on limiting strategic arms,

Mindful of their obligations under Article VI of the Treaty on the Non-Proliferation of Nuclear Weapons,

Declaring their intention to achieve at the earliest possible date the cessation of the nuclear arms race and to take effective measures toward reductions in strategic arms, nuclear disarmament, and general and complete disarmament,

Desiring to contribute to the relaxation of international tension and the strengthening of trust between States,

Have agreed as follows:

### Article I

1. Each party undertakes to limit anti-ballistic missile (ABM) systems and to adopt other measures in accordance with the provisions of this Treaty.

2. Each Party undertakes not to deploy ABM systems for a defense of the territory of its country and not to provide a base for such a defense, and not to deploy ABM systems for defense of an individual region except as provided for in Article III of this Treaty.

### Article II

1. For the purpose of this Treaty an ABM system is a system to counter strategic ballistic missiles or their elements in flight trajectory, currently consisting of:

(a) ABM interceptor missiles, which are interceptor missiles constructed and deployed for an ABM role, or of a type tested in an ABM mode;

ARMS CONTROL AND DISARMAMENT AGREEMENTS

(b) ABM launchers, which are launchers constructed and deployed for launching ABM interceptor missiles; and

(c) ABM radars, which are radars constructed and deployed for an ABM role, or of a type tested in an ABM mode.

2. The ABM system components listed in paragraph 1 of this Article include those which are:

(a) operational;
(b) under construction;
(c) undergoing testing;
(d) undergoing overhaul, repair or conversion; or
(e) mothballed.

### Article III

Each Party undertakes not to deploy ABM systems or their components except that:

(a) within one ABM system deployment area having a radius of one hundred and fifty kilometers and centered on the Party's national capital, a Party may deploy: (1) no more than one hundred ABM launchers and no more than one hundred ABM interceptor missiles at launch sites, and (2) ABM radars within no more than six ABM radar complexes, the area of each complex being circular and having a diameter of no more than three kilometers; and

(b) within one ABM system deployment area having a radius of one hundred and fifty kilometers and containing ICBM silo launchers, a Party may deploy: (1) no more than one hundred ABM launchers and no more than one hundred ABM interceptor missiles at launch sites, (2) two large phased-array ABM radars comparable in potential to corresponding ABM radars operational or under construction on the date of signature of the Treaty in an ABM system deployment area containing ICBM silo launchers, and (3) no more than eighteen ABM radars each having a potential less than the potential of the smaller of the above-mentioned two large phased-array ABM radars.

### Article IV

The limitations provided for in Article III shall not apply to ABM systems or their components used for development or testing, and located within current or additionally agreed test ranges. Each Party may have no more than a total of fifteen ABM launchers at test ranges.

### Article V

1. Each Party undertakes not to develop, test, or deploy ABM systems or components which are sea-based, air-based, space-based, or mobile land-based.

2. Each Party undertakes not to develop, test, or deploy ABM launchers for launching more than one ABM interceptor missile at a time from each launcher, not to modify deployed launchers to provide them with such a capability, not to develop, test, or deploy automatic or semi-automatic or other similar systems for rapid reload of ABM launchers.

### Article VI

To enhance assurance of the effectiveness of the limitations on ABM systems and their components provided by the Treaty, each Party undertakes:

SALT ONE — ABM TREATY

(a) not to give missiles, launchers, or radars, other than ABM interceptor missiles, ABM launchers, or ABM radars, capabilities to counter strategic ballistic missiles or their elements in flight trajectory, and not to test them in an ABM mode; and

(b) not to deploy in the future radars for early warning of strategic ballistic missile attack except at locations along the periphery of its national territory and oriented outward.

### Article VII

Subject to the provisions of this Treaty, modernization and replacement of ABM systems or their components may be carried out.

### Article VIII

ABM systems or their components in excess of the numbers or outside the areas specified in this Treaty, as well as ABM systems or their components prohibited by this Treaty, shall be destroyed or dismantled under agreed procedures within the shortest possible agreed period of time.

### Article IX

To assure the viability and effectiveness of this Treaty, each Party undertakes not to transfer to other States, and not to deploy outside its national territory, ABM systems or their components limited by this Treaty.

### Article X

Each Party undertakes not to assume any international obligations which would conflict with this Treaty.

### Article XI

The Parties undertake to continue active negotiations for limitations on strategic offensive arms.

### Article XII

1. For the purpose of providing assurance of compliance with the provisions of this Treaty, each Party shall use national technical means of verification at its disposal in a manner consistent with generally recognized principles of international law.

2. Each Party undertakes not to interfere with the national technical means of verification of the other Party operating in accordance with paragraph 1 of this Article.

3. Each Party undertakes not to use deliberate concealment measures which impede verification by national technical means of compliance with the provisions of this Treaty. This obligation shall not require changes in current construction, assembly, conversion, or overhaul practices.

### Article XIII

1. To promote the objectives and implementation of the provisions of this Treaty, the Parties shall establish promptly a Standing Consultative Commission, within the framework of which they will:

(a) consider questions concerning compliance with the obligations assumed and related situations which may be considered ambiguous;

ARMS CONTROL AND DISARMAMENT AGREEMENTS

(b) provide on a voluntary basis such information as either Party considers necessary to assure confidence in compliance with the obligations assumed;

(c) consider questions involving unintended interference with national technical means of verification;

(d) consider possible changes in the strategic situation which have a bearing on the provisions of this Treaty;

(e) agree upon procedures and dates for destruction or dismantling of ABM systems or their components in cases provided for by the provisions of this Treaty;

(f) consider, as appropriate, possible proposals for further increasing the viability of this Treaty; including proposals for amendments in accordance with the provisions of this Treaty;

(g) consider, as appropriate, proposals for further measures aimed at limiting strategic arms.

2. The Parties through consultation shall establish, and may amend as appropriate, Regulations for the Standing Consultative Commission governing procedures, composition and other relevant matters.

Article XIV

1. Each Party may propose amendments to this Treaty. Agreed amendments shall enter into force in accordance with the procedures governing the entry into force of this Treaty.

2. Five years after entry into force of this Treaty, and at five-year intervals thereafter, the Parties shall together conduct a review of this Treaty.

Article XV

1. This Treaty shall be of unlimited duration.

2. Each Party shall, in exercising its national sovereignty, have the right to withdraw from this Treaty if it decides that extraordinary events related to the subject matter of this Treaty have jeopardized its supreme interests. It shall give notice of its decision to the other Party six months prior to withdrawal from the Treaty. Such notice shall include a statement of the extraordinary events the notifying Party regards as having jeopardized its supreme interests.

Article XVI

1. This Treaty shall be subject to ratification in accordance with the constitutional procedures of each Party. The Treaty shall enter into force on the day of the exchange of instruments of ratification.

2. This Treaty shall be registered pursuant to Article 102 of the Charter of the United Nations.

**DONE** at Moscow on May 26, 1972, in two copies, each in the English and Russian languages, both texts being equally authentic.

**FOR THE UNITED STATES OF AMERICA**

**FOR THE UNION OF SOVIET SOCIALIST REPUBLICS**

*President of the United States of America*

*General Secretary of the Central Committee of the CPSU*

## Agreed Statements, Common Understandings, and Unilateral Statements Regarding the Treaty Between the United States of America and the Union of Soviet Socialist Republics on the Limitation of Anti-Ballistic Missiles

### 1. Agreed Statements

The document set forth below was agreed upon and initialed by the Heads of the Delegations on May 26, 1972 (letter designations added):

**AGREED STATEMENTS REGARDING THE TREATY BETWEEN THE UNITED STATES OF AMERICA AND THE UNION OF SOVIET SOCIALIST REPUBLICS ON THE LIMITATION OF ANTI-BALLISTIC MISSILE SYTEMS**

[A]

The Parties understand that, in addition to the ABM radars which may be deployed in accordance with subparagraph (a) of Article III of the Treaty, those non-phased-array ABM radars operational on the date of signature of the Treaty within the ABM system deployment area for defense of the national capital may be retained.

[B]

The Parties understand that the potential (the product of mean emitted power in watts and antenna area in square meters) of the smaller of the two large phased-array ABM radars referred to in subparagraph (b) of Article III of the Treaty is considered for purposes of the Treaty to be three million.

[C]

The Parties understand that the center of the ABM system deployment area centered on the national capital and the center of the ABM system deployment area containing ICBM silo launchers for each Party shall be separated by no less than thirteen hundred kilometers.

[D]

In order to insure fulfillment of the obligation not to deploy ABM systems and their components except as provided in Article III of the Treaty, the Parties agree that in the event ABM systems based on other physical principles and including components capable of substituting for ABM interceptor missiles, ABM launchers, or ABM radars are created in the future, specific limitations on such systems and their components would be subject to discussion in accordance with Article XIII and agreement in accordance with Article XIV of the Treaty.

ARMS CONTROL AND DISARMAMENT AGREEMENTS

[E]

The Parties understand that Article V of the Treaty includes obligations not to develop, test or deploy ABM interceptor missiles for the delivery by each ABM interceptor missile of more than one independently guided warhead.

[F]

The Parties agree not to deploy phased-array radars having a potential (the product of mean emitted power in watts and antenna area in square meters) exceeding three million, except as provided for in Articles III, IV and VI of the Treaty, or except for the purposes of tracking objects in outer space or for use as national technical means of verification.

[G]

The Parties understand that Article IX of the Treaty includes the obligation of the US and the USSR not to provide to other States technical descriptions or blue prints specially worked out for the construction of ABM systems and their components limited by the Treaty.

## 2. Common Understandings

Common understanding of the Parties on the following matters was reached during the negotiations:

### A. Location of ICBM Defenses

The U.S. Delegation made the following statement on May 26, 1972:

Article III of the ABM Treaty provides for each side one ABM system deployment area centered on its national capital and one ABM system deployment area containing ICBM silo launchers. The two sides have registered agreement on the following statement: "The Parties understand that the center of the ABM system deployment area centered on the national capital and the center of the ABM system deployment area containing ICBM silo launchers for each Party shall be separated by no less than thirteen hundred kilometers." In this connection, the U.S. side notes that its ABM system deployment area for defense of ICBM silo launchers, located west of the Mississippi River, will be centered in the Grand Forks ICBM silo launcher deployment area. (See Agreed Statement [C].)

### B. ABM Test Ranges

The U.S. Delegation made the following statement on April 26, 1972:

Article IV of the ABM Treaty provides that "the limitations provided for in Article III shall not apply to ABM systems or their components used for development or testing, and located within current or additionally agreed test ranges." We believe it would be useful to assure that there is no misunderstanding as to current ABM test ranges. It is our understanding that ABM test ranges encompass the area within which ABM components are located for test purposes. The current U.S. ABM test ranges are at White Sands, New Mexico, and at Kwajalein Atoll, and the current Soviet ABM test range is near Sary Shagan in Kazakhstan. We consider that non-phased array radars of types used for range safety or instrumentation purposes may be located outside of ABM test ranges. We interpret the reference in Article IV to "additionally agreed test

SALT ONE—AGREED STATEMENTS

ranges" to mean that ABM components will not be located at any other test ranges without prior agreement between our Governments that there will be such additional ABM test ranges.

On May 5, 1972, the Soviet Delegation stated that there was a common understanding on what ABM test ranges were, that the use of the types of non-ABM radars for range safety or instrumentation was not limited under the Treaty, that the reference in Article IV to "additionally agreed" test ranges was sufficiently clear, and that national means permitted identifying current test ranges.

### C. Mobile ABM Systems

On January 29, 1972, the U.S. Delegation made the following statement:

Article V(1) of the Joint Draft Text of the ABM Treaty includes an undertaking not to develop, test, or deploy mobile land-based ABM systems and their components. On May 5, 1971, the U.S. side indicated that, in its view, a prohibition on deployment of mobile ABM systems and components would rule out the deployment of ABM launchers and radars which were not permanent fixed types. At that time, we asked for the Soviet view of this interpretation. Does the Soviet side agree with the U.S. side's interpretation put forward on May 5, 1971?

On April 13, 1972, the Soviet Delegation said there is a general common understanding on this matter.

### D. Standing Consultative Commission

Ambassador Smith made the following statement on May 22, 1972:

The United States proposes that the sides agree that, with regard to initial implementation of the ABM Treaty's Article XIII on the Standing Consultative Commission (SCC) and of the consultation Articles to the Interim Agreement on offensive arms and the Accidents Agreement,[1] agreement establishing the SCC will be worked out early in the follow-on SALT negotiations; until that is completed, the following arrangements will prevail: when SALT is in session, any consultation desired by either side under these Articles can be carried out by the two SALT Delegations; when SALT is not in session, *ad hoc* arrangements for any desired consultations under these Articles may be made through diplomatic channels.

Minister Semenov replied that, on an *ad referendum* basis, he could agree that the U.S. statement corresponded to the Soviet understanding.

### E. Standstill

On May 6, 1972, Minister Semenov made the following statement:

In an effort to accommodate the wishes of the U.S. side, the Soviet Delegation is prepared to proceed on the basis that the two sides will in fact observe the obligations of both the Interim Agreement and the ABM Treaty beginning from the date of signature of these two documents.

In reply, the U.S. Delegation made the following statement on May 20, 1972:

---

[1] See Article 7 of Agreement to Reduce the Risk of Outbreak of Nuclear War Between the United States of America and the Union of Soviet Socialist Republics, signed Sept. 30, 1971.

ARMS CONTROL AND DISARMAMENT AGREEMENTS

The U.S. agrees in principle with the Soviet statement made on May 6 concerning observance of obligations beginning from date of signature but we would like to make clear our understanding that this means that, pending ratification and acceptance, neither side would take any action prohibited by the agreements after they had entered into force. This understanding would continue to apply in the absence of notification by either signatory of its intention not to proceed with ratification or approval.

The Soviet Delegation indicated agreement with the U.S. statement.

### 3. Unilateral Statements

The following noteworthy unilateral statements were made during the negotiations by the United States Delegation:

#### A. Withdrawal from the ABM Treaty

On May 9, 1972, Ambassador Smith made the following statement:

The U.S. Delegation has stressed the importance the U.S. Government attaches to achieving agreement on more complete limitations on strategic offensive arms, following agreement on an ABM Treaty and on an Interim Agreement on certain measures with respect to the limitation of strategic offensive arms. The U.S. Delegation believes that an objective of the follow-on negotiations should be to constrain and reduce on a long-term basis threats to the survivability of our respective strategic retaliatory forces. The USSR Delegation has also indicated that the objectives of SALT would remain unfulfilled without the achievement of an agreement providing for more complete limitations on strategic offensive arms. Both sides recognize that the initial agreements would be steps toward the achievement of more complete limitations on strategic arms. If an agreement providing for more complete strategic offensive arms limitations were not achieved within five years, U.S. supreme interests could be jeopardized. Should that occur, it would constitute a basis for withdrawal from the ABM Treaty. The U.S. does not wish to see such a situation occur, nor do we believe that the USSR does. It is because we wish to prevent such a situation that we emphasize the importance the U.S. Government attaches to achievement of more complete limitations on strategic offensive arms. The U.S. Executive will inform the Congress, in connection with Congressional consideration of the ABM Treaty and the Interim Agreement, of this statement of the U.S. position.

#### B. Tested in ABM Mode

On April 7, 1972, the U.S. Delegation made the following statement:

Article II of the Joint Text Draft uses the term "tested in an ABM mode," in defining ABM components, and Article VI includes certain obligations concerning such testing. We believe that the sides should have a common understanding of this phrase. First, we would note that the testing provisions of the ABM Treaty are intended to apply to testing which occurs after the date of signature of the Treaty, and not to any testing which may have occurred in the past. Next, we would amplify the remarks we have made on this subject during the previous Helsinki phase by setting forth the objectives which govern the U.S. view on the subject, namely, while prohibiting testing of non-ABM components for ABM purposes: not to prevent testing of ABM components, and not to prevent testing of non-ABM components for

SALT ONE—AGREED STATEMENTS

non-ABM purposes. To clarify our interpretation of "tested in an ABM mode," we note that we would consider a launcher, missile or radar to be "tested in an ABM mode" if, for example, any of the following events occur: (1) a launcher is used to launch an ABM interceptor missile, (2) an interceptor missile is flight tested against a target vehicle which has a flight trajectory with characteristics of a strategic ballistic missile flight trajectory, or is flight tested in conjunction with the test of an ABM interceptor missile or an ABM radar at the same test range, or is flight tested to an altitude inconsistent with interception of targets against which air defenses are deployed, (3) a radar makes measurements on a cooperative target vehicle of the kind referred to in item (2) above during the reentry portion of its trajectory or makes measurements in conjunction with the test of an ABM interceptor missile or an ABM radar at the same test range. Radars used for purposes such as range safety or instrumentation would be exempt from application of these criteria.

### C.  No-Transfer Article of ABM Treaty

On April 18, 1972, the U.S. Delegation made the following statement:

In regard to this Article [IX], I have a brief and I believe self-explanatory statement to make. The U.S. side wishes to make clear that the provisions of this Article do not set a precedent for whatever provision may be considered for a Treaty on Limiting Strategic Offensive Arms. The question of transfer of strategic offensive arms is a far more complex issue, which may require a different solution.

### D.  No Increase in Defense of Early Warning Radars

On July 28, 1970, the U.S. Delegation made the following statement:

Since Hen House radars [Soviet ballistic missile early warning radars] can detect and track ballistic missile warheads at great distances, they have a significant ABM potential. Accordingly, the U.S. would regard any increase in the defenses of such radars by surface-to-air missiles as inconsistent with an agreement.

# APPENDIX C: OTHER APPLICATIONS
# OF DIRECTED ENERGY WEAPONS

This Background Paper treats only one—and probably one of the most difficult—military application of directed energy. Many other applications of widely varying plausibility vie for funding and attention. An assessment of all these schemes is well beyond the scope of this Paper, but the list below is provided for reference. Mention of a scheme does not imply that it has any technical or military promise; this question would have to be properly studied.

*Anti-satellite (ASAT)*. Directed energy attack on satellites from space-, air-, or ground-based weapons is substantially easier than boost phase BMD. A satellite's orbit is completely predictable, making it in effect a fixed target. Long dwell times and low fluences suffice for ASAT attack on unshielded satellites. For instance, long illumination at just a few watts/cm² (several times the sun's normal irradiance in space) could upset the thermal control systems that allow spacecraft to endure the extremes of heat and cold in outer space. Substantial hardening of large and complex satellites (including sensors) to directed energy weapons from all directions at all times is impractical. Unlike BMD, which must handle thousands of boosters in a few minutes, ASATs would have fewer targets and longer attack times. Last, BMDs must operate under the most hostile circumstances imaginable, whereas the superpowers might use ASATs in scenarios short of nuclear war.

This Background Paper has stressed (see Section 5.1) that maturation of the same technologies involved in boost phase BMD virtually assures potent ASATs. The so-called "Star Wars" systems could well be their own worst threats. Besides the intrinsic ease of ASAT over BMD, a Soviet defense suppression ASAT attack on U.S. defensive battle stations would have three key factors working in its favor: 1) The Soviets would pick the time and sequence of attack on the U.S. BMD system and launch of their ICBMs; 2) The Soviets need not destroy the entire defensive constellation, but only "punch a hole" for their ICBMs to pass through; 3) The attack would take place over Soviet territory.

Ground-based laser ASATs, presumably using excimer or free-electron lasers for best atmospheric propagation, would have the advantages of large size and power supplies. Airborne lasers could avoid some of the propagation disturbances introduced by denser air at low altitudes, but turbulence around the airplane skin could require adaptive beam compensation.

Space-based directed energy ASATs are the most interesting category of all, since they would be, in effect, long-range space mines. Rather than positioning itself next to its quarry like an ordinary space mine, a laser could be thousands of kilometers away and still be able to strike within milliseconds upon receipt of a radio signal from the ground.

*Strategic offense*. If they mature, the directed energy devices discussed for BMD might turn out to have been better termed "offensive breakthroughs" than "defensive breakthroughs." Consider, for example, a fleet of Soviet x-ray lasers launched simultaneously with (or minutes before) a Soviet first-strike ICBM attack. The pop-up x-ray lasers' job would be to intercept any U.S. ICBMs launched before arrival of Soviet silo-killing RVs. The Soviet x-ray lasers would therefore deprive the U.S. of its option for launch under attack. Microwave generators might be used for EMP-like attack on the U.S. command and control system. Another example of offensive use of beam weapons would be Soviet ASAT attack on U.S. warning, communications, nuclear detonation detection, or navigation satellites important to the U.S. rataliatory capability. Yet another example would of course be suppression of any U.S. BMD that used space-based weapons or sensors.

*Bus intercept*. This Background Paper has focused on intercept of ICBMs before booster burnout. Intercept of the bus or post boost vehicle poses a rather different challenge. Post-boost phase for today's ICBMs is rather long (several minutes) but could be shortened drastically on future ICBMs. Bus tracking requires a different sensor than booster tracking, since the bus plume is much less conspicuous, and the bus rocket motor may not operate continuously. The bus is a target of declining value as it dispenses its RVs. Interruption of bus operation would not prevent the bus and its contents from continuing their ballistic flight to the target country, though the aim might be very wide of the target. Operating above the atmosphere, the bus can deploy light-weight shields, decoys, and sensor countermeasures (e.g., corner reflectors). On the other hand, x-ray lasers and neutral particle beams that cannot penetrate the atmosphere can attack the bus in space.

*Anti-SLBM.* A number of schemes have been suggested for using directed energy weapons against SLBMs, besides the obvious extensions of ICBM defense. Thus pop-up x-ray lasers could be positioned on U.S. coasts or ships at sea to intercept SLBMs launched from nearby Soviet submarines. Aircraft patrolling coastal waters and carrying lasers could attack ascending SLBMs in their area.

*Anti-IRBM.* Intermediate range ballistic missiles (IRBMs) have short boost phases and potentially low trajectories, making anti-IRBM defense rather different from anti-ICBM defense and perhaps better accomplished with ground-based terminal BMD systems deployed in the theater.

*Defense of satellites (DSAT).* Low-power wide-divergence (small optics) laser satellites (perhaps HF for high specific energy) could serve as "escorts" for other satellites, defending the other satellites from hostile objects—mines, ASAT missiles —approaching within a given range.

*Anti-aircraft.* At least four schemes have been broached for using directed energy weapons against aircraft or cruise missiles. The most ambitious would involve a worldwide constellation of trackers (possibly LWIR) and beam weapons (possibly DF or short wavelength lasers) to attack Soviet Blackjack strategic bombers, Backfire bombers attacking U.S. aircraft carriers, Soviet

airborne command post "Doomsday planes," Soviet AWACS radar planes, and so on. In a second scheme, B-1 or B-52 bombers would be outfitted with lasers (possibly DF) to protect them from Soviet fighters, surface-to-air missiles (SAMs), and air-to-air missiles. A third scheme equips carrier battle groups with lasers or particle beams to defend themselves against cruise missile attack. Fourth and last, ground-based beam weapons might replace surface-to-air missiles for local air defense.

*Midcourse and terminal BMD.* Intense electron beams have long been studied as replacements for interceptors in reentry BMD. In midcourse BMD, beam weapons might not only destroy RVs, but aid discrimination of RVs from lightweight decoys: lasers, particle beams, or x-ray lasers would illuminate approaching objects, and sensors would use the response of each object as an extra piece of data to judge whether it was a true RV (see Section 6).

*Submarine communications.* This scheme would use a blue-green laser to communicate with submerged submarines. Seawater is opaque to all but VLF and ELF radio frequencies, used for submarine communications today, and to the blue-green portion of the visible light spectrum. A blue-green laser beam originating on a satellite, reflected from a space-based mirror, or carried by an airplane would be modulated in accordance with the message to be transmitted and directed at a given spot on the ocean. After transmission of the full message, the beam would dwell on a neighboring spot and transmit again, and so on, eventually covering all submarine patrol areas. Optical sensors on the submarine hull would detect the message.

*Blinding sensors and seekers.* Analysts have studied a wide range of tactical applications for lasers, involving blinding of battlefield sensors, missile seekers, and even human beings.

# Office of Technology Assessment

The Office of Technology Assessment (OTA) was created in 1972 as an analytical arm of Congress. OTA's basic function is to help legislative policy-makers anticipate and plan for the consequences of technological changes and to examine the many ways, expected and unexpected, in which technology affects people's lives. The assessment of technology calls for exploration of the physical, biological, economic, social, and political impacts that can result from applications of scientific knowledge. OTA provides Congress with independent and timely information about the potential effects—both beneficial and harmful—of technological applications.

Requests for studies are made by chairmen of standing committees of the House of Representatives or Senate; by the Technology Assessment Board, the governing body of OTA; or by the Director of OTA in consultation with the Board.

The Technology Assessment Board is composed of six members of the House, six members of the Senate, and the OTA Director, who is a non-voting member.

OTA has studies under way in nine program areas: energy and materials; industry, technology, and employment; international security and commerce; biological applications; food and renewable resources; health; communication and information technologies; oceans and environment; and science, transportation, and innovation.

# BALLISTIC MISSILE DEFENSES AND U.S. NATIONAL SECURITY

## SUMMARY REPORT

Fred S. Hoffman, *Study Director*

October 1983

*Prepared for the*

# FUTURE SECURITY STRATEGY STUDY

# BALLISTIC MISSILE DEFENSES AND U.S. NATIONAL SECURITY

## SUMMARY REPORT

Fred S. Hoffman, *Study Director*

October 1983

.

*Prepared for the*

# FUTURE SECURITY STRATEGY STUDY

# Acknowledgments

This report is a summary of work performed by a Study Team whose members were: Mr. Fred S. Hoffman, Director; Mr. Leon Sloss, Deputy Director; Mr. Fritz Ermarth; Mr. Craig Hartsell; Mr. Frank Hoeber; Dr. Marvin King; Mr. Paul Kozemchak; Lt. Gen. C. J. LeVan, USA (Ret.); Dr. James J. Martin; Mr. Marc Millot; Mr. Lawrence O'Neill; and Dr. Harry Sauerwein. The work of the Study Team has been reviewed by a Senior Policy Review Group consisting of Professor John Deutch; Dr. Charles Herzfeld; Mr. Andrew W. Marshall; Dr. Michael May; Professor Henry S. Rowen; General John Vogt, USAF (Ret.); Ambassador Seymour Weiss; Mr. Albert Wohlstetter; and Mr. James Woolsey. Supporting papers have been contributed by Mr. Craig Hartsell, Dr. James J. Martin, Mr. John Baker, Lt. Gen. C. J. LeVan, Mr. Douglas Hart, Mr. Marc Millot, Dr. David S. Yost, Mr. Leon Sloss, and Mr. Frank Hoeber.

The Study also benefitted from comments and suggestions by Dr. Thomas Brown, Dr. Ashton Carter, and Dr. Thomas Rona.

The Panel also has had the invaluable cooperation of Lt. Col. Irving Schuetze, USA.

Responsibility for the views expressed herein rests with the Study Team.

# Preface

President Reagan has directed an "effort to define a long-term research and development program...to achieve our ultimate goal of eliminating the threat posed by strategic nuclear missiles...." The President noted that the achievement of the ultimate goal was a "formidable technical task" that would probably take decades, and that "as we proceed we must remain constant in preserving the nuclear deterrent...maintaining a solid capability for flexible response...pursue real reductions in nuclear arms...(and) reduce the risk of a conventional military conflict escalating to nuclear war by improving our nonnuclear capabilities."

Two studies assisted in that effort: (1) the Defensive Technologies Study (DTS) to review the technologies relevant to defenses against ballistic missiles and recommend a specific set of long-term programs to make the necessary technological advances, and (2) the Future Security Strategy Study (FSSS) to assess the role of defensive systems in our future security strategy. The implications for defense policy, strategy, and arms control were addressed by two FSSS teams: an interagency team led by Mr. Franklin C. Miller, and a team of outside experts led by Mr. Fred S. Hoffman. This is a report on the results of the work of the team of outside experts. The work was done under the auspices of the Institute for Defense Analyses at the request of the Office of the Under Secretary of Defense for Policy to assist the interagency team.

This report and its conclusions do not necessarily represent the views of the Department of Defense or the Institute for Defense Analyses.

## CONTENTS

# SUMMARY REPORT

## A. MAJOR CONCLUSIONS AND RECOMMENDATIONS

### The Strategic Need for Defensive Systems

1. *U.S. national security requires vigorous development of technical opportunities for advanced ballistic missile defense systems.*

- Effective U.S. defensive systems can play an essential role in reducing reliance on threats of massive destruction that are increasingly hollow and morally unacceptable. A strategy that places increased reliance on defensive systems can offer a new basis for managing our long-term relationship with the Soviet Union. It can open new opportunities for pursuing a prudent defense of Western security through both unilateral measures and agreements. The Soviets have often used arms negotiations to pursue competitive military advantage. The Soviet Union is likely to cooperate in pursuing agreements that are mutually beneficial *only* if it concludes that it cannot accomplish its present political goals because it faces Western firmness and ability to resist coercion.

- Technologies for ballistic missile defenses, together with those for precise, effective, and discriminate nuclear and nonnuclear offensive systems, are advancing rapidly. They can present opportunities for resisting aggression and deterring conflict that are safer and more humane than exclusive reliance on the threat of nuclear retaliation.

- A satisfactory deterrent requires a combination of more discriminating and effective offensive systems to respond to enemy attacks plus defensive systems to deny the achievement of enemy attack objectives. Such a deterrent can counter the erosion of confidence in our alliance guarantees caused by the adverse shifts in the military balance since the 1960s.

- Readiness to deploy advanced ballistic missile defense systems is a necessary part of a U.S. hedge against the increasingly ominous possibility of one-sided Soviet deployment of such systems. Such a Soviet deployment, superimposed on the present nuclear balance, would have disastrous consequences for U.S. and allied security. Clearly this possibility, especially in the near term, also requires precautionary measures to enhance the ability of our offensive forces to penetrate defenses.

**The Preferred Path to the President's Goal: Intermediate Options**

2. *The new technologies offer the possibility of a multilayered defense system able to intercept offensive missiles in each phase of their trajectories.* In the long term, such systems might provide a nearly leakproof defense against large ballistic missile attacks. However, their components vary substantially in technical risk, development lead time, and cost, and in the policy issues they raise. Consequently, partial systems, or systems with more modest technical goals, may be feasible earlier than the full system.

3. *Such "intermediate" systems may offer useful capabilities.* The assessment in this study of the utility of intermediate systems is necessarily tentative, owing to the current lack of specificity in systems design, effectiveness and costs. Nevertheless, it indicates that, given a reasonable degree of success in our R&D efforts, intermediate systems can strengthen deterrence. They will greatly complicate Soviet attack plans and reduce Soviet confidence in a successful outcome at various levels of conflict and attack sizes, both nuclear *and nonnuclear.* Even U.S. defenses of limited capability can deny Soviet planners confidence in their ability to destroy a sufficient set of military targets to satisfy enemy attack objectives, thereby strengthening deterrence. Intermediate defenses can also reduce damage if conflict occurs. The combined effects of these intermediate capabilities could help to reassure our allies about the credibility of our guarantees.

4. *A flexible research and development (R&D) program designed to offer early options for the deployment of intermediate systems, while proceeding toward the President's ultimate goal, is preferable to one that defers the availability of components having a shorter development lead time* in order to optimize the allocation of R&D resources for development of the "full system."

- Intermediate defense systems can help to ameliorate our security problems in the interim while full systems are being developed.

- The full-system approach involves higher technical risk and higher cost. On the other hand, an approach explicitly addressing the utility of intermediate systems offers a hedge against the possibility that nearly leakproof defenses may take a very long time, or may prove to be unattainable in a practical sense against a Soviet effort to counter the defense.

- The deployment of intermediate systems would also provide operational experience with some components of later, more comprehensive, and more advanced defense systems, increasing the effectiveness of the development effort.

5. We have considered several possible intermediate options:

- *Anti-Tactical Missile (ATM) Options*

   Deployment of an anti-tactical missile (ATM) system is an intermediate option that might be available relatively early. The system might combine some advanced midcourse and terminal components identified by the Defensive Technologies Study with a terminal underlay. The advanced components, though developed initially in an ATM mode, might later play a role in continental United States (CONUS) defense. Such an option addresses the pressing military need to protect allied forces as well as our own, in theaters of operations, from either nonnuclear or nuclear attack. It would directly benefit our allies as well as ourselves. Inclusion of such an option in our long-range R&D program on ballistic missile defenses should reduce allied anxieties that our increased emphasis on defenses might indicate a weakening in our commitment to the defense of Europe. We can pursue such a program option *within ABM Treaty constraints*. Such a course is therefore consistent with a policy of deferring decisions on modifying or withdrawing from the treaty.

- *Intermediate CONUS Options*

Intermediate capabilities may also have important applications in CONUS, initially to defend critical installations such as $C^3I$ nodes. As the defense system is thickened, it also will add to Soviet uncertainties in targeting, even in large-scale attacks, thereby enhancing deterrence. Depending on rates of progress in the R&D program, a two-phase defense of high effectiveness against moderate threats might comprise both endoatmospheric and exoatmospheric components employing space-based sensors and ground-based interceptors. These intermediate components would be the lower tiers in a full multilayered system.

- *Limited Boost-Phase Intercept Options*

Some intermediate options may provide useful near-term leverage on Soviet plans and programs even if they prove unable to meet fully sophisticated Soviet responses. An early boost-phase intercept system with capability against large rockets similar to those that are an important part of Soviet forces may be one example. Such an option could impose costs on the Soviets and increase their incentive to move toward an offensive posture that is more stable and less threatening. A definitive assessment of the utility of such options must specify their technological and political feasibility, timing, and cost, and the ease with which they can be countered.

6. Pursuit of the President's goal, especially if it is interpreted solely in terms of the full, nearly leakproof system, will raise questions about our readiness to defend against other threats, notably that of air attack by possible advanced bombers and cruise missiles. An appropriate response to such questions will require an early and comprehensive review of air defense technologies, leading to the development of useful systems concepts.

**Defensive Systems and Stability of Deterrence**

7. Deployment of defensive systems can increase stability, but to attain this goal we must design our offensive and defensive forces properly; especially, we must not allow them to be vulnerable. In combination with other measures, defenses can contribute to reducing the prelaunch vulnerability of our offensive forces. To increase stability, defenses must themselves avoid high vulnerability, must be robust in the face of enemy technical or tactical countermeasures, and must compete favorably in cost with expansion of the Soviet offensive force.

8. As currently assessed, some boost-phase intercept systems and other space-based components pose serious policy problems, because of engagement time constraints. Space-based components may also be highly vulnerable to Soviet boost-phase intercept systems, or anti-satellite (ASAT) systems. It will be imperative to design systems which are not themselves subject to rapid attack. Alternative approaches need to be developed in the R&D program that permit safe arrangements for the operation of the defensive system.

### Soviet Policies, Initiative, and Responses

9. *The common assumption that the decision to initiate widespread deployment of ballistic missile defense systems rests with the United States alone is completely unjustified.* Soviet history, doctrine, and programs all indicate that the Soviets are likely (and better prepared than we) to initiate a widespread antiballistic missile (ABM) deployment whenever they deem it to their advantage.

10. The long-term course of Soviet military policy plans and programs is uncertain in detail, but unless there is a major change in their political goals, the Soviets are highly likely to continue to aim at being able to defeat any combination of external enemies.

- The Soviets will almost certainly continue to maintain and upgrade their large air defenses and to conduct programs for R&D and modernization of their ballistic missile defenses. These activities will increasingly create uncertainty about the ability of U.S. missile forces to penetrate without countermeasures, and about the possibility of a sudden (open) or gradual (clandestine) Soviet breakout from the ABM Treaty constraints. The importance of such uncertainty is intensified because of the substantial Soviet investments in air defense and passive defenses of elements of the Soviet military and government. Even without violating ABM Treaty constraints, the Soviets will probably deploy a substantial ATM defense, exacerbating our problems in theaters of operations and making them more difficult to correct.

- On the other hand, if the Soviets believe that a Western deployment of defenses will substantially improve the West's capability to resist attack or coercion, they will try to prevent a Western deployment through political means or arms negotiations.

- If the United States deploys defensive systems, the Soviets will probably seek to maintain their offensive threat through a set of measures that will depend on their assessment of the defenses and their own technological options. Depending on the defense effectiveness and leverage, such a response may not fully restore Soviet offensive capabilities.

- If, over time, the Soviets become convinced that the West has the resolve and ability to block Soviet achievement of their long-term goals of destabilization and domination of other states, they may move from their present political/military policies to become more willing to agree to reducing the nuclear threat, through a combination of mutual restrictions on offensive forces and deployment of defensive systems.

# B. SUPPORTING RATIONALE

President Reagan's directive to assess the role of defensive systems has required the FSSS to consider the relation of these systems to our strategic objectives and to Soviet programs and policy. The role of intermediate defensive systems has been a major focus of our study.

### 1. *The Need for Defensive Systems in our Security Strategy*

There is a broad consensus that reliance on nuclear retaliatory threats raises serious political and moral problems, particularly in contingencies where the enemy use of force has been constrained. Technologies for defensive systems and those for extremely precise and discriminating attacks on strategic targets have been advancing very rapidly. (Many technologies are common to both functions.) Together they offer substantial promise of a basis for protecting our national security interests, and those of our allies, that is more humane and more prudent than sole reliance on threats of nuclear response. The case for increasing the emphasis on defensive programs in our national security strategy rests on several grounds, in addition to the broad, long-term objectives mentioned by the President in his March 23 speech:

- The massive increase in Soviet power at all levels of conflict is eroding confidence in the threat of U.S. nuclear response to Soviet attacks against our allies. A continuation of this erosion could ultimately undermine our traditional alliance structure.

- If the Soviet Union persists in the buildup of nuclear offensive forces, for the next decade and beyond the United States may not wish to restore, by offensive means alone, a military balance consistent with our strategic needs. Soviet willingness and ability to match or overmatch increases in U.S. nuclear forces suggest that while additions to our forces are needed to maintain the continued viability of our nuclear deterrent, such additions alone may not preserve confidence in our alliance guarantees.

- The public in the United States and other Western countries is increasingly anxious about the danger of nuclear war and the prospects for a supposedly unending nuclear arms race. Those expressing this anxiety, however, frequently ignore the fact that the U.S. nuclear stockpile has been declining, both in numbers and in megatons, while Soviet forces have increased massively in both. A U.S. counter to the Soviet buildup that emphasized increases in U.S. nuclear stockpiles would exacerbate public anxieties.

- Arms agreements, despite widespread Western hopes for them, have to date failed to prevent growing instability in the balance—and the deterioration—in the Western position relative to the East. Offensive force limitation agreements, originally associated in the U.S. arms control strategy with the ABM Treaty, have failed to restrain the Soviet offensive buildup; *de facto* reductions in the explosive yield and size of U.S. strategic nuclear stocks have not prevented vast increases in the size and destructiveness of the Soviet stockpile.

- Rapidly advancing technologies offer new opportunities for active defense deployment against ballistic missile attack that did not exist when, over a decade ago, the United States abandoned plans for defense deployments against nuclear attack. Technologies for sensing and discrimination of targets, directing the means of intercept, and destroying targets have created the possibility of a system of layered defenses that would pose successive, independent barriers to penetrating missiles. There has been improvement in some (not all) aspects of defense vulnerability. Given successful outcomes to development programs and robustness in the face of Soviet countermeasures, such defenses would permit only a very small proportion of even a very large attacking ballistic missile force to reach target. Such defenses might also offer high leverage in competing with offensive responses.

2. *Ballistic Missile Defenses in the Soviet Union*

The Soviets maintain a high level of activity in programs relevant to defenses against nuclear attack including:

- Active programs for modernizing deployed air and ballistic missile defense systems which together give them the basis for a very rapid deployment of widespread ballistic missile defenses, if they decide to ignore ABM Treaty obligations completely and openly.

- Large and diverse R&D programs in areas of technology for advanced ballistic missile and air defense systems.

- A space launch capacity significantly greater than our own, if not as sophisticated.

A substantial Soviet lead in deployed defensive systems, superimposed on their growing offensive threat against our nuclear offensive forces, could destroy the stability of the strategic balance.

*The decision to initiate widespread deployment of ballistic missile defenses does not rest with the United States alone. The common assumption that it does is completely unjustified.* The Soviets give every appearance of preparing for such a deployment whenever they believe they will derive significant strategic advantage from doing so. Their activities include some that are questionable under the ABM Treaty. Unless the public is aware and kept aware of Soviet activities in this area, the United States will probably be blamed for initiating "another round in the arms race." The state of U.S. preparedness to deploy capable defenses will be an important element in the Soviets' assessment of their own options. Active U.S. R&D programs on advanced defensive systems can assist in deterring a Soviet deployment designed to exploit an asymmetry in their favor.

### 3. *Alternative Paths to the President's Objective*

The path to the President's ultimate objective may be designed to go directly toward the ultimate objective of a full, multilayered system that offers nearly leakproof defenses against very large offensive forces. Under some conditions such a path might be an optimal use of limited R&D resources, concentrating first on those technologies that present the greatest difficulty and require the greatest lead times.

Alternatively, R&D programs might be designed to provide earlier options for the deployment of intermediate systems, based on technologies that can contribute to the ultimate objective, as such systems become technically feasible and offer useful capabilities. Such a path toward the President's ultimate goal might generate earlier funding demands to support deployment of intermediate systems and would require early treatment of some of the policy issues. Also, at least one variant considered in our report, an ATM deployment for theaters of operations, could be undertaken without modification of the ABM Treaty.

The principal benefits of an R&D path providing options for earlier, partial deployments are:

- Possibilities for an early contribution to improving the deteriorating military balance.

- Its explicit provision of a hedge against the risks inherent in a program where each of a large number of demanding technological goals must be met in order to realize any useful result at all.

- The likelihood that early deployments of parts of the ultimate system may also prove to be the most effective path to achieving such a system; early operational experience with some system elements can contribute useful feedback to the development process.

### 4. *Intermediate Defensive Systems, Soviet Strategy, and Deterrence*

Fundamentally, the choice between the two paths depends on the utility of intermediate systems in meeting our national security objectives. In the discussion of ballistic missile defenses that preceded the U.S. proposal of the ABM Treaty, opponents of such defenses argued that the utility of widespread defense deployments should be judged in terms of their ability to protect population from large attacks aimed primarily at urban-industrial areas. Because of the destructiveness of nuclear weapons, nearly leakproof defenses are required to provide a high level of protection for population against such attacks. Moreover, opponents at that time also divided our strategic objectives into two categories: deterrence of war and limiting damage if deterrence failed. They relegated defenses exclusively to the second objective and ignored the essential complementarity between the two objectives. Consequently, they assigned defenses no role in deterrence.

We have reexamined this issue, and we conclude that defenses of intermediate levels of capability can make critically important contributions to our national security objectives. *In particular, they can reinforce or help maintain deterrence by denying the Soviets confidence in their ability to achieve the strategic objectives of their contemplated attacks as they assess a decision to go to war.* By strengthening deterrence *at various levels of conflict*, defenses can also contribute valuable reassurance to our allies.

Deterrence rests on the Soviets' assessment of their political/military alternatives. This, in turn, depends on their objectives and style in planning for and using military force. It also depends on their estimates of the effectiveness of weapons and forces on both sides. Soviet assessments on these matters may differ sharply from our own. Specifically, the past behavior of the Soviets suggests they credit defensive systems with greater capability than we do. If true, this will increase the contribution of defensive systems to deterrence.

Because of the long lead times, assessment of the strategic role of defenses also requires very long-term projections about the nature of the Soviet state. While such projections cannot be made with confidence, there is no current basis for projecting a fundamental change in the Soviet attitude toward external relations. We consider below the possibility that appropriate management by the West of its long-term relations with the Soviets might induce a fundamental change. Desirable as this goal is, the most probable projection for the foreseeable future is that they will continue to set a high priority on their ability to control, subvert, or coerce other states as the basis for their foreign relations. In this case, military power will continue to play a major role for the Soviets, and many present elements of style in the application of that power can be expected to persist:

- Domination of the Eurasian periphery is a primary strategic objective. The Soviets' preferred mode in exploiting their military power is to apply it to deter, influence, coerce—in short, to control—other states, if possible without combat. But the ability to so apply this power depends on strength in actual combat.

- The Soviet objective in combat is victory, defined as survival of the Soviet state and military power (with as little damage as possible) and the imposition of the Soviet will on opponents. Soviet doctrine and practice contemplate limited war, viewed in terms of Soviet ability to impose limitations on opponents for Soviet strategic advantage.

- Soviet plans unite the roles of various elements of military forces in a coherent strategic architecture, embracing offense, defense, and combined arms in various theaters of operations. Destruction of an enemy is subordinate to the achievement of the goal of victory. The Soviets' concept for use of strategic offensive and defensive capability is, consequently, to deter attacks by U.S. intercontinental forces, to separate the United States from its allies in the Eurasian periphery, and to limit damage in the event that U.S. offensive forces are used against the Soviet Union.

- Uncertainty is a dominant factor in all combat, creating an unlimited demand for superiority in forces. Soviet planners seek ways to control uncertainty but, faced with uncertainty over which they cannot exercise a high degree of control, Soviet military action may be deterred. Uncertainties are particularly important in technically complex interactions between offense and defense.

Such a view of military force and its political applications may appear inconsistent with Soviet threats of inevitable apocalyptic destruction in the event of war at any level—but such threats are intended to play on the fears of the Western public. While very great destruction might in fact result from Soviet attacks, the discussion above suggests that the Soviets give priority to military targets. In the absence of defenses, their massive offensive forces make it possible for them to attack large numbers of targets, including urban-industrial targets as well as high-priority military targets.

Whether they would conduct such attacks from the outset or withhold attacks against urban-industrial targets to deter U.S. retaliation must be a matter of conjecture. In any case, intermediate levels of defense capability might deny them the ability to destroy with high confidence all of their high-priority targets and force them to concentrate their attack on such targets, diverting weapons that might otherwise be directed against cities. Moreover, if defenses can deny the Soviets confidence in achievement of their military attack objectives, this will strengthen deterrence of such attacks. Thus, to the extent that such attacks are necessary to overall Soviet plans, defenses can help deter lower levels of conflict.

### 5. The Military Utility of Intermediate Defensive Systems

Defensive systems affect attack planning in a variety of ways, depending on the characteristics and effectiveness of the defenses, the objectives of the attack, and the responses of the defense and offense to the measures adopted by the other side.

Any defense system can be overcome by an attack large enough to exhaust the intercept capability of the defense. The size of attack against which the defense is designed is therefore one major characteristic of a defensive system. The cost of expanding the defense to deal with a given increase in the size and cost of the offense is a measure of the leverage of the defense. Another characteristic is its effectiveness—its probability of destroying an offensive missile.

If the defense has sufficiently high capacity, effectiveness, and leverage, it can of course essentially preclude attacks. Such defenses may result from the R&D programs pursuant to the President's goal, but it is more likely that the results will be more modest. Even a modest level of effectiveness—for example, a kill probability of 0.5 for each layer of a four-layer defense—yields an overall "leakage" rate of only about 6 percent for an attack size that does not exceed the total intercept capacity of the various layers. Such a leakage rate is, of course, sufficient to create catastrophic damage in an attack of, say, 5,000 reentry vehicles (RVs) aimed at cities. It would mean 300 RVs arriving at targets—sufficient to destroy a very large part of our urban structure and population even if distributed in a nonoptimal fashion from the point of view of the offense.

Against an extensive military target system, however, with an attack objective of destroying large fractions of specific target sets (such as critical $C^3I$ facilities) with high confidence, such a leakage rate would be totally inadequate for the offense. The more specific the attack objectives and the higher the confidence required by the offense, the greater the leverage exacted by the defense. For example, in the previous four-layer case, if the defense required a high-confidence penetration against a specific target, it would need to fire at least 30 RVs to a single target since the defense firing doctrine is unknown to the attacker. As these are expected-value calculations, an attacker would have to double or triple the above values to attain high confidence in killing a specific target. Clearly an attacking force of 5,000 RVs that could destroy a very large military target system in the absence of defenses would be totally inadequate to achieve high confidence of destruction of a large fraction of a defended target set amounting to hundreds of targets. Yet, this is precisely what is required to achieve the strategic objectives of a large-scale nuclear attack.

The situation is even more dramatic in the case of limited attacks on restricted target systems, intended to achieve a decisive strategic advantage while continuing to deter further escalation of the level of nuclear attack. Such attacks would be precluded entirely by defenses of the sort discussed, would deny the attacker's confidence in the outcome, or would require a level of force inconsistent with limiting the level of violence, while depleting the attacker's inventory available for other tasks.

Offense and defense have a rich menu of responses from which they can choose. These include fractionation of payload to increase the number of warheads for a given missile force, the use of decoys, and the use of preferential offense or defense tactics. The outcome of the contest is likely to be uncertain to both sides so long as the defense keeps pace with additions to offensive force size by expanding its intercept capacity and upgrading its critical subsystems. Uncertainty about the offense-defense engagement itself contributes to deterrence of attack by denying confidence in the attack outcome.

We have considered the effect of introducing defenses in hypothetical representative military situations, taking account of what we know of Soviet objectives and operational style in combat. In their doctrine, the Soviets stress operations designed to bring large-scale conflict to a quick and decisive end, at as low a level of violence as is consistent with achievement of Soviet strategic aims. To achieve this objective in a conflict involving NATO, a major aspect of their operations is intense initial attacks on critical NATO military targets in the rear, particularly those relevant to NATO's theater nuclear capabilities and air power. Such attacks (including those in the nonnuclear phase of combat) are intended to contribute to Soviet goals at that level, to reduce NATO's ability and resolve to initiate nuclear attacks if the nonnuclear defense fails to hold, and to assist in nuclear preemption of a NATO nuclear attack. High confidence in degrading NATO air power is also essential to support utilization of Soviet operational maneuver groups designed to disrupt NATO rear areas.

The Soviets plan to use a wide variety of means to accomplish this task. Tactical ballistic missiles (TBMs) are taking an increasing role in this mission during the initial stages of either nuclear or nonnuclear combat as their accuracy increases and the sophistication of high-explosive warheads increases. Inability to destroy critical target systems would cast doubt on the feasibility of the entire Soviet attack plan, and so contribute to deterrence of theater combat, nuclear or nonnuclear.

In the event of imminent or actual large-scale conflict in Europe, another high-priority Soviet task would be to prevent quick reinforcement and resupply from the United States.

Early and obvious success in this respect, by demonstrating the hopelessness of resistance, might abort European resistance altogether or end a conflict in its very early stages. In the absence of defenses, the Soviets might attempt this task by nonnuclear tactical ballistic missile attacks on reception facilities in Europe. The Soviets could also accomplish this task with higher confidence by means of quite limited nuclear attacks on such facilities in Europe and on a restricted set of force projection targets in CONUS.

While the risk of provoking large-scale U.S. response to nuclear attacks on CONUS might be unacceptable to the Soviets, they might also feel that—given the stakes, the risks of escalation if conflict in Europe is prolonged, and the strength of their deterrent to U.S. initiation of a large-scale nuclear exchange—the *relative* risks might be acceptable if the attack size were small enough and their confidence of success sufficiently high. Without defenses, very small numbers of ballistic missiles could in fact achieve high confidence in such an attack. However, an intermediate ballistic missile defense deployment of moderate capabilities could force the Soviets to increase their attack size radically. This would reduce or eliminate the Soviets' confidence that they could achieve their attack objectives while controlling the risks of a large-scale nuclear exchange. The role of intermediate defenses in large-scale nuclear attacks has already been discussed at the beginning of this section.

Soviet response to prospective or actual defense deployments by the United States also will have longer-run aspects. The Soviets' initial reaction will be to assess the nature, effects, and likelihood of a U.S. defense deployment. Barring fundamental changes in their conception of their relations to other states and their security needs, they will seek to prevent such a deployment through manipulation of public opinion or negotiations over arms agreements. (We consider the possibility of a fundamental change in Soviet political/military objectives in the discussion of arms agreements below.)

If the Soviets fail to prevent the deployment of defenses, they will assess their alternative responses in the light of the strategic architecture discussed above, the effectiveness and leverage of the U.S. ballistic missile defenses, and other relevant U.S. offensive and defensive capabilities (e.g., air defense). If the new defensive technologies offer sufficient leverage against the offense and they cannot prevent the West from deploying defensive systems, the Soviets may accept a reduction in their long-range offensive threat against the West, which might be reflected in arms agreements. In this case, they would probably seek to compensate by increasing their relative strength in other areas of military capability. Their current program emphases suggest that they would be more likely to respond with a continuing buildup in their long-range offensive forces. However, such a buildup would not necessarily be sufficient to maintain their current level of confidence in the achievement of the strategic objectives of those forces.

### 6. *Managing the Long-Term Competition with the Soviet Union*

Current Soviet policy on arms agreements is dominated by the Soviet Union's attempt to derive unilateral advantage from arms negotiations and agreements, by accepting only arrangements that permit continued Soviet increases in military strength while using the negotiation process to inhibit Western increases in military strength. There is no evidence that Soviet emphasis on competitive advantage over mutual benefit will change in the near future, unless a fundamental change occurs in the Soviet Union's underlying foreign policy objectives. Such a change might be induced in the long run by a conviction among Soviet leaders that the West was able and resolved to block the Soviet Union's attempts to extend its power and influence by reliance on military strength. If such a change occurred, the possibilities for reaching much more substantial arms agreements might increase. In that event, it might also be possible to reach agreements restricting offensive forces so as to permit defensive systems to diminish the nuclear threat. Soviet belief in the seriousness of U.S. resolve to deploy such defenses might itself contribute to such a change.

### 7. *Defenses and Stability*

Deployment of defensive systems can increase stability, but to attain this we must design our offensive and defensive forces properly—and, especially, we must not allow them to be vulnerable. In combination with other measures, defenses can contribute to reducing the prelaunch vulnerability of our offensive forces. To increase stability, defenses must themselves avoid high vulnerability, must be robust in the face of enemy technical or tactical countermeasures, and must compete favorably in cost terms with expansion of the Soviet offensive force. A defense that was highly effective for an attack below some threshold but lost effectiveness very rapidly for larger attacks might decrease stability if superimposed on vulnerable offensive systems. Boost-phase and midcourse layers may present problems of both vulnerability and high sensitivity to attack size. Nevertheless, if this vulnerability can be limited through technical and tactical measures, these layers may constitute very useful elements of properly designed multilayered systems where their sensitivity is compensated by the capabilities of other system components.

### 8. *A Perspective on Costs*

We do not yet have a basis for estimating the full cost of the necessary research program nor the cost of systems development or various possible defensive deployment options. It is clear, however, that costs and the tradeoffs they require would present important issues for defense policy. While not insignificant, total systems costs would be spread over many years. There is no reason at present to assume that the potential contributions of defensive systems to our security would not prove sufficient to warrant the costs of deploying the systems when we are in a better situation to assess their costs and benefits.

# THE STRATEGIC DEFENSE INITIATIVE

## Defensive Technologies Study

# THE STRATEGIC DEFENSE INITIATIVE

## Defensive Technologies Study

**Department of Defense**

**April 1984**

On March 23, 1983, President Reagan challenged the
scientific community of the United States to investigate
whether new technologies could provide the means for countering the awesome threat of nuclear ballistic missiles.

Following his historic speech, the President directed an
intensive study to define the technologies necessary for
defending the United States and our allies from ballistic
missile attack. We collected over 50 of our nation's top
scientists and engineers and asked them to assess the feasibility of achieving this goal and to structure a research program
to develop the technologies that could provide an effective
defense against ballistic missiles. This report summarizes the
results of their effort, the Defensive Technologies Study.

The principal finding of the Defensive Technologies Study
Team was that, despite the uncertainties, new technologies hold
great promise for achieving the President's goal of eliminating
the threat of ballistic missiles to ourselves and our allies.
Based on the technical recommendations of this study the United
States has structured a focused research and technology program
of the highest priority to pursue these new technologies. This
Strategic Defense Initiative will provide future Presidents with
an option to enhance our deterrence capability by basing it on a
mix of offensive and defensive forces. The Strategic Defense
Initiative will have three aspects as its hallmark: innovation,
focused technology programs, and technical demonstration milestones.

Our scientists and engineers are aware, like the President,
that we face significant technical challenges and uncertainties.
Yet, as we move into the next decade, I am confident that our
greatest asset, our people's ingenuity and creativity, will make
the President's vision a reality. We will pursue the Strategic
Defense Initiative with utmost vigor. I believe that within the
technologies reviewed by the Defensive Technologies Study are
the seeds of a safer world. We owe it to ourselves, our allies,
and most of all to our children to meet the President's
challenge.

## CONTENTS

*PREFACE*

In March 1983 President Reagan established as a long-term national goal an end to the threat of ballistic missiles. He said that "we must thoroughly examine every opportunity for reducing tensions and for introducing stability into the strategic calculus on both sides." He asked the scientific community to give the United States "the means of rendering" the ballistic missile threat "impotent and obsolete."

Shortly after his address to the Nation, the President directed that an intensive analysis be conducted, to include a Defensive Technologies Study to identify the most promising approaches to effective defense against ballistic missiles and to describe a technically feasible research and development program. A study team was formed and worked under the leadership of Dr. James C. Fletcher. The team's report is summarized here.

## SUMMARY AND CONCLUSIONS OF THE DEFENSIVE TECHNOLOGIES STUDY

The Defensive Technologies Study analyzed the technological feasibility of developing an effective defense against ballistic missiles and proposed programs in the areas of

- surveillance, acquisition, and tracking;
- directed energy weapons;
- conventional weapons;
- battle management, communications, and data processing;
- systems concepts;
- countermeasures and tactics.

Classified reports for each area and a *Summary, Defense Technology Plan* have been issued. Presented here is an unclassified overview of the summary report, with its principal findings.

The Study Team identified a long-term, technically feasible research and development plan. The goal of the study was to provide the basis for selecting the technology paths to follow when a specific defensive strategy is chosen. At the same time, near-term demonstrations of some system components were identified that could provide options for early deployment and meaningful levels of effectiveness against constrained threats. The plan also incorporates ideas for enhancing the defense of NATO and other allies.

The study reviewed, evaluated, and placed priorities on the technological issues underlying the ballistic missile defense of the United States and its allies. Also reviewed was a set of strategic defense system concepts and supporting technologies in various states of development. In addition, the study considered system concepts where technological attributes were not preeminent, for example, concepts constrained by fiscal considerations. The study did not consider defenses against threats other than ballistic missiles, such as bombers and cruise missiles or conventional forces; these issues are dealt with in other Department of Defense studies.

The Defensive Technologies Study Team identified a research and development program to allow knowledgeable decisions on whether, several years from now, to begin an engineering validation phase that, in turn, could lead to an effective defensive capability in the 21st century. Similarly, intermediate deployments could be feasible that would provide meaningful levels of defense, especially against constrained threats.

The Defensive Technologies Study concluded that

- powerful new technologies are becoming available that justify a major technology development effort offering future technical options to implement a defensive strategy;
- focused development of technologies for a comprehensive ballistic missile defense will require strong central management;
- the most effective systems have multiple layers, or tiers;
- survivability of the system components is a critical issue whose resolution requires a combination of technologies and tactics that remain to be worked out;
- significant demonstrations of developing technologies for critical ballistic missile defense functions can be performed over the next ten years that will provide visible evidence of progress in developing the technical capabilities required of an effective in-depth defense system.

## ADVANCES IN DEFENSIVE TECHNOLOGIES

The ballistic missile threat has increased significantly over the past twenty years, so an appropriate question is: "What has happened to justify another evaluation of ballistic missile defense as a basis for a major change in strategy?" Advances in defensive technologies warrant such a reevaluation.

Two decades ago there were no reliable approaches to the problem of boost-phase intercept; however, multiple approaches now exist based on directed energy concepts such as particle beams and lasers and kinetic-energy target destruction mechanisms.

Intercept in midcourse was difficult twenty years ago because of no credible concepts for decoy discrimination, the intercept cost, and the collateral effects of nuclear weapons used for the interceptor warheads. Today, multispectral sensing of discriminants with laser imaging and millimeter-wave radar, birth-to-death tracking, and direct-impact projectiles that have promise as inexpensive interceptors appear to eliminate the difficulties of midcourse intercept.

In the 1960s an inability to discriminate penetration aids at high altitudes and limited interceptor performance resulted in very small defended areas for each terminal site and required an unacceptably high number of interceptors for effective defense. Now, technological advances may offer ways to discriminate among incoming objects and to allow intercepts at high altitudes. When these improvements are coupled with the potential for boost-phase and midcourse intercepts to disrupt pattern attacks, the effectiveness of terminal defenses is significantly increased.

Likewise, 1960s technology in computer hardware and software and signal processing was incapable of supporting battle management of the multitiered defense. Because of technological advances, the needed command, control, and communications facilities in all likelihood will be realized.

Several new technologies and concepts emerged from the work of the Defensive Technologies Study Team that, considered with those already well known, illustrate how far defensive technology has progressed over the past two decades. For example, throughout the phases of a ballistic missile trajectory, there are many observables, and by using both active and passive sensors, a selection of them can be measured. That is, it is likely that discrimination can be done between a warhead and a decoy or debris as threatening objects proceed toward their targets. An active sensor works on the same principle as radar; a passive sensor relies on radiation emanating from the target. Some possible technologies the study identified for surveillance, acquisition, and tracking were active techniques such as thermal response of a target to a continuous-wave laser and passive techniques such as imaging with infrared sensors. Although any one sensor can be defeated, it is very difficult to defeat several operating simultaneously.

The study also identified several concepts for the intercept and destruction of targets. Kinetic-energy, or impact, devices include exoatmospheric and high endoatmospheric, nonnuclear, rocket-propelled projectiles and hypervelocity guns. Directed energy concepts with significant potential include ground- or space-based particle beams. Also identified were potential concepts for enhanced battle management and command, control, and communications as well as several different ways to ensure space systems survivability.

## THE THREAT

Various potential threats were considered, ranging from an attack with fewer than 100 ballistic missiles and a few hundred warheads to a simultaneous launch attack with more than 3,000 missiles and over 30,000 warheads. The Study Team selected a defense-in-depth approach because of the stress imposed by a maximum, unconstrained ballistic missile offense. The critical technologies highlighted later are best understood in the context of this threat.

## PROGRAM MANAGEMENT

The study concluded that a high priority should be placed on central management of the research and development program and there should be streamlined budgeting and contracting and effective security.

## THE BALLISTIC MISSILE DEFENSE ENVIRONMENT

The four phases of a typical ballistic missile trajectory are shown in Figure 1. First, there is a boost phase when the first- and second-stage engines are burning and offering intense, highly specific observables. A post-boost, or bus deployment, phase occurs next, during which multiple warheads and penetration aids are released from a post-boost vehicle. Then, there is a midcourse phase when warheads and penetration aids travel on ballistic trajectories above the atmosphere. Finally, there is a terminal phase in which the warheads and penetration aids reenter the atmosphere and are affected by atmospheric drag.

A ballistic missile defense capable of engaging the target all along its flight path must perform certain key functions:

- *Rapid and reliable warning of an attack and initiation of the engagement.* This requires global, full-time surveillance of ballistic missile launch areas to detect an attack and define its destination and intensity, determine likely targeted areas, and provide data for hand-off to boost-phase intercept and post-boost vehicle tracking systems.

- *Efficient intercept and destruction of the booster and post-boost vehicle.* The defense must be capable of dealing with attacks ranging from a few tens of missiles to a massive, simultaneous launch. In attacking post-boost vehicles, the defense prefers to attack as early as possible to minimize the number of penetration aids deployed.
- *Efficient discrimination through bulk filtering of lightweight penetration aids.* The price to the offense in mass, volume, and investment for credible decoys should be high.
- *Enduring birth-to-death tracking of all threatening objects.* This enables unambiguous hand-over, with few errors, of reentry vehicles to designated interceptors.
- *Low-cost target intercept and destruction in midcourse.* There should be recognition of the assigned target in the midst of a large array of penetration aids and debris. The cost to the defense for interceptors should be less than the cost to the offense for warheads.
- *High endoatmospheric terminal intercept and destruction.* This involves relatively short-range intercept of each reentering warhead.
- *Battle management, communications, and data processing.* These elements coordinate the system components for effectiveness and economy of force.

It is generally accepted, on the basis of many years of ballistic missile defense studies and associated experiments, that an efficient defense against a high-level threat would be a multitiered defense-in-depth requiring all the capabilities listed above. For each tier there will be leakage, that is, threatening objects that have not been intercepted and hence move on to the next phase. For example, three tiers, each of which allows 10 percent leakage, yielding an overall leakage of 0.1 percent, are likely to be less costly than a single layer that is 99.9 percent effective. In addition, a multitiered defense is the optimum counter to structured attacks; any given offense response affects only one phase.

The defended area of a terminal-defense interceptor is determined, working backward in a ballistic missile trajectory, by how fast the

interceptor can fly and how early it can be launched. Terminal-
defense interceptors fly within the atmosphere, and their velocity is
limited. How early they can be launched depends on the requirements
for discrimination of the target from penetration aids and accompanying debris. Because the terminal defense of a large area requires
many interceptor launch sites, the defense is vulnerable to saturation
tactics.

It is desirable, therefore, to complement the terminal defense with
area defenses that intercept at long ranges. Such a complement is
found in a system for exoatmospheric intercepts in the midcourse
phase.

Intercept outside the atmosphere requires the defense to cope with
decoys designed to attract interceptors and exhaust the defending
force prematurely. Fortunately, available engagement times in midcourse are longer than in other phases. The midcourse defensive
system must provide both early filtering, or discrimination, of
nonthreatening objects and continuing attrition of threatening objects
if the defense is to minimize the pressure on the terminal system.
Intercept before midcourse is attractive because starting the defense at
midcourse accepts the potential of a large increase in targets from
multiple independently targeted reentry vehicle and decoy deployment.

The ability to respond effectively to an unconstrained threat is
strongly dependent on the viability of a boost-phase intercept system.
For every booster destroyed, the number of objects to be identified
and sorted out by the remaining elements of a layered ballistic missile
defense system is reduced significantly. Because each future booster
could be capable of deploying tens of reentry vehicles and hundreds of
decoys, the leverage, or the advantage gained by the defense, may be
100 to 1 or more. A boost-phase system is itself constrained by the
relatively short engagement times and the potentially large number of
targets. Because of these constraints, an efficient surveillance and
battle-management system is needed.

That phase of flight in which post-boost vehicle operations occur is
a transition from boost phase to midcourse. In this phase the leverage
gained by the defense decreases with time as decoys and reentry
vehicles are deployed. On the other hand, the post-boost phase offers
additional time for intercept by boost-phase weapons, and above all

an opportunity to discriminate between warheads and deception objects as they are deployed.

The phenomenology and required technology for each of these phases of a ballistic missile trajectory are quite different. In each phase of a ballistic missile flight, a defensive system must perform the basic functions of (1) surveillance, acquisition, and tracking and (2) intercept and target destruction.

## SURVEILLANCE, ACQUISITION, AND TRACKING

Just as there are many tiers to the overall ballistic missile defense system, there can be more than one tier in each of the phases. These space-based surveillance, acquisition, and tracking components perform different tasks because the nature of a structured attack changes as the threatening objects proceed along their trajectories. To illustrate this point and also to indicate how the components of one phase may interact with those of another phase, two potential technologies will be described—(1) infrared sensors and laser designators for the midcourse phase and (2) infrared sensors and laser trackers for the terminal phase.

The surveillance, acquisition, and tracking function includes sensing information for battle management and processing signals and data for discrimination of threatening reentry vehicles from other objects. As each potential reentry vehicle is released from its postboost vehicle, it begins ballistic midcourse flight accompanied by deployment hardware and possibly by decoys. Each credible object must be accounted for in a birth-to-death track, even if the price is many decoy false alarms. Interceptor vehicles of the defense must also be tracked.

The midcourse sensors must be able to discriminate between the threatening reentry vehicles that have survived through the postboost deployment phase and nonthreatening objects such as decoys and debris. They must also provide reentry vehicle position and trajectory data for firing interceptors and assessing target destruction. Most reentry vehicles must be recognized, even if again there are many false alarms. Requirements are to track all objects designated as reentry vehicles and other objects that may be confusing to later tiers.

Space-based, passive, infrared sensors could provide a way to meet these requirements. They could permit long-range detection of cold bodies against the space background, rejection of simple lightweight objects, and birth-to-death tracking of designated objects. Laser trackers could provide imaging to determine if targets had been destroyed and precision tracking of objects as they continue through midcourse. As the objects proceed along their trajectories, data on them are handed off from sensor to sensor and track files on threatening objects are progressively improved.

The terminal phase is the final line of defense. The tasks of surveillance are to acquire and sort all objects that have leaked through early defense layers and to identify the remaining reentry vehicles. Such actions will, where possible, be based on hand-overs from the midcourse engagement. Objects include reentry vehicles shot at but not destroyed, reentry vehicles never detected, and decoys and other objects that were neither discriminated nor destroyed. These credible objects must be handed off to terminal-phase interceptors.

An innovative concept for the terminal phase is the airborne optical adjunct—a long-endurance platform that would be put into position on warning of attack—that would detect arriving reentry vehicles using infrared sensors, as those space-based sensors had done in midcourse, tracking those that were not previously selected. The airborne sensors would also provide the data necessary for additional discrimination. They could acquire and track objects in late exoatmospheric flight and observe interactions with the atmosphere from the beginning of reentry. Then, a laser or radar would precisely measure the position of each object and refine its track just before committing the interceptors.

## INTERCEPT AND TARGET DESTRUCTION

A variety of mechanisms, including directed energy, can destroy a target at any point along its trajectory. The study identified several promising ones. An excimer laser, for example, can be configured to produce a single giant pulse that delivers a resulting shock wave to a target. The shock causes structural collapse. A continuous-wave or repetitively pulsed laser delivers radiant thermal energy to the target. Contact is maintained until a hole is burned through the target or the temperature of the entire target is raised to a damaging level. Examples included in this category are free-electron lasers, chemical lasers (hydrogen fluoride or deuterium fluoride), and repetitively pulsed excimer lasers. Another way to destroy a target is with a neutral-particle beam, which deposits sufficient energy within a target to destroy its internal components. Guns and missiles destroy their targets through kinetic-energy impact. Here, homing projectiles are propelled by chemical rockets or by hypervelocity guns, such as the electromagnetic gun based on the idea of an open solenoid.

Figures 2, 3, and 4 show ballistic missile defense during boost, midcourse, and terminal phases.

## BATTLE MANAGEMENT

The purpose of battle management is to optimize the use of defense resources—it is a data-processing and communication system that includes the command, control, and communication facilities. Its tasks are situation monitoring, resource accounting, resource allocation, and reporting.

A layered battle-management system would correspond to the different layers of the ballistic missile defense system, with each layer being semiautonomous with its own processing resources, rules of engagement, sensor inputs, and weapons. During an engagement, data would be handed over from one phase to the next. Its exact architecture would be highly dependent on the mix of sensors and weapons and the geographical scope of the defensive system that it manages.

Sensors survey the field of battle, and their raw data are filtered to reduce the volume. Later processes organize these data according to the size of the object; information specific enough to determine its orbital parameters and positions as a function of time; and a listing of other data that bear on the identity, classification, and threat status of the object being tracked. In principle, all objects in the field of view of the sensors are candidates, and all objects that cannot readily be rejected as nonthreatening will appear in the file, which is the representation of the total battle situation.

The resources of the defense system include the sensors and weapons, the data-processing and communication gear, and the platforms or stations on which these and other components reside.

The allocation of defense system resources, both sensor and weapon, is a dynamic process that must be repeated with each significant change in the situation. Sensors must be assigned to sectors or to targets of interest at appropriate times to acquire necessary data, and weapons must be assigned to targets within a framework implemented by rules of engagement. An optimum allocation of resources involves extrapolating the present situation into the future and selecting a course of action that optimizes some quantity, for example, the number of targets destroyed. In each phase there are options available to the commander depending on the nature of the threat. The options also differ because events happen within different time frames.

Ultimately, data must be distributed to authorities external to the defense system to infer or sense the development of hostilities, determine a defense condition level and take appropriate actions with respect to weapons release, assist in inferring the attacker's intent, and evaluate the effectiveness of the defense and anticipate damage.

Developing hardware will not be as difficult as developing appropriate software. Very large (order of 10 million lines of code) software that operates reliably, safely, and predictably will have to be deployed. Fault-tolerant, high-performance computing will be necessary. It must be maintenance-free for ten years, radiation-hardened, able to withstand single-event upset, and designed to degrade gracefully. The main problem of network communication is managing networks of space-, air-, and ground-based resources. Other problems are real-time protocols and dynamic reconfiguration. In addition, specific ballistic missile defense algorithms, for example, target assignment, as well as a simulation environment for evaluating architectures will have to be developed.

## SURVIVABILITY

Survivability is potentially a serious problem for the space-based components. The most likely threats to the components of a defense system are direct-ascent anti-satellite weapons; ground- or air-based lasers; orbital anti-satellites, both conventional and directed energy; space mines; and fragment clouds.

The approaches to enhancing survivability against a determined attacker are the classic ones that have been used to enhance the survivability of aircraft and surface ships: hardening, evasion, proliferation, deception, and active defense. Applying these functions in combinations will be necessary to counter the spectrum of potential attacks.

Ideally, the defense system should be designed to withstand an attack meant to saturate the system, that is, to survive an attack requiring the commitment of all defense system resources.

## OFFENSIVE RESPONSES

In all considerations of offense versus defense, there is a continuing dynamic interaction. Each action can stimulate a countermeasure. In response to the development of a ballistic missile defense system, history indicates that a potential opponent will, in general, proceed in a straightforward manner with the lowest level of countering technology judged adequate. There would be continual work on possible technical responses, and it should be noted that each projected response involves a trade-off; for example, hardening of booster rockets means a reduced payload or range.

## CRITICAL TECHNOLOGIES

The Defensive Technologies Study Team concentrated on critical technologies, that is, the technologies basic to the longest lead-time items in a multitiered, four-phase ballistic missile defense system capable of defending against a massive and responsive threat. The concern was primarily with the technologies that are paramount—the concepts whose feasibility will determine whether an effective defense is possible.

There are several critical technological issues that will probably require research programs of ten to twenty years:

- *Boost and post-boost phases.* As mentioned earlier, the ability to effectively respond to an unconstrained threat is strongly dependent on meeting it appropriately during the boost and bus deployment, or post-boost, phases. This is especially important for a responsive threat.

- *Threat clouds.* Large threat clouds—that is, dense concentrations of reentry vehicles, decoys, and debris in great numbers—must be identified and sorted out during the midcourse phase and high reentry.
- *Survivability.* It will be necessary to develop a combination of tactics and mechanisms ensuring the survival of the system's space-based components.
- *Interceptors.* By having inexpensive interceptors in the midcourse phase and in early reentry, intercept can be economical enough to permit attacks on threatening objects that cannot be discriminated.
- *Battle Management.* Tools are needed for developing battle-management software.

The study also identified five- to ten-year research programs dealing with other issues. One category is space logistics. In order of priority within this category, it is desirable to have

(1) a heavy-lift launch vehicle for space-based platforms of up to 100 metric tons;
(2) a capability to service the space components;
(3) a capability to make available, on orbit, sufficient materials for space-component shielding against attack;
(4) an ability to transfer items from one orbit to another.

In addition to these items, multimegawatt power sources for space applications would be required.


## NEAR-TERM DEMONSTRATIONS AND DEPLOYMENTS

An informed decision on system development cannot be made before the end of the decade, but there may be reasons for near-term feasibility demonstrations that could be developed into elements of a total ballistic missile defense system. Unlike the boost and post-boost phases, the trade-offs between competing technological approaches for the midcourse and terminal phases are relatively well understood. Although we cannot yet pick detailed designs for the major components of the midcourse and terminal-phase defenses, the best generic approaches are known and the set of competing technologies is narrow. A number of near-term demonstrations could be done before the end of the decade that typify technological milestones. Such demonstrations could include, among others,

- a space-based acquisition, tracking, and pointing experiment;
- a megawatt-class, visible-light, ground-based laser demonstration;
- an airborne optical adjunct demonstration;
- a high-speed, endoatmospheric, nonnuclear interceptor missile demonstration.

In the next five years, there are decision points that will affect the technologies available by 1990. Between 1990 and 2000 the United States may decide to provide increasing protection for its allies and itself by deploying portions of the complete four-phase system. Such deployments might be evolutionary, leading to the final, low-leakage system.

The members of the Defensive Technologies Study Team finished their work with a sense of optimism. The technological challenges of a strategic defense initiative are great but not insurmountable. By pursuing the long-term, technically feasible research and development plan identified by the Study Team and presented in this report, the United States will reach that point where knowledgeable decisions concerning an engineering validation phase can be made with confidence. The scientific community may indeed give the United States "the means of rendering" the ballistic missile threat "impotent and obsolete."

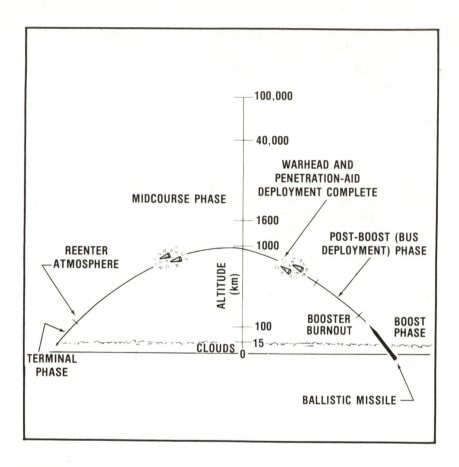

**Figure 1.** Phases of a typical ballistic missile trajectory. During the boost phase, the rocket engines accelerate the missile payload through and out of the atmosphere and provide intense, highly specific observables. A post-boost, or bus deployment, phase occurs next, during which multiple warheads and penetration aids are released from a post-boost vehicle. In the midcourse phase, the warheads and penetration aids travel on trajectories above the atmosphere, and they reenter it in the terminal phase, where they are affected by atmospheric drag.

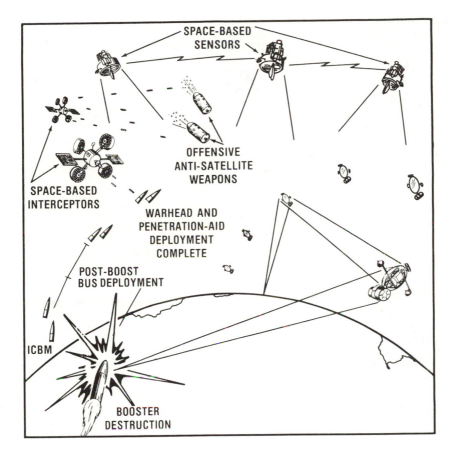

**SPACE-BASED SENSORS**

**OFFENSIVE ANTI-SATELLITE WEAPONS**

**SPACE-BASED INTERCEPTORS**

**WARHEAD AND PENETRATION-AID DEPLOYMENT COMPLETE**

**POST-BOOST BUS DEPLOYMENT**

**ICBM**

**BOOSTER DESTRUCTION**

**Figure 2.** Strawman concept for ballistic missile defense during the boost phase. An essential requirement is a global, full-time surveillance capability to detect an attack and define its destination and intensity, determine targeted areas, and provide data to guide boost-phase intercept and post-boost vehicle tracking systems. Attacks may range from a few missiles to a massive, simultaneous launch. For every booster destroyed, the number of objects to be identified and sorted out by the remaining elements of a multitiered defense system will be reduced significantly. An early defensive response will minimize the numbers of deployed penetration aids. The transition (post-boost phase) from boost phase to midcourse allows additional time for intercept by boost-phase weapons and for discrimination between warheads and deception objects. Space-based sensors detect and define the attack. Space-based interceptors protect the sensors from offensive anti-satellite weapons and, as a secondary mission, attack the missiles. In this depiction nonnuclear, direct-impact projectiles are used against the offensive weapons.

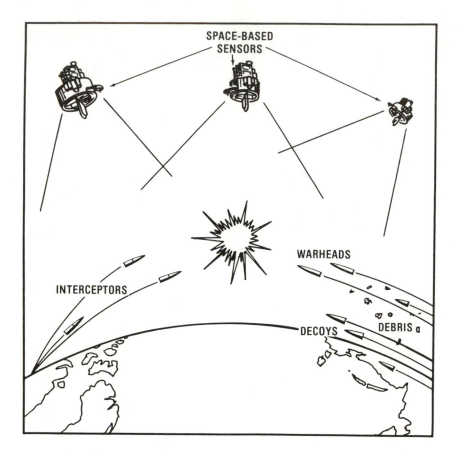

**Figure 3.** Strawman concept for ballistic missile defense during the midcourse phase. Intercept outside the atmosphere during the midcourse phase requires the defense to cope with decoys designed to attract interceptors and exhaust the defending force. Continuing discrimination of nonthreatening objects and continuing attrition of reentry vehicles will reduce the pressure on the terminal-phase system. Engagement times are longer here than in other phases. The figure shows space-based sensors that discriminate among the warheads, decoys, and debris and the interceptors that the defense has committed. The nonnuclear, direct-impact projectiles speed toward warheads that the sensors have identified.

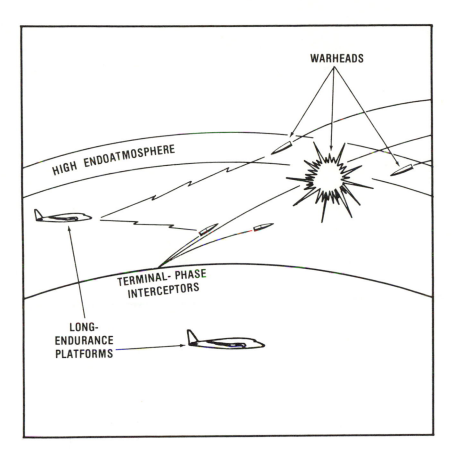

**Figure 4.** Strawman concept for ballistic missile defense during the terminal phase. This phase is the final line of defense. Threatening objects include warheads shot at but not destroyed, objects never detected, and decoys neither discriminated nor destroyed. These objects must be dealt with by terminal-phase interceptors. An airborne optical adjunct is shown here. Reentry vehicles are detected in late exoatmospheric flight with sensors on these long-endurance platforms. The interceptors—nonnuclear, direct-impact projectiles—are guided to the warheads that survived the engagements in previous phases.

# ORGANIZATION

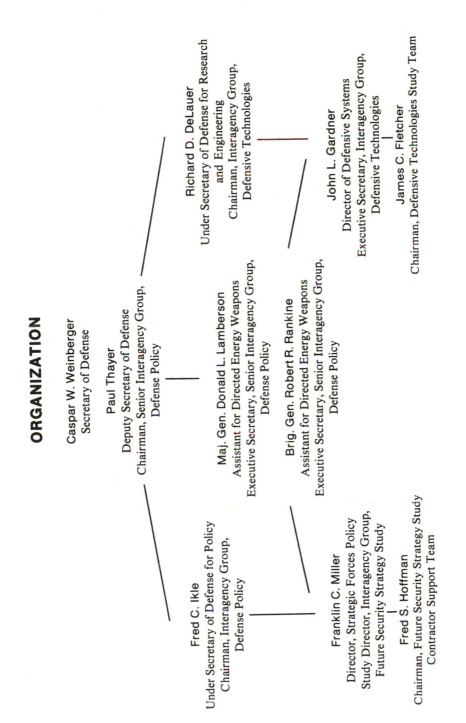

Caspar W. Weinberger
Secretary of Defense

Paul Thayer
Deputy Secretary of Defense
Chairman, Senior Interagency Group,
Defense Policy

Richard D. DeLauer
Under Secretary of Defense for Research
and Engineering
Chairman, Interagency Group,
Defensive Technologies

Maj. Gen. Donald L. Lamberson
Assistant for Directed Energy Weapons
Executive Secretary, Senior Interagency Group,
Defense Policy

Brig. Gen. Robert R. Rankine
Assistant for Directed Energy Weapons
Executive Secretary, Senior Interagency Group,
Defense Policy

John L. Gardner
Director of Defensive Systems
Executive Secretary, Interagency Group,
Defensive Technologies

James C. Fletcher
Chairman, Defensive Technologies Study Team

Fred C. Ikle
Under Secretary of Defense for Policy
Chairman, Interagency Group,
Defense Policy

Franklin C. Miller
Director, Strategic Forces Policy
Study Director, Interagency Group,
Future Security Strategy Study

Fred S. Hoffman
Chairman, Future Security Strategy Study
Contractor Support Team

# DEFENSIVE TECHNOLOGIES STUDY TEAM MEMBERSHIP

## Study Team Leadership

Dr. James C. Fletcher,
Study Chairman
The University of Pittsburgh

Dr. Harold M. Agnew,
Vice Chairman
General Atomic Technologies, Inc.

Major General John C. Toomay,
Deputy Chairman
United States Air Force (Retired)

Dr. Alexander H. Flax,
Deputy Chairman
Institute for Defense Analyses

Mr. John L. Gardner,
Executive Secretary
Office of the Secretary of Defense

Major Simon P. Worden,
Executive Military Assistant
Office of the Secretary of Defense

## Conventional Weapons Panel

Dr. Julian Davidson,
Panel Chairman
Science Applications, Inc.

Dr. Delmar Bergen
Los Alamos National Laboratory

Mr. T. Jeff Coleman
Coleman Research, Inc.

Dr. Clarke DeJonge
Science Applications, Inc.

Dr. Harry D. Fair
Defense Advanced Research Projects
Agency

Dr. James Katechis
U.S. Army Ballistic Missile Defense
System Command

Dr. Joseph R. Mayersak
USAF Armament Division

Lt Col Miles Clements,
Military Assistant
Headquarters, Department of the
Army

Lt Col Peter E. Gleszer,
Military Assistant
Headquarters, Department of the
Army

## Systems Concepts Panel

Dr. Edward T. Gerry,
Panel Chairman (Boost-Phase
Systems)
W. J. Schafer Associates, Inc.

Dr. Wayne R. Winton,
Panel Chairman (Midcourse
Systems)
Sparta, Inc.

Mr. Charles R. Wieser,
Panel Chairman (Terminal-
Phase Systems)
Physical Dynamics, Inc.

Dr. J. E. Lowder
Sparta, Inc.

Dr. James R. Fisher — U.S. Army Ballistic Missile Defense Systems Command

Dr. Louis C. Marquet — Defense Advanced Research Projects Agency

## Directed Energy Weapons Panel

Dr. Gerold Yonas, Panel Chairman — Sandia National Laboratories

Dr. J. Richard Airey — Science Applications, Inc.

Dr. Robert T. Andrews — Lawrence Livermore National Laboratory

Dr. Richard J. Briggs — Lawrence Livermore National Laboratory

Dr. Gregory H. Canavan — Los Alamos National Laboratory

Dr. Robert W. Selden — Los Alamos National Laboratory

Dr. Petras Avizonis — USAF Weapons Laboratory

Captain Richard J. Joseph — Defense Nuclear Agency

Dr. Joseph A. Mangano — Defense Advanced Research Projects Agency

Dr. Robert C. Sepucha — Defense Advanced Research Projects Agency

Captain Alan Evans, Military Assistant — USAF Systems Command

## Systems Integration Panel

Lieutenant General Kenneth B. Cooper, Panel Chairman — United States Army (Retired) Systems Planning Corporation

Mr. Wallace D. Henderson — Braddock, Dunn and McDonald

Mr. Robert T. Poppe — General Research Corporation

Mr. John M. Bachkosky — Office of the Secretary of Defense

## Surveillance, Acquisition, Tracking, and Kill Assessment Panel

Dr. John L. Allen, Panel Chairman — John Allen Associates, Inc.

Dr. George F. Aroyan — Hughes Aircraft Company

Dr. John A. Jamieson — John Jamieson, Inc.

Mr. William Z. Lemnios — Massachusetts Institute of Technology

Mr. Dennis P. Murray — Research and Development Associates

Mr. Robert G. Richards — Aerojet Electronics Systems

Mr. Fritz Steudel — Raytheon Corporation

Lt (jg) Patricia A. O'Rourke, Military Assistant — Office of the Chief of Naval Operations

## Countermeasures and Tactics Panel

| | |
|---|---|
| Dr. Alexander H. Flax, Panel Chairman | Institute for Defense Analyses |
| Dr. Robert B. Barker | Lawrence Livermore National Laboratory |
| Dr. Robert G. Clem | Sandia National Laboratories |
| Dr. Walter R. Sooy | Lawrence Livermore National Laboratory |
| Dr. James W. Somers | Defense Nuclear Agency |

## Battle Management, $C^3$, and Data Processing Panel

| | |
|---|---|
| Dr. Brockway McMillan, Panel Chairman | Private Consultant |
| Dr. Duane A. Adams | Private Consultant |
| Dr. Harry I. Davis | Private Consultant |
| Mr. J. R. Logie | Private Consultant |
| Dr. Robert E. Nicholls | Massachusetts Institute of Technology/Lincoln Laboratory |
| Mr. Robert Yost | Science Applications, Inc. |

## Executive Scientific Review Group

| | |
|---|---|
| Dr. Edward Frieman, Group Chairman | Science Applications, Inc. |
| Lt Col Michael Havey | Office of Science and Technology Policy |
| Dr. Solomon J. Buchsbaum | Bell Laboratories |
| Mr. Daniel Fink | Private Consultant |
| Mr. Bert Fowler | Mitre Corporation |
| Dr. Eugene Fubini | Private Consultant |
| Admiral Bobby R. Inman | United States Navy (Retired) Microelectronics and Computer Technology Corporation |
| Dr. Michael May | Lawrence Livermore National Laboratory |
| General E. C. Meyer | United States Army (Retired) |
| Professor William A. Nierenberg | Scripps Institute of Oceanography |
| Dr. David Packard | Hewlett Packard Company |

# GLOSSARY

**active sensor**    A system that includes both a detector and a source of illumination. A camera with a flash attachment is an active sensor.

**airborne optical adjunct**    A set of sensors designed to detect, track, and discriminate an incoming warhead. The sensors are typically optical or infrared devices flown in an aircraft stationed above clouds.

**algorithm**    Rules for solving a problem using computer language.

**architecture**    The physical structure of a computer system, which can include both hardware and software (programs).

**birth-to-death tracking**    The ability to track a missile and its payload from launch until it is intercepted or reaches its target.

**boost phase**    The portion of a missile flight during which the payload is accelerated by the large rocket motors. For a multiple-stage rocket, boost phase involves all motor stages.

**booster**    The rocket that "boosts" the payload to accelerate it from the earth's surface into a ballistic trajectory, during which no additional force is applied to the payload.

**bus deployment phase**    The portion of a missile flight during which multiple warheads are deployed on different paths to different targets (also referred to as the post-boost phase). The warheads on a single missile are carried on a platform, or "bus" (also referred to as a post-boost vehicle), which has small rocket motors to move the bus slightly from its original path.

**chemical laser**    A laser in which chemical action is used to produce the pulses of coherent light.

*323*

**coherent light**    The state in which light waves are in phase over the time scale of interest. Light travels in discrete bundles of energy called photons. Each photon may be treated like an ocean wave. If all the waves are in phase, they are said to be coherent. When light is coherent, the effects of each photon build on top of the others. A laser produces coherent light and therefore can concentrate energy.

**cold bodies**    Objects at or near low ambient temperature, which radiate infrared radiation. All objects radiate electromagnetic energy, and if the object is hot enough, this energy is visible light.

**constrained threat**    A situation where opponents are limited in the number of warheads or types of missiles, for example, by arms control agreement.

**continuous-wave laser**    A laser in which the coherent light is generated continuously rather than at fixed time intervals.

**decoy**    A device that is constructed to look and behave like a nuclear-weapon-carrying warhead, but which is far less costly, much less massive, and can be deployed in large numbers to complicate defenses.

**directed energy**    Energy in the form of particle or laser beams that can be sent long distances at nearly the speed of light.

**discriminate**    The process of observing a set of attacking objects and determining which are the real warheads and which are decoys and other nonthreatening objects.

**dynamic reconfiguration**    A means whereby a battle-management system can change its condition during a battle to respond to changing circumstances, such as the destruction of some defensive components.

**electromagnetic gun**    A gun based on the idea of an open solenoid. The projectile is accelerated by electromagnetic forces rather than by an explosion, as in a conventional gun.

**endoatmospheric**    When all activities take place within the earth's atmosphere, generally considered as occurring at altitudes below 100 kilometers.

| | |
|---|---|
| excimer laser | A chemical laser that uses noble gases. |
| exoatmospheric | When all activities take place outside the earth's atmosphere, generally considered as occurring at altitudes above 100 kilometers. |
| fragment clouds | Clusters of small objects placed in front of a target in space. This is a simple way to destroy the target. |
| free-electron laser | A laser in which electrons are converted to coherent light. The electrons are supplied by an accelerator and power for the laser by electrical energy. |
| hypervelocity gun | A gun that can accelerate projectiles to 5 kilometers per second or more, for example, an electromagnetic, or rail, gun. |
| imaging | The process of identifying an object by obtaining a high-quality image of it. |
| infrared sensor | A sensor to detect the infrared radiation from a cold body such as a missile reentry vehicle. |
| Interagency Groups | Two groups, one for Defensive Technologies and one for Defense Policy, set up to monitor the work of each study team. |
| intercept | The act of destroying a target. |
| kinetic energy | The energy from the momentum of an object. |
| laser | A device for generating coherent visible or infrared light. |
| laser designator | The use of a low-power laser to illuminate a target so that a weapon equipped with a special tracker can home in on the designated target. |
| laser imaging | A new technology where a laser beam can be used in a way similar to a radar beam to produce a high-quality image of an object. |
| laser tracker | The process of using a laser to illuminate a target so that specialized sensors can detect the reflected laser light and track the target. |

| | |
|---|---|
| leakage | The percentage of warheads that get through a defensive system intact and operational. |
| midcourse phase | The long period of a warhead's flight to its target after it has been dispensed from the post-boost vehicle until it reenters the atmosphere over its target. |
| multispectral sensing | A method of using many different bands of the spectrum to sense a target, for example, visible and infrared light. If several bands are used, deceptive measures become much more difficult. |
| neutral-particle beam | An energetic beam of neutral atoms (no net electric charge). A particle accelerator moves the particle to nearly the speed of light. |
| particle beam | A stream of atoms or subatomic particles (electrons, protons, or neutrons) accelerated to nearly the speed of light. |
| passive sensor | A sensor that only detects radiation naturally emitted (infrared radiation) or reflected (sunlight) from a target. |
| penetration aids | Methods to defeat defenses by camouflage, deception, decoys, and countermeasures. |
| post-boost vehicle | The portion of a rocket payload that carries the multiple warheads and has maneuvering capability to place each warhead on its final trajectory to a target (also referred to as a "bus"). |
| radiant energy | The energy from radiation such as electrons, protons, or alpha particles. |
| real-time protocols | Computer programs capable of making decisions as rapidly as input information is received. |
| repetitively pulsed laser | A laser that fires its beam in sequential short bursts, as opposed to a continuous beam or a single pulse. |
| responsive threat | Offensive forces that have been modified to defeat a defensive system. |

Senior Interagency Group
Set up in response to the President's directive to study the ballistic missile defense problem. The group reported to the National Security Council and consisted of senior representatives from the Department of Defense, the Department of State, the Arms Control and Disarmament Agency, the Central Intelligence Agency, the National Aeronautics and Space Administration, the National Security Council, and the Office of Science and Technology Policy.

signal processing
A computer system's capability to organize the raw data received from many different sources.

single-event upset
Electronic components of a battle-management system performing abnormally because of radiation.

structured attack
Timing the arrival of a sequence of warheads at their targets to create maximum destructive effects.

terminal phase
The final phase of a ballistic missile trajectory, during which warheads and penetration aids reenter the atmosphere.

threat clouds
Dense concentrations of both threatening and nonthreatening objects. The defense must distinguish between them.

*Library of Congress Cataloging-in-Publication Data*

The Star wars controversy.

  Includes bibliographies.
  1. Strategic Defense Initiative.   2. Ballistic missile defenses.
I. Miller, Steven E.   II. Van Evera, Stephen.   III. International security.
UG743.S7   1986      358′.1754      86-4280
ISBN 0-691-07713-4 (alk. paper)
ISBN 0-691-02253-4 (pbk.)